AF847235

Digital Innovations in Architecture, Engineering and Construction

Series Editors

Diogo Ribeiro, Department of Civil Engineering, Polytechnic Institute of Porto, Porto, Portugal

M. Z. Naser, Glenn Department of Civil Engineering, Clemson University, Clemson, USA

Rudi Stouffs, Department of Architecture, National University of Singapore, Singapore, Singapore

Marzia Bolpagni, Mace Group, London, UK

The Architecture, Engineering and Construction (AEC) industry is experiencing an unprecedented transformation from conventional labor-intensive activities to automation using innovative digital technologies and processes. This new paradigm also requires systemic changes focused on social, economic and sustainability aspects. Within the scope of Industry 4.0, digital technologies are a key factor in interconnecting information between the physical built environment and the digital virtual ecosystem. The most advanced virtual ecosystems allow to simulate the built to enable a real-time data-driven decision-making. This Book Series promotes and expedites the dissemination of recent research, advances, and applications in the field of digital innovations in the AEC industry. Topics of interest include but are not limited to:

- Industrialization: digital fabrication, modularization, cobotics, lean.
- Material innovations: bio-inspired, nano and recycled materials.
- Reality capture: computer vision, photogrammetry, laser scanning, drones.
- Extended reality: augmented, virtual and mixed reality.
- Sustainability and circular building economy.
- Interoperability: building/city information modeling.
- Interactive and adaptive architecture.
- Computational design: data-driven, generative and performance-based design.
- Simulation and analysis: digital twins, virtual cities.
- Data analytics: artificial intelligence, machine/deep learning.
- Health and safety: mobile and wearable devices, QR codes, RFID.
- Big data: GIS, IoT, sensors, cloud computing.
- Smart transactions, cybersecurity, gamification, blockchain.
- Quality and project management, business models, legal prospective.
- Risk and disaster management.

Cristina Cantagallo · Valentino Sangiorgio ·
Humberto Varum · Francesco Fiorito ·
Fabio Fatiguso

Digitization of Built Heritage

Approaches and Methods for Data Acquisition, Analysis, and Intervention

Springer

Cristina Cantagallo
Department of Engineering and Geology (INGEO)
University "G. d'Annunzio" of Chieti-Pescara
Pescara, Italy

Valentino Sangiorgio
Department of Engineering and Geology (INGEO)
University "G. d'Annunzio" of Chieti-Pescara
Pescara, Italy

Humberto Varum
Department of Civil and Geo Resources Engineering (CONSTRUCT-LESE)
University of Porto
Porto, Portugal

Francesco Fiorito
Department of Civil, Environmental, Land, Construction and Chemistry (DICATECh)
Polytechnic University of Bari
Bari, Italy

Fabio Fatiguso
Department of Civil, Environmental, Land, Construction and Chemistry (DICATECh)
Polytechnic University of Bari
Bari, Italy

ISSN 2731-7269 ISSN 2731-7277 (electronic)
Digital Innovations in Architecture, Engineering and Construction
ISBN 978-3-031-96597-5 ISBN 978-3-031-96598-2 (eBook)
https://doi.org/10.1007/978-3-031-96598-2

© The Editor(s) (if applicable) and The Author(s), under exclusive license to Springer Nature Switzerland AG 2025

This work is subject to copyright. All rights are solely and exclusively licensed by the Publisher, whether the whole or part of the material is concerned, specifically the rights of translation, reprinting, reuse of illustrations, recitation, broadcasting, reproduction on microfilms or in any other physical way, and transmission or information storage and retrieval, electronic adaptation, computer software, or by similar or dissimilar methodology now known or hereafter developed.
The use of general descriptive names, registered names, trademarks, service marks, etc. in this publication does not imply, even in the absence of a specific statement, that such names are exempt from the relevant protective laws and regulations and therefore free for general use.
The publisher, the authors and the editors are safe to assume that the advice and information in this book are believed to be true and accurate at the date of publication. Neither the publisher nor the authors or the editors give a warranty, expressed or implied, with respect to the material contained herein or for any errors or omissions that may have been made. The publisher remains neutral with regard to jurisdictional claims in published maps and institutional affiliations.

This Springer imprint is published by the registered company Springer Nature Switzerland AG
The registered company address is: Gewerbestrasse 11, 6330 Cham, Switzerland

If disposing of this product, please recycle the paper.

Preface

The preservation of built heritage is a critical issue in the face of increasing threats from both natural disasters and the accelerating impacts of climate change. The cultural and historical value of these structures requires innovative approaches to ensure their protection and conservation for future generations. This monograph responds to this urgent need by exploring state-of-the-art methods for **digitizing built heritage** and the corresponding processes of **acquisition, analysis, and definition of risk mitigation strategies**.

Targeted at graduate and postgraduate students, researchers, conservation practitioners, architects, engineers, and policymakers, this work presents a comprehensive framework for applying digital technologies to heritage conservation. A detailed exploration of the knowledge path of cultural heritage buildings is provided, employing tools such as 3D laser scanning, photogrammetry, and Heritage Building Information Modelling (HBIM), followed by an examination of how these data can be utilized for the assessment and management of both **rapid- and slow-onset disasters**. By integrating accurate assessments of **seismic and climate change risks**, this work supports **practical mitigation strategies** and intervention plans, particularly in areas vulnerable to seismic events or climate-related impacts. The proposed digital platform consolidates data from diverse sources, enabling stakeholders to make informed decisions on risk mitigation and heritage conservation.

The monograph **initially** focuses on the **digital data acquisition** phase and the strategies for creating a **Digital Identity Card** for heritage buildings. The monograph emphasizes the need for a comprehensive and integrated **digital strategy** in the context of cultural heritage management that can offer a solution for systematically collecting, processing, and storing data from various sources. This Digital Identity Card consolidates all relevant data, including geometric features, material properties, historical context, and current condition, providing a comprehensive understanding of the building's vulnerabilities and guiding mitigation strategies. This digital representation is essential for understanding a building's vulnerabilities and devising strategies to address potential threats. Incorporating technologies such as photogrammetry, 3D laser scanning, and UAV-based surveys, this framework not only facilitates precise architectural documentation but also enables the identification of structural weaknesses and the monitoring of long-term conditions. Particular emphasis is placed on the **knowledge path of historic buildings**, which is thoroughly explored in all its phases and components. **Traditional and innovative methods** are integrated to ensure that every aspect of the building's history, structure, and condition is meticulously documented. Through a combination of **conventional surveying techniques** and **cutting-edge digital tools**, the monograph provides a holistic framework for understanding the building, from its historical significance to its material properties and structural integrity.

Secondly, the **operational phase** addresses **multi-scale and multi-risk analyses**, with particular attention to seismic risk as a Rapid-Onset Disaster (ROD) and climate change as a Slow-Onset Disaster (SOD). Seismic risk, a major threat to built heritage, is one of the central themes in this work. The loss of heritage buildings during earthquakes is not only a physical disaster but also a cultural one, severing the link between communities and their past. The monograph addresses the need for a **multi-scale approach to seismic risk assessment**, considering both territorial and individual building scales. Large-scale analyses are conducted directly from the platform, using algorithms based on parameters that can be easily derived from rapid building inspections or targeted interviews with local experts. The goal is to produce valuable assessments for spatial planning and resource allocation. In contrast, individual heritage buildings are evaluated through advanced techniques, such as Finite Element Modeling (FEM) and non-linear analysis, essential for identifying both local and global collapse mechanisms, enabling **targeted strengthening interventions**. The slow-onset risks posed by **climate change** represent another significant challenge for the conservation of built heritage. Rising global temperatures, increased humidity, extreme weather events, and sea level rise all have a detrimental effect on the materials and structural integrity of historical buildings. Moreover, the increasingly frequent occurrence of heat waves makes global warming particularly alarming for people exposed to heat stress, which is worsened by the Urban Heat Island (UHI) effect in cities. The monograph dedicates significant attention to these risks, offering strategies for long-term monitoring, **preventive conservation**, and adaptation. On one hand, the effects of climate change—such as thermal stress, freeze-thaw cycles, and salt crystallization in coastal areas—are examined; on the other hand, particular attention is also given to the intensity of the UHI. By employing climate models and vulnerability mapping tools, this work provides methodologies for predicting the impact of climate change and UHI on built heritage and integrating these predictions into regional planning and conservation efforts.

Thirdly, the implementation phase explains the process for **automating multi-scale and multi-risk analyses** by proposing state-of-the-art techniques. To ensure a dynamic and comprehensive digitization of the built heritage, all data related to the knowledge phase of the heritage building have to be integrated into an IT platform that can store, organize, and manage all data, as well as **calculate large-scale vulnerability and risks**, with a dynamic system capable of continuous updates.

A distinctive feature of the proposed approach is the focus on public engagement and dissemination of knowledge. Digital tools like 3D visualizations, Virtual Reality (VR), and Augmented Reality (AR) applications are explored as means of making cultural heritage more accessible and engaging for the public. These tools are not only educational but also vital for raising awareness of the importance of heritage conservation. By offering interactive experiences, these digital innovations help foster a deeper connection between communities and their cultural heritage, encouraging active participation in its preservation.

The contribution of this monograph is found in its **multidisciplinary, multi-scale approach to heritage conservation**. It bridges the gap between architectural knowledge, structural analysis, and environmental risk management by offering a comprehensive framework for digitizing the built heritage. This combination provides heritage conservation professionals with a practical, scalable model that merges traditional and innovative methods, ensuring the preservation of historical and cultural value while addressing the evolving challenges of modern conservation. The use of advanced data acquisition technologies, coupled with rigorous risk assessment methodologies, **risk mitigation strategies**, and public engagement tools, creates a robust framework for protecting heritage buildings from both immediate and long-term threats. Through the development of a centralized digital platform, this work empowers heritage professionals, policymakers, and communities to make informed decisions, safeguarding these irreplaceable assets for future generations.

Pescara, Italy — Cristina Cantagallo
Pescara, Italy — Valentino Sangiorgio
Porto, Portugal — Humberto Varum
Bari, Italy — Francesco Fiorito
Bari, Italy — Fabio Fatiguso

Acknowledgments The study presented in this article was partially funded by the Projects:

- GENESIS: *Seismic risk manaGEmeNt for the touristic valorisation of thE hiStorIcal centers of Southern Italy* with a consortium of 23 partners, founded by the National Operative Programme. Overall budget: 8 873 812 € (UNICH 840 000 €). PON MIUR "Research and Innovation" 2014–2020 and FSC. D.D. 13/07/2017 n. 1735. Industrial research and experimental development projects in the 12 Smart Specialization areas. Specialization area: Cultural Heritage. Project Code ARS01_00883. Seismic risk Management for tourist valorization of the historical center of southern Italy.
- INTERCONNECTING: *Data, Analysis, and Digital Immersive Models for Sustainable Conservation of the Built Heritage: risk assessment and proactive strategies.* PNRR-Next Generation EU—1.3 Extended partnerships—CHANGES Spoke 7—Protection and conservation of cultural heritage against climate changes, natural and anthropic risks. Leader: University of Florence. Topic: Data processing and digital modeling.
- MULTI-TWIN: *Intelligenza artificiale a supporto di analisi Multi-rischio mediante digital-Twin.* PNRR-Next Generation EU—1.4 National Centers. Mission 4, "Education and Research"—Component 2, "From Research to Enterprise"—Investment Line 1.4, Project 'National Centre for HPC, Big Data and Quantum Computing—HPC' CUP H93C22000450007. Topic: Application of AI, data analysis and virtual reality for monitoring and management of infrastructure and resources as well as multi-hazard analysis at a regional scale with a focus on historic centers, critical buildings and environments (i.e., forests).

The opinions and conclusions presented by the authors do not necessarily reflect those of the funding agencies.

Competing Interests The authors have no competing interests to declare that are relevant to the content of this manuscript.

Contents

1 Introduction 1
1.1 Safeguarding Built Heritage Through Digital Innovation 1
1.2 Main Objectives 4
1.3 Innovative Aspects 6
1.4 Structure and Main Contents of the Book 8
References 13

2 Creating a Digital Identity Card of the Building Heritage: Data Collection, Digitalization, and Processing 15
2.1 Digitalization in the Construction Sector 15
2.2 Definition of Digital Heritage 18
2.3 Storage, Digitization, Organization, and Processing of Building Heritage Data 20
References 22

3 Knowledge of the Built Heritage 23
3.1 Knowledge Path 23
3.2 Identification of the Building 24
3.3 Geometric Survey 29
3.3.1 General Aspects 29
3.3.2 Traditional Technologies 30
3.3.3 Reality Data Capture Technologies 35
3.4 Critical-Historical Analysis 48
3.4.1 Documentary Research and Interpretation 48
3.4.2 Construction Phases 51
3.5 Building Techniques and Construction Details 56
3.5.1 General Objectives 56
3.5.2 Material Survey 57
3.5.3 Non-destructive Survey Techniques 64
3.5.4 Decay and Crack Survey 72
3.5.5 Construction Techniques 78

3.6 Material Properties 83
3.7 Soil and Foundations 90
3.8 Knowledge Levels and Confidence Factors 91
3.9 Heritage Building Information Modelling 93
3.9.1 Introduction to BIM 93
3.9.2 Evolution from BIM to HBIM 95
3.9.3 Components of HBIM 97
3.9.4 Applications of HBIM 101
3.9.5 Technical Challenges and Solutions in HBIM 103
3.9.6 Future Directions in HBIM 105
References 107

4 Multi-scale and Multi-risk Analysis of Built Heritage 111
4.1 Seismic Vulnerability Levels: Four-Level Approach 111
4.2 Preliminary Vulnerability Assessment and Masonry Quality Index 114
4.2.1 Masonry Quality Index 114
4.2.2 Preliminary Vulnerability Assessment and Masonry Disaggregation 121
4.3 Seismic Vulnerability and Risk for Large-Scale Analyses 121
4.3.1 Empirical Methods 123
4.3.2 Analytical/Mechanical Methods 126
4.3.3 Vulnerability Index Methods 129
4.3.4 Seismic Risk Assessment for Large-Scale Analyses 150
4.4 Seismic Vulnerability and Risk for Individual Heritage Buildings 161
4.4.1 Seismic Behavior of Masonry Heritage Buildings 161
4.4.2 Vulnerability Assessment for First-Mode Mechanisms (LV2) 162
4.4.3 Vulnerability Assessment for Second-Mode Mechanisms (LV3) 166
4.4.4 Risk Assessment for Individual Heritage Buildings 174
4.5 Climate Change Vulnerability and Risk for Large-Scale Analyses 180
4.5.1 Climate Change and Urban Risk: An Introductory Overview 180
4.5.2 Urban Heat Island Intensity Evaluation 182
4.5.3 A Quantitative Index for Assessing Urban Heat Island Intensity 184
4.5.4 Microclimate Analyses and Simulations 187
4.6 Climate Change Vulnerability and Risk for Individual Heritage Buildings 193
4.6.1 Introduction 193
4.6.2 Temperature 194
4.6.3 Moisture 196

4.6.4 Precipitation 198
4.6.5 Wind 200
4.6.6 Solar Radiation 201
References 203

5 Risk Mitigation Strategies: Conservation, Maintenance and Intervention 211
5.1 Conservation and Restoration Principles 211
5.2 Monitoring for Preventive Maintenance in Building Heritage 219
5.2.1 Key Concepts of Structural Monitoring Systems 219
5.2.2 Classification of Monitoring Systems 219
5.2.3 Static Monitoring Systems 220
5.2.4 Dynamic Monitoring Systems 225
5.3 Strengthening Interventions 229
5.3.1 Overview 229
5.3.2 Interventions for Local Collapse Mechanisms 232
5.3.3 Interventions for Global Collapse Mechanisms 234
5.3.4 Local Repairs 241
5.4 Climate Change Actions for Heritage Buildings 246
5.4.1 Urban Scale Strategies for UHII Mitigation 246
5.4.2 Temperature 247
5.4.3 Moisture 250
5.4.4 Precipitation 253
5.4.5 Wind 256
5.4.6 Solar Radiation 258
References 261

6 Automation of the Risk Analysis for Built Heritage 269
6.1 Objectives and Ambition 269
6.2 Overview of Existing Platforms for Risk Assessment 270
6.3 From Datalake to Database: Structuring and Interoperability of Data for Built Heritage Digitalization 275
6.4 Multiscale Digitalization: From Data Acquisition to Management 276
6.4.1 The Role of Multiscale Digitalization in Heritage Risk Management 276
6.4.2 Expeditious Approach: Large-Scale Vulnerability and Risk Assessment 277
6.4.3 Detailed Approach: Single-Building Vulnerability and Risk Assessment 278
References 279

7 Conclusions 281

Authors' Contributions

This book is written by *Cristina Cantagallo, Valentino Sangiorgio, Humberto Varum, Francesco Fiorito,* and *Fabio Fatiguso.*

Cristina Cantagallo (corresponding and main author) dealt with conceptualization, writing (original draft preparation), data curation, formal analysis, investigation, and visualization. *Valentino Sangiorgio* dealt with conceptualization, visualization, review, editing, and supervision. *Humberto Varum, Fiorito Francesco, and Fabio Fatiguso* dealt with review, editing and supervision.

In particular, the contribution of each author is reported for every chapter of the book in the following.

Chapter 1: Introduction—*Cristina Cantagallo, Valentino Sangiorgio, Humberto Varum, Francesco Fiorito,* and *Fabio Fatiguso.*

Chapter 2: Creating a Digital Identity Card of the Building Heritage: Data Collection, Digitalization, and Processing—*Cristina Cantagallo and Fabio Fatiguso.*

Chapter 3: Knowledge of the Built Heritage—*Cristina Cantagallo, Humberto Varum,* and *Valentino Sangiorgio.*

Chapter 4: Multi-scale and Multi-risk Analysis of Built Heritage—*Cristina Cantagallo, Valentino Sangiorgio, Humberto Varum, Francesco Fiorito,* and *Fabio Fatiguso.*

Chapter 5: Risk Mitigation Strategies: Conservation, Maintenance and Intervention—*Cristina Cantagallo, Valentino Sangiorgio,* and *Francesco Fiorito.*

Chapter 6: Automation of the Risk Analysis for Built Heritage—*Cristina Cantagallo, Valentino Sangiorgio, and Fabio Fatiguso.*

Chapter 7: Conclusions—*Cristina Cantagallo* and *Valentino Sangiorgio.*

Chapter 1
Introduction

Abstract This chapter introduces the need to safeguard built heritage through digital innovation, emphasizing the threats posed by both rapid-onset disasters, such as earthquakes, and slow-onset processes linked to climate change and urban heat island. It presents the main objectives of the monograph, focusing on developing an integrated digital system capable of combining multi-scale data, vulnerability assessments, and risk mitigation strategies. The chapter highlights key innovative aspects, including the comprehensive approach to digitization—not limited to geometric and material knowledge, but extended to vulnerability and risk analyses—and the creation of a dynamic digital platform for heritage management. Finally, it outlines the structure and main contents of the book, offering readers a clear roadmap through the methodologies, tools, and conservation strategies discussed in the following chapters.

1.1 Safeguarding Built Heritage Through Digital Innovation

Cultural heritage represents a testament to our past, a tangible link to the stories, traditions, and identities that shaped our societies. It embodies the artistic and technological advancements of its era, serving as tangible link to our collective history. However, this legacy faces a constant struggle against time and the relentless forces of nature. Natural disasters, particularly earthquakes, pose a significant threat, potentially causing catastrophic damage, especially to architectural heritage. The devastating consequences of earthquakes, as witnessed in numerous historical events worldwide, underscore the urgent need for effective measures to protect our built heritage from seismic hazards. In fact, the loss of a historical building is not merely a physical destruction; it represents the severing of a vital link to our collective memory.

The devastating impact of **earthquakes** on cultural heritage is well documented. Throughout history, these seismic events have caused irreparable damage to countless structures, leaving behind scars on our collective memory. The partial collapse of the Colosseum in Rome during an earthquake in the fourteenth century, the partial

© The Author(s), under exclusive license to Springer Nature Switzerland AG 2025

C. Cantagallo et al., *Digitization of Built Heritage*, Digital Innovations in Architecture, Engineering and Construction, https://doi.org/10.1007/978-3-031-96598-2_1

Fig. 1.1 **a** San Agustin Church, Manila, after the 1880 earthquake (source: https://picryl.com, http://commons.wikimedia.org/. License: public domain); **b** Ruins of City Hall after the 1906 San Francisco earthquake, California (source: https://picryl.com, http://www.loc.gov/. Contributor: Detroit Publishing Co. License: public domain); **c** First Unitarian Church damaged by the 1906 San Francisco earthquake, California (source: https://picryl.com, http://www.loc.gov/. Contributor: Detroit Publishing Co. License: public domain)

collapse of a bell tower of the San Agustin Church in Manila (Fig. 1.1a) after the 1880 earthquake and the demolition of the San Francisco City Hall (Fig. 1.1b), together with and the severe damage sustained by the First Unitarian Church (Fig. 1.1c) during the 1906 San Francisco earthquake in California served as stark reminders of the vulnerability of our cultural heritage to seismic forces.

In addition to earthquakes, which are rapid-onset disasters, cultural heritage is also threatened by slow-onset disasters, such as **climate change**. Disasters can generally be categorized as either rapid or slow in onset, each presenting distinct challenges for risk management and heritage preservation.

Rapid-onset disasters are those that occur suddenly and with little warning. Earthquakes, tsunamis, floods, and volcanic eruptions fall under this category. These events cause immediate and often catastrophic damage. They require quick responses and immediate assessments of structural damage, often leaving little time for preventive action.

On the other hand, **slow-onset disasters** emerge gradually over time, often as the result of long-term environmental changes. These include phenomena such as sea level rise, desertification, changes in water availability, and coastal erosion, all of which are linked to climate change. Although their impact may not be as immediately visible as that of rapid-onset events, the cumulative effect of these disasters can lead to significant structural degradation and vulnerability of heritage buildings over time. Addressing these threats requires a different approach, focusing on long-term monitoring, adaptation, and resilience strategies.

In the following chapters, both types of disasters—earthquakes and climate change—will be explored in relation to their impact on cultural heritage, along with the specific methodologies required for their mitigation and prevention.

The **protection** of cultural heritage from earthquakes and climate change requires a multifaceted approach that encompasses prevention, preparedness, and response. At the heart of this endeavour lies a deep understanding of the built heritage itself, encompassing its structural integrity, material properties, and historical context.

This knowledge is crucial for identifying vulnerabilities, assessing seismic risk, and developing effective mitigation strategies (Olivero et al. 2019).

Traditionally, the **management** of cultural heritage has been hampered by the sheer volume and complexity of historical structures, each with unique characteristics and construction methods. This vast array of data, often scattered across different sources and formats, presents a significant challenge for comprehensive analysis and risk assessment. Moreover, the dynamic nature of the built environment, with ongoing modifications and deterioration over time, further complicates the task of maintaining accurate and up-to-date information.

The advent of **digital technologies** has revolutionized the approach to cultural heritage preservation. Digitization offers a powerful toolkit for documenting, analysing, and visualizing cultural assets, enabling the creation of comprehensive knowledge networks and fostering interdisciplinary research (Marra 2022). This digital transformation has opened up new avenues for understanding the built heritage, providing insights that were previously inaccessible.

Digital documentation techniques, such as 3D laser scanning and photogrammetry, allow for the creation of detailed and accurate representations of historical structures. These digital models can be analysed to identify structural defects, assess material properties, and simulate the behaviour of structures under seismic loads. This information is essential for developing targeted interventions aimed at strengthening vulnerable elements and enhancing overall seismic resilience. The integration of geospatial data, such as satellite imagery and topographic maps, provides a broader context for assessing seismic risk.

Digital technologies also play a crucial role in **disseminating knowledge** and promoting **public awareness** about cultural heritage preservation. Interactive 3D visualizations, virtual reality experiences, and augmented reality applications can bring cultural heritage to life, fostering engagement and appreciation among the public. These innovative tools can also be used to educate communities about seismic risk and encourage preparedness measures.

The digitization of cultural heritage is not merely a technological endeavour; it is a cultural imperative. By harnessing the power of digital technologies, we can safeguard our cultural heritage for future generations, ensuring that these tangible links to our past continue to inspire, educate, and enrich our lives.

Digitalization ensures accessibility and longevity of the stored information through the digitalization process, allowing the passage of information and the entire documentary apparatus to future generations. In fact, digital tools facilitate the swift storage, manipulation, sharing, and transfer of information globally. However, an effective digitalization can be obtained only by standardized documentation protocols that integrate efficiently into decision-making processes for preventive conservation. Despite the growing support for open cultural heritage data and its reuse in Italy and in various countries, there is still a lack of structured approaches to data collection and registration, devising a sustainable strategy for protecting cultural heritage becomes challenging (Masciotta et al. 2021). Many institutions maintain their data in isolated silos, hindering comprehensive analysis and hindering the development of effective risk mitigation strategies. There is a critical need for a centralized platform that integrates various data sources, encompassing a building's knowledge (historical context, construction details, material properties), vulnerability (susceptibility to specific types of damage), and risk levels (likelihood of damage occurring). This integrated system would provide a holistic view of cultural heritage, empowering proactive measures to safeguard it from natural disasters.

This monograph aims to address this critical gap. It explores innovative approaches and methods for digitizing the built heritage, focusing on the creation of a dynamic and flexible digital platform capable of accumulating and analysing comprehensive data. This data will generally come from building knowledge, multi-scale vulnerability assessments and risk mitigation strategies. The user can use existing building data either to carry out their own vulnerability and risk assessment when such assessments does not exist, or to update existing assessments. Furthermore, the collected data can also serve to implement automatic procedures aimed at identifying vulnerability and risk on large scale. Ultimately, this information will support the development of effective risk mitigation strategies, allowing policy makers to take proactive measures to protect our cultural heritage from natural disasters.

1.2 Main Objectives

This monograph delves into traditional and innovative methods for creating an integrated digital system that encompasses all data involving built heritage, organized according to a multi-scale approach ranging from the territorial scale to that of the individual building.

The data digitalization process is both integrated and innovative. It includes not only data derived from the geometrical knowledge and damage assessments of the built heritage, but also the results of analyses conducted for various types of disasters, different types of constructions, and at different scales and accuracy levels, along with indications of possible risk mitigation strategies. The data system derived from the building knowledge and the hazard of the site where it is located makes possible the definition of its vulnerability and the associated risk, both at the territorial scale and at the scale of the individual building.

The key objectives of the book and associated research are as follows:

Develop a comprehensive data framework: To establish a robust and structured data framework that can effectively capture and organize all relevant information related to built heritage. This framework should encompass data from various sources, including geometric surveys, damage assessments, historical records, material properties, geospatial data, and seismic and climate change data.

Implement a multi-scale approach: To adopt a multi-scale approach that considers the seismic risk of built heritage at different levels, ranging from the territorial scale to the individual building scale. This approach will enable a more comprehensive understanding of risk factors and facilitate targeted mitigation strategies (Fig. 1.2).

(a) (b)

Fig. 1.2 Different types of urban context representing the different scales used in the framework presented in this book. **a** Representative example of the territorial scale: historic centre of Matera (MT, Italy); **b** representative example of the building scale "D. Ridola" Museum (MT, Italy)

Integrate innovative data acquisition techniques: To adopt innovative data acquisition techniques, such as 3D laser scanning, photogrammetry, and advanced imaging technologies, to create detailed and accurate digital representations of built heritage assets. These digital models will serve as a foundation for further analysis and risk assessment.

Utilize advanced seismic analysis methods: To employ advanced methods, such as finite element modelling and non-linear analyses, to assess the vulnerability of different types of built heritage structures. This includes both seismic analysis for rapid-onset disasters and assessments of the impacts of climate change for slow-onset disasters, providing insights into the potential behaviour of these structures under various stress conditions.

Develop risk mitigation strategies: To formulate risk mitigation strategies tailored to the specific needs of different built heritage assets and their surrounding environments. These strategies may include structural strengthening, retrofitting, non-structural measures, and emergency preparedness plans.

Create an integrated digital platform: To design and develop a digital platform that seamlessly integrates all the collected data, analysis results, and risk mitigation strategies. This platform will serve as a central repository for information and a powerful tool for decision-making and risk management (Fig. 1.3).

Promote knowledge dissemination and public awareness: To disseminate knowledge and raise public awareness about the importance of cultural heritage preservation and risk mitigation through digital innovation.

By achieving these objectives, this work aims to contribute significantly to the advancement of built heritage preservation and risk management practices with a specific focus on seismic risks and those related to climate change. The integrated digital platform and the associated data-driven insights will enable heritage professionals, policymakers, and communities to make informed decisions that safeguard our cultural heritage for future generations. In addition, students and researchers will be provided with a solid guide to achieve heritage-aware preservation, including (a)

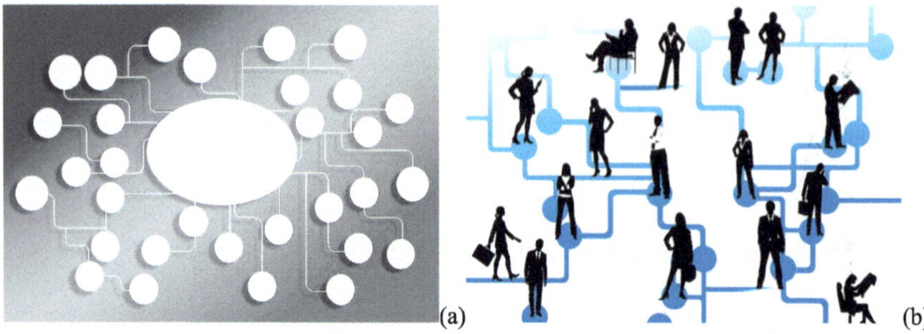

Fig. 1.3 The digital platform as a tool to gather multiple information from different sources and different sectors and as means to systematise the work done by different types of professionals working in the field of cultural heritage. *Source* https://picryl.com. License: public domain

traditional and innovative approaches, (b) the use of new digital data acquisition and storage technologies, and (c) a multi-scale approach involving both architectural and structural aspects.

1.3 Innovative Aspects

The framework proposed in this book presents significant innovative aspects, which can be summarized in the following points:

1. **Focus on a Specific Cultural Heritage Domain**: The work is dedicated to a specific area of cultural heritage, "Built heritage." This allows for the identification of specific processes and methods for the conservation and risk mitigation of architectural heritage.
2. **Comprehensive Digitalization of Built Heritage**: Most works on the digitization of cultural heritage and built heritage focus on the knowledge of the artefact, indicating and describing innovative and digitally informed methods for producing the geometric survey, the crack pattern, the assessment of degradation, and the material survey (Shen et al. 2010; Vähä et al. 2013; Ibem and Laryea 2014; Valero et al. 2015; Mcgibbon 2020). In this work, the digitization of built heritage is understood in an integrated and comprehensive sense and includes not only the aspects related to the knowledge of the building but also, and above all, the aspects related to the vulnerability and risk of the protected asset.
3. **Integrated Digital Platform for Heritage Management**: The set of integrated data described in this text can be visualized on a dynamic, flexible, and updatable computer platform, which contains an identity card for each analysed historical building. The implementation of the platform is currently underway within the **GENESIS** Project: Seismic Risk Management for the Touristic Valorisation of the Historical Centers of Southern Italy. PON MIUR "Research and Innovation" 2014–2020 and FSC. D.D. 13/07/2017 n. 1735. Industrial research

and experimental development projects in the 12 Smart Specialization areas. Specialization area: Cultural Heritage. Project Code ARS01_00883, and within the **INTERCONNECTING** Project, published by Executive Decree no. 210/2024—Prot. 31,283 of 9 February 2024. Cascade Call for Proposals F for the Project 'Cultural Heritage Active Innovation for Next-Gen Sustainable Society (CHANGES). SPOKE 7' Project code PE000020.

4. **Multi-Scale Approach for Differentiated Analysis**: The data system and the corresponding analyses follow a multi-scale approach differentiated by level of detail and analysis type (Fig. 1.4). In particular, in the field of built heritage, the approach followed in this framework distinguishes two levels of analysis: the territorial (or large-scale) level, which aims to analyse all buildings contained within a city or an area characterized by some homogeneous construction types (e.g., historical centre, expansion zone, commercial zone, etc.), and the building level, which analyses the building according to an in-depth and detailed approach. This subdivision represents the key to having a systematic and codified approach to the protection of protected assets and digitization, capable of capturing its various aspects and meeting the different needs of stakeholders.
5. **Support for Stakeholders across Different Sectors**: The approaches and methods for data acquisition, analysis, and interventions and the implementation of a digital platform according to a systematic and integrated framework represent significant support for (a) students, doctoral students, and researchers, (b) designers and cultural heritage operators, and (c) enterprises. The approaches and methodologies described in the book allow the reader to define, calculate, and implement the vulnerability and risk of a property or an area according to an integrated approach. The analysis starts with the knowledge of the building and then defines the different methods for calculating vulnerability and risk and for different types of buildings. Subsequently, different risk mitigation strategies are identified, specifically studied for built heritage as a function of the identified vulnerability level. Finally, the methods and approaches to implement all the data within a dynamic and flexible digital platform are identified. This platform can represent an effective support for public administrations that, thanks to the data provided, are able to effectively plan interventions through an appropriate allocation of resources, and for designers and enterprises operating in the field of

Fig. 1.4 Diagram representing the multi-scale approach used in this work

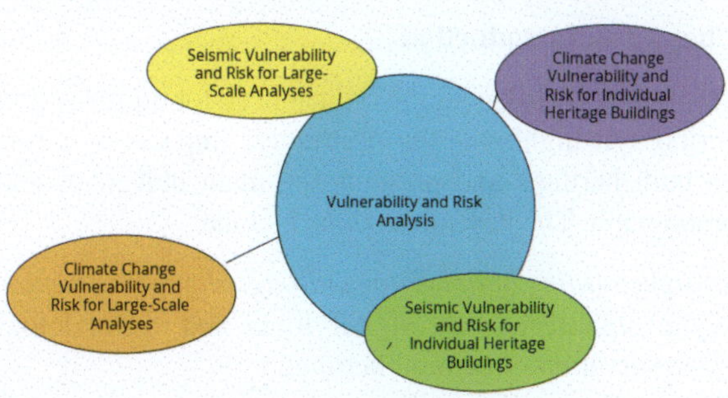

built heritage conservation that, thanks to the platform, have a solid and reliable database for the design and implementation of future consolidation interventions.

These innovative aspects set the proposed framework apart from existing approaches to built heritage digitization and risk assessment. The comprehensive nature of the data collection, the multi-scale analysis approach, and the integrated digital platform provide a powerful tool for managing and protecting built heritage assets from different hazards. The framework's potential to support a wide range of stakeholders, from researchers to practitioners, further enhances its value and contribution to the field of cultural heritage preservation.

1.4 Structure and Main Contents of the Book

This book is structured to provide a systematic exploration of digital innovations in the safeguarding of built heritage, with a specific focus on data acquisition, analysis, and intervention strategies to enhance resilience, particularly in the face of seismic threats and climate change. Each section contains both **theoretical and practical components**, allowing readers to grasp the general aspects of the topic and the primary references in the literature, as well as the necessary methods for applying this knowledge to real-world cases.

> ***Digitization Notes***
> In order to create a comprehensive guide to the methods and tools required for the digitization of built heritage, each chapter and subsection is designed with clarity and coherence. Moreover, where needed, a grey box will summarize the operational steps and recommendations for the digitalization of the respective content, offering practical advice and key points for implementation.

The structure and contents are meticulously crafted to meet the needs of both academic scholars and professionals in the field of cultural heritage preservation. They are summarized in the following list and in Fig. 1.5.

Chapter 1: Introduction

This chapter introduces the critical need for digital approaches in preserving cultural heritage. It highlights the destructive impacts of earthquakes and climate change on built heritage and presents the main objectives and innovative aspects of the monograph. The key subsections include:

- *Safeguarding built heritage through digital innovation.* This subsection discusses the significance of digital tools and methodologies in the protection and preservation of cultural heritage.

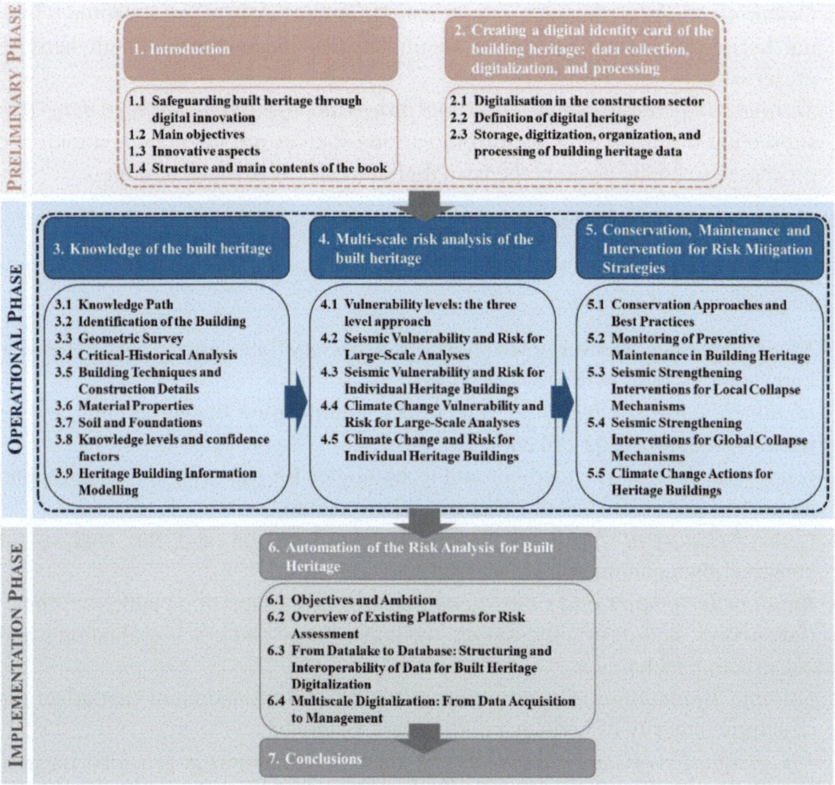

Fig. 1.5 Structure and chapters of the book

- *Main objectives*. The primary goals of the monograph are outlined, emphasizing the role of digitalization in risk assessment, monitoring, and intervention planning.
- *Innovative aspects*. The novel approaches presented in the book are highlighted, particularly the use of automation and digital technologies for heritage conservation.
- *Structure and main contents of the book*. An overview of the book's layout and main themes is provided, helping the reader understand the logical flow of topics and their interconnections.

Chapter 2: Creating a Digital Identity Card for Building Heritage

This chapter explores the methodologies for collecting, digitizing, and processing data to create a "**Digital Identity Card**" for heritage buildings. The key subsections include:

- *Digitalization in the construction sector*. This subsection provides an overview of the role of digital technologies in the modern construction sector and their application to heritage preservation.

- *Definition of digital heritage.* A conceptual framework is offered, defining "digital heritage" and explaining its significance in the context of built heritage preservation.
- *Storage, digitization, organization, and processing of building heritage data.* This subsection discusses how data from various sources are integrated, stored, and processed to create a comprehensive digital model of heritage buildings.

Chapter 3: The Knowledge of the Built Heritage

This chapter provides an in-depth exploration of the various aspects involved in understanding built heritage:

- *Knowledge Path.* An overview of the steps necessary for acquiring comprehensive knowledge of built heritage is provided.
- *Identification of the building.* The correct identification of the building and its territorial context is described.
- *Geometric Survey.* The methods and technologies for capturing the physical and geometric characteristics of structures are explained.
- *Critical-Historical Analysis.* Historical documentation and the analysis of construction techniques are discussed.
- *Building Techniques and Construction Details.* This subsection addresses material surveys, non-destructive survey methods, decay and crack evaluations, and construction techniques.
- *Soil and Foundations.* The properties of the soil and foundations that affect the structural integrity of heritage buildings are examined.
- *Knowledge Levels and Confidence Factors.* A discussion is provided on how increased knowledge of the building reduces the confidence factors used in safety analyses, thus minimizing interventions.
- *Heritage Building Information Modelling (HBIM).* The digitalization of collected data into an HBIM model for enhanced conservation and management of heritage buildings is explored.

Each section concludes with a grey box offering practical guidance on digitalization, outlining key steps for digitizing specific components of the knowledge path.

Chapter 4: The Multi-Scale and Multi-Risk Analysis of the Built Heritage

This chapter addresses the various approaches for assessing vulnerability and risk across different scales:

- *Seismic Vulnerability Levels: Four-Level Approach.* The multi-scale approach for evaluating vulnerability in built heritage is introduced, including preliminary, first, second, and third-level assessments, depending on the depth of analysis required for both large-scale and individual buildings.
- *Preliminary Vulnerability Assessment and Masonry Quality Index.* In the case of poor quality masonry, the disintegration of the material often occurs before the activation of the other damage mechanisms. Therefore, it is necessary first

to check the quality of the masonry, and then, in case of poor quality, to carry out interventions on the masonry. If this preliminary vulnerability assessment is ignored, the heritage building vulnerability can be underestimated.

- *Seismic Vulnerability and Risk for Large-Scale Analyses.* This subsection focuses on the assessment of seismic risk on a large scale, which often includes compartments or even entire urban areas. The analysis allows for prioritization in terms of intervention strategies, identifying areas and structures that are most vulnerable to seismic events.
- *Seismic Vulnerability and Risk for Individual Heritage Buildings.* For individual heritage buildings, the risk assessment is more detailed, incorporating local factors such as material properties, construction techniques, and the building's historical modifications. This subsection delves into the use of finite element models (FEM) and nonlinear analyses to assess the structural behaviour, identifying local and global collapse mechanisms. Beyond the structural aspects, particular attention is given to the functional risk associated with the loss of artistic assets, such as frescoes, sculptures, and other decorative elements
- *Climate Change Vulnerability and Risk for Large-Scale Analyses.* Changing environmental conditions (such as rising temperatures, increased frequency of extreme weather events, and sea-level rise) impact not only the structural integrity and preservation of heritage sites, but also the safety and comfort of their users. The section presents methodologies for assessing the vulnerability of heritage buildings across regions by using climate models and vulnerability mapping tools. It highlights the importance of integrating these analyses into regional planning initiatives, ensuring that large-scale heritage conservation strategies consider climate change as a central risk factor.
- *Climate Change Vulnerability and Risk for Individual Heritage Buildings.* This subsection covers how factors such as thermal stress, humidity variations, and freeze–thaw cycles can accelerate the degradation of materials and lead to building vulnerabilities. It also addresses the specific risks posed to non-structural and artistic assets within the building, as changing environmental conditions can cause deterioration to delicate materials, such as wood, plaster, and pigments. Functional risk here refers to the potential loss or damage to these artistic components, which, though not affecting the building's stability, significantly impact its cultural value.

Where needed, **grey boxes** provide operational guidelines for the digitalization of multi-risk assessment content at these different scales, as well as for automating vulnerability and risk assessments at the territorial scale.

Chapter 5: Risk Mitigation Strategies: Conservation, Maintenance and Intervention

This chapter presents an analysis of the most common risk mitigation strategies and approaches for heritage buildings. The main subsections include:

- *Conservation and Restoration Principles.* This section focuses on the conservation principles outlined in international documents such as the Venice Charter (1964), the ICOMOS/ISCARSAH Recommendations for the Analysis and Restoration of

Architectural Heritage (ICOMOS/ISCARSAH 2005), and the ISO/FDIS 13822 guidelines (ISO/TC96/SC2 2010).
- *Monitoring for Preventive Maintenance in Building Heritage.* Monitoring heritage buildings is an important support for the preservation of cultural property. For this reason, this section discusses the main aspects of this powerful tool.
- *Strengthening Interventions.* This chapter introduces a series of advanced and traditional techniques aimed at improving the behavior of masonry structures, preventing different types of collapse mechanisms.
- *Climate Change Actions for Heritage Buildings.* This section discusses the main intervention strategies aimed at mitigating both the effects of the aggressive environment related to climate change and the Urban Heat Island Intensity (UHII).

Each section concludes with grey boxes providing recommendations and key steps for the digitalization of interventions proposed in the literature, as well as for their automatic or semi-automatic planning, aimed at defining tailored maintenance plans or intervention programs based on vulnerability and risk assessments.

Chapter 6: Automation of the Risk Analysis for Built Heritage

This chapter delves into the technological advancements that facilitate the digitalization and automation of vulnerability and risk analysis processes. It includes the following subsections:

- *Objectives and Tools.* The main objectives and the tools available for digitalizing risk analysis are introduced.
- *Overview of Existing Platforms for Risk Assessment.* This section describes a brief overview of some of the most important platforms currently used for risk assessment.
- *From Datalake to Database: Structuring and Interoperability of Data for Built Heritage Digitalization.* The transition from datalake to database is described in order to ensure interoperability among datasets and to enable advanced computational processes, such as vulnerability modeling, derivation of fragility curves, and exposure-based risk assessments.
- *Multiscale Digitalization: From Data Acquisition to Management.* The Role of Multiscale Digitalization in Heritage Risk Management and how this instrument can be implemented in a digital platform is described in detail.

Chapter 7: Conclusions

The final chapter summarizes the key insights and contributions of the monograph, reinforcing the importance of digital innovation for the preservation and risk management of built heritage.

This structured approach not only provides a logical progression through the topic but also facilitates a deep understanding of each component involved in the digitization and risk management of built heritage. This layout ensures that the monograph serves as a comprehensive guide for future research and practical applications in the field.

References

Ibem EO, Laryea S (2014) Survey of digital technologies in procurement of construction projects. Autom Constr 46:11–21

ICOMOS/ISCARSAH (2005) Recommendations for the analysis, conservation and structural restoration of architectural heritage

ISO/TC96/SC2 (2010) Bases for design of structures—assessment of existing structures

Marra A (2022) Know4CARE. Publica, Alghero

Masciotta MG, Morais MJ, Ramos LF, Oliveira DV, Sánchez-Aparicio LJ, González-Aguilera D (2021) A digital-based integrated methodology for the preventive conservation of cultural heritage: the experience of heritagecare project. Int J Architectural Heritage 15:844–863. https://doi.org/10.1080/15583058.2019.1668985

Mcgibbon S (2020) A Common Structured Integrated Collaborative Digitised (CrOsS) Framework for the Historic Building Repair and Maintenance (R&M) Sector. PhD Thesis in Construction Project Management, Heriot-Watt University

Olivero S, Migliorini M, Moretti F, Lingua A, Matrone F, Colucci E, Chiabrando F, Spanò A, Bottero M, Lombardi P, Assumma V, Datola G, Noardo F, Campostrini P, Rinaldi E, Appiotti F, Hempel L (2019) The RESCULT project: a new European interoperable database for improving the resilience of Cultural Heritage subject to disasters

Shen W, Hao Q, Mak H, Neelamkavil J, Xie H, Dickinson J, Thomas R, Pardasani A, Xue H (2010) Systems integration and collaboration in architecture, engineering, construction, and facilities management: a review. Adv Eng Inform 24:196–207. https://doi.org/10.1016/j.aei.2009.09.001

Vähä P, Heikkilä T, Kilpeläinen P, Järviluoma M, Gambao E (2013) Extending automation of building construction—survey on potential sensor technologies and robotic applications. Autom Constr 36:168–178

Valero E, Adán A, Cerrada C (2015) Evolution of RFID applications in construction: a literature review. Sensors (Switzerland) 15:15988–16008. https://doi.org/10.3390/s150715988

Venice Charter (1964) International charter for the conservation and restoration of monuments and sites. Venice Charter

Chapter 2
Creating a Digital Identity Card of the Building Heritage: Data Collection, Digitalization, and Processing

Abstract This chapter examines the creation of a Digital Identity Card for built heritage, focusing on data collection, digitalization, and processing. It begins by discussing the evolution and impact of digitalization in the construction sector, emphasizing how technologies such as Building Information Modelling (BIM) and cloud-based tools enhance efficiency, collaboration, and sustainability. It then explores the definition and scope of digital heritage, highlighting its role in documenting, preserving, and interpreting historic buildings through advanced recording, management, and visualization techniques. Finally, the chapter presents methodologies for the storage, organization, and processing of heritage data, detailing the integration of diverse datasets within datalakes and structured databases to support informed decision-making in conservation and risk management.

2.1 Digitalization in the Construction Sector

Digitization in the construction industry marks a revolutionary shift from traditional methods to a more technologically integrated approach. This transformation is driven by the adoption of digital tools that enhance efficiency, accuracy, and collaboration across all stages of construction projects. Innovative technologies such as Building Information Modelling (BIM), cloud computing, and artificial intelligence have become fundamental in reshaping construction processes. These technologies enable better project visualization, improved precision in material usage, and optimized workflow management. Additionally, the integration of IoT devices and advanced data analytics supports real-time monitoring and management of construction sites, thereby enhancing decision-making and reducing operational risks. As a result, the construction sector is witnessing a significant reduction in costs and project delivery times, while increasing safety and environmental sustainability.

The digital transformation in the construction sector started when designers began transitioning from physical drawing boards to utilizing computer-aided design (CAD) technologies several decades ago. This evolution laid the groundwork for the integration of Building Information Modelling (BIM), a pivotal digital management

© The Author(s), under exclusive license to Springer Nature Switzerland AG 2025

C. Cantagallo et al., *Digitization of Built Heritage*, Digital Innovations in Architecture, Engineering and Construction, https://doi.org/10.1007/978-3-031-96598-2_2

system that has since revolutionized the industry. BIM optimizes productivity and quality in building and infrastructure projects by incorporating a 3D object-oriented model that contains embedded data accessible to all stakeholders at every project phase, from design through to maintenance. In this digital ecosystem, each participant maintains control over their data, yet project managers benefit from a consolidated data source to enhance decision-making processes. The adoption of BIM is expected to streamline resource utilization and bolster communication and collaboration across design teams, leading to enhanced flexibility and strategic foresight in project management—advantages that ultimately benefit clients, project managers, and contractors alike. BIM also facilitates the identification and resolution of potential design and construction conflicts, supports detailed maintenance schedules, and enhances property and asset management through performance-based contracting.

Puolitaival et al. (2018) provide a deeper exploration into the evolution of the digitalisation in the construction management sector through a systematic literature review of the articles published in construction specific journals containing the phrase 'digital technology'. The study reveals that digital technology, as a concept in construction, didn't gain significant traction until the 1980s, with a more noticeable increase in academic discussions only starting in the 2000s. Most of the articles analysed in the study primarily discuss the advent and implications of Building Information Modeling (BIM), along with newer technologies like digital imaging, augmented and mixed reality, monitoring and control technology and digital data processing. A limited number of articles instead analyse mobile solutions, internet, intranet, extranet, advanced internet (e.g. IoT, semantic web), automated and digital fabrication, digital project management, additive manufacturing or sensor technology.

Digitalization within the construction industry offers extensive benefits, particularly in improving energy efficiency throughout various building lifecycle phases and contributing to the overall reduction of carbon emissions in buildings. When integrated with innovations like mobile devices, 3D cameras, cloud technology, and IoT, BIM helps dissolve traditional barriers between different phases of construction, centering data at the heart of the construction process to minimize risks and enhance operational predictability. These benefits can be summarized in the following list and in Fig. 2.1.

- **Increased Efficiency**: Automation and digital tools streamline various construction processes, reducing the time and labour required for project completion.
- **Improved Accuracy and Quality**: Digital technologies, like BIM, enhance the accuracy of construction drawings and specifications, leading to fewer errors and reworks.
- **Enhanced Collaboration**: Cloud-based platforms enable seamless communication and data sharing among stakeholders, regardless of their location, which improves project coordination.
- **Cost Reduction**: Efficient resource management and reduced waste contribute to significant cost savings over the project lifecycle.

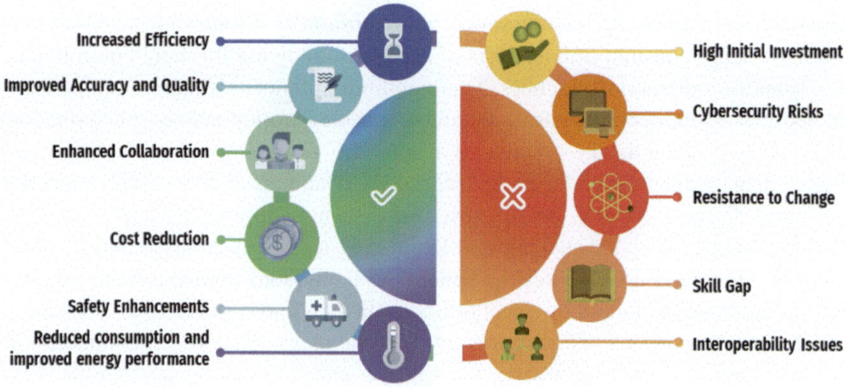

Fig. 2.1 Benefits and limits of the digitalisation in the construction sector

- **Safety Enhancements**: Digitization provides better monitoring and management of safety practices, reducing the risk of accidents on site.
- **Reduced consumption and improved energy performance**: Digitization enhances sustainability by enabling advanced simulations and analyses of building energy consumption, which facilitates more eco-friendly construction methods.

However, there are also limits to this process of change, which future research must try to overcome. The following list and Fig. 2.1 summarize some of them.

- **High Initial Investment**: Implementing digital technologies requires substantial upfront investment in software, training, and system integration.
- **Cybersecurity Risks**: Increased use of digital tools exposes construction projects to potential cybersecurity threats, which can compromise sensitive data.
- **Resistance to Change**: The construction industry often exhibits a hesitancy to adopt new technologies due to cultural resistance and the traditional nature of the business.
- **Skill Gap**: There is a significant skill gap as current workforce may lack the digital skills required to operate new technologies effectively.
- **Interoperability Issues**: Different construction software systems often have compatibility issues, which can hinder the seamless exchange of data.

Literature studies underscore the benefits and the transformative potential of digitalization in the construction sector, suggesting that continued technological advancements will be pivotal in shaping more sustainable, efficient, and collaborative construction practices. For example, Erdogan et al. (2009) explore a future-oriented approach to construction, using scenario planning to predict how technological advancements could shape the construction industry and bring benefits by 2030. Research predicts a significant evolution in the way information is managed and shared, transitioning the industry from a document-driven to a **model-driven**

approach. Moreover, digitization supports **sustainability** through advanced simulation and energy consumption analysis of buildings, allowing for more environmentally friendly construction practices. This paradigm shift not only increases efficiency but also fosters more collaborative and interconnected work practices. Such integration reflects in the industry's increased adaptability to market needs and improved change management capabilities. As technology continues to evolve, it is expected that these digital tools will become even more central to construction processes, enabling smarter decision-making and facilitating a more responsive and resilient industry. This trend underscores the benefits of continuous innovation and adaptation in construction practices to meet future challenges and opportunities. Shen et al. (2010) complement this vision by reviewing systems integration and collaboration within the architecture, engineering, construction, and facilities management sectors. Their analysis highlights the critical role of web-based technologies and Building Information Modelling (BIM) in enhancing project management and operational efficiency. They advocate for further research into integration technologies that span the complete lifecycle of construction projects, emphasizing that future developments should continue to focus on creating a more seamless and collaborative construction industry environment.

The European Commission actively supports digitalization in the construction sector by funding for research projects like HIT2GAP (https://cordis.europa.eu/project/id/680708) and Built2Spec (https://built2spec-project.eu/), which focus on innovating construction practices for enhanced quality control and compliance with energy efficiency standards. Moreover, projects like SPHERE (https://sphere-project.eu/) and BIM-SPEED (https://www.bim-speed.eu/en) further highlight efforts to advance building stock performance and integrate digital tools across the construction sector, encouraging inclusive and collaborative approaches to innovation that align with the Open Innovation 2.0 paradigm, fostering broad structural changes across the industry.

2.2 Definition of Digital Heritage

The concept of "digital heritage," as recognized by UNESCO in its 2003 "Charter on the Preservation of Digital Heritage," defines digital heritage as **computer-based materials of enduring value, originating from diverse communities and sectors, and necessitating active preservation to maintain continuity**. This foundational definition underscores the evolving importance of digital heritage in the preservation and appreciation of global cultural resources. The agreement on this definition is critical as it sets the standard for how digital heritage is approached and managed worldwide.

From the early applications in the 1990s, digital technology has increasingly been adopted for the documentation, analysis, and dissemination of cultural heritage. This has achieved remarkable results in enhancing the accessibility and understanding of heritage resources. By the early 2010s, international acknowledgment of digital heritage was evident in symposiums and conferences organized by institutions like UNESCO. Parallel to the broader concept of digital heritage, **Historic Building Digitisation** focuses on the conversion of data from physical structures into digital formats. This process facilitates the detailed documentation, interpretation, and public access of historical buildings. Technologies such as 3D laser scanning and photogrammetry have revolutionized how historic buildings are preserved and experienced. As noted by McGibbon (2020), this process is crucial for digitally documenting and visually accessing these structures, enhancing both academic research and public engagement.

The current scope of work in digital cultural heritage can be broadly categorized into three distinct yet interrelated segments, each retaining a focus that could be described as aspects of "mechanical" digitization: digital documentation, research management, and display/visualization along with interpretation.

1. **Digital Documentation**. This facet involves advanced recording technologies such as remote sensing, aerial radar, photogrammetry, and image-based modelling, 3D laser scanning, and infrared/multispectral image exploration. The practitioners in this area typically have specialized training in surveying and mapping disciplines. These technologies enable the precise capture and recreation of cultural artifacts and sites, ensuring that every detail is preserved digitally for future generations.
2. **Research Management**. The research management aspect focuses on the application of digital management technologies to safeguard cultural heritage. This includes disaster prevention, ongoing monitoring and protection, and the utilization of computer simulations and analysis databases. Information management platforms such as Building Information Modelling (BIM) are central to this effort, providing robust tools for managing complex data and enhancing the preservation strategies. Professionals in this segment often have a strong background in computer graphics and data management, enabling them to innovate and improve preservation methodologies through technology.
3. **Display, Visualization, and Interpretation**. This segment leverages digital technologies to create engaging and interactive experiences for educational and cultural dissemination purposes. Techniques involved include website creation for interactive networks, 3D modelling, processing and animation, as well as innovative imaging solutions like light projections and holographic displays. Virtual and augmented reality technologies play a crucial role in bringing cultural heritage to life, allowing people to experience history in immersive and interactive ways. Specialists in multimedia design and development spearhead these initiatives, blending creativity with technological expertise to expand the ways in which cultural narratives are shared and understood.

Each of these segments, while functioning independently with its own set of experts and technological tools, contributes to a cohesive whole, enhancing the preservation, management, and dissemination of cultural heritage in the digital realm. This integrative approach not only safeguards heritage but also makes it accessible and relatable to a global audience, bridging the gap between the past and the future through innovative digital practices.

2.3 Storage, Digitization, Organization, and Processing of Building Heritage Data

In today's digital era, the preservation and management of built heritage demand innovative solutions that leverage the capabilities of modern information technology. **Built heritage**, encompassing historical buildings, monuments, and archaeological sites, represents a vital link to our past. However, effectively managing the vast amounts of data associated with these assets poses significant challenges. To address this, the development of a **specialized IT platform is essential**. This platform will be meticulously designed to collect, organize, and process information from various sources, ensuring comprehensive coverage and accessibility of heritage data.

The methodology for developing such a platform involves a multi-disciplinary approach that integrates advanced technologies with the principles of heritage conservation. This book proposes a structured process to implement this platform, that includes the identification of data sources, the deployment of data collection tools, the establishment of a robust data management system, and the implementation of sophisticated data processing and analysis techniques. By adhering to this methodology, the **IT platform** aims to preserve valuable historical information, provide a comprehensive digital repository for heritage management, and enhance the accessibility and usability of this data for researchers, policymakers, and the public at large. Moreover, the proposed IT platform for the built heritage is pivotal for informed decision-making in heritage conservation, providing a dynamic tool that adapts to the evolving needs of heritage management.

The construction of the comprehensive digital repository for heritage management should be carried out according to the following steps:

- **Data Integration into a Heritage Data Lake**. The initial stage in building a comprehensive digital repository for heritage management involves the collection of data into a data lake. A data lake is a centralized repository that allows for the storage of all types of data, both structured and unstructured, at any scale. In the context of built heritage, this data can range from high-resolution images and 3D scans of structures to historical documents, and metadata about each site.
- **Collection and Aggregation**. Data is collected from a variety of sources. These include digital archives, sensor outputs from monitoring devices, satellite and aerial imagery, and direct inputs from field research such as photographs, notes, and video recordings. Modern tools like drones and LiDAR (Light Detection and Ranging) are particularly useful for capturing detailed spatial data of inaccessible areas. Once collected, data is aggregated into the data lake, providing a raw, but comprehensive dataset that serves as the foundation for further analysis and processing.

- **Parameterization and Preliminary Organization**. After the data is stored in the data lake, it undergoes initial processing to parameterize and categorize the information based on predefined criteria. For instance, data might be tagged with metadata such as the date of acquisition, type of data (e.g., visual, textual, spatial), source, and geographical coordinates. This parameterization facilitates the initial organization of the data, making it easier to retrieve and manipulate for specific purposes.
- **Transfer to a Structured Database**. Following parameterization, selected datasets are transferred from the data lake to a more structured database environment. This transition involves a transformation process, where data is normalized and schematized according to the database's schema. This structured database allows for efficient querying and more complex analyses. It is designed to support ongoing heritage management tasks, enabling quick access to organized data and facilitating integration with other digital tools and platforms.
- **Processing and Analysis**. In the database, data can be further processed using advanced data analysis techniques such as machine learning algorithms to identify patterns, predict structural issues, or automate the classification of heritage elements. These processes enhance the data's usability, turning raw data into actionable insights that can inform conservation policies and practices.
- **Implementation of Data Governance**. Data governance policies are crucial at this stage to ensure data integrity, security, and compliance with international standards and local regulations. Establishing clear guidelines on data usage, access controls, and regular audits helps in maintaining the reliability and confidentiality of sensitive heritage data.

By leveraging a data lake for initial data collection and a structured database for organized storage and analysis, heritage managers can create a dynamic and scalable digital ecosystem. This system not only preserves invaluable historical information but also enhances the accessibility and practical utility of the data for various stakeholders involved in heritage conservation.

An example of the application of the methodologies for creating a Digital Identity Card of a heritage building is provided in Fig. 2.2. This Digital Identity Card integrates various types of data, including technical features, risk assessments, and proposed mitigation strategies. The visual representation offers a clear and concise summary of the key information gathered during the documentation and analysis phases, outlining the building's current condition and the necessary interventions for its preservation.

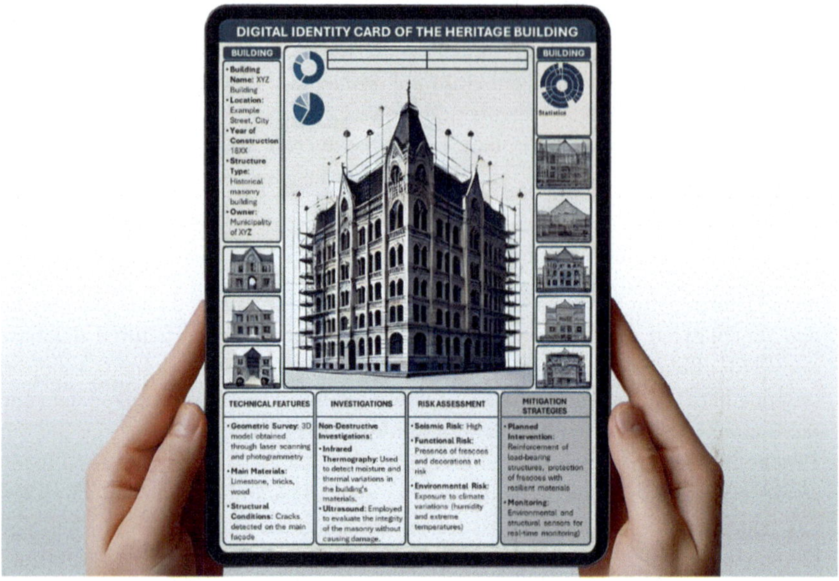

Fig. 2.2 Digital Identity Card for the heritage building. This figure consolidates essential information such as the building's name, location, construction details, technical features, risk assessments, and planned mitigation strategies. It serves as a comprehensive tool for heritage conservation and risk management

References

Erdogan B, Abbott C, Aouad G, Kazi AS (Sami) (2009) Construction IT in 2030: a scenario planning approach. J Inf Technol Constr 14

McGibbon S (2020) A Common Structured Integrated Collaborative Digitised (CrOsS) Framework for the Historic Building Repair and Maintenance (R&M) Sector. PhD Thesis, Heriot-Watt University

Puolitaival T, Kestle L, Kähkönen K (2018) What's the real story around digital technologies in construction management? In: Do K, Sutrisna M, Hammad A, Ramanayaka CE (eds) 42nd Australasian Universities Building Education Association (AUBEA) conference 2018: educating building professionals for the future of the globalised world. Singapore, pp 251–258

Shen W, Hao Q, Mak H, Neelamkavil J, Xie H, Dickinson J, Thomas R, Pardasani A, Xue H (2010) Systems integration and collaboration in architecture, engineering, construction, and facilities management: a review. Adv Eng Inform 24:196–207. https://doi.org/10.1016/j.aei.2009.09.001

Chapter 3
Knowledge of the Built Heritage

Abstract This chapter provides a comprehensive overview of the knowledge acquisition process for built heritage, structured through multiple levels of investigation. It begins by outlining the knowledge path, followed by the identification of the building and the geometric survey, comparing traditional techniques with reality data capture technologies such as photogrammetry, terrestrial laser scanning, and UAVs. The chapter then addresses the critical-historical analysis, including documentary research, interpretation, and the study of construction phases. Detailed attention is given to building techniques and construction details, as well as to material surveys, non-destructive investigation methods (including infrared thermography, endoscopy, and ultrasound), decay analysis, crack pattern analysis and foundation conditions. The final section is dedicated to Heritage Building Information Modelling (HBIM), presenting its evolution, components, and applications in conservation, restoration, facility management, and documentation. The chapter also discusses technical challenges related to data accuracy, interoperability, and heterogeneous data management in HBIM, while outlining future directions for integrating emerging technologies and sustainable conservation practices.

3.1 Knowledge Path

Knowledge of a historic building is a fundamental prerequisite for a reliable seismic safety assessment and consequently for the choice of an effective improvement intervention. The problems of a historic building are those common to all existing buildings. However, in the case of protected cultural heritage, it is even more important to know the original characteristics of the structure, the changes that have occurred over time due to the phenomena of damage resulting from anthropogenic transformations, the aging of materials and the calamitous events. Moreover, considering the need to prevent irreparable losses, the execution of a complete investigation campaign may be too invasive on the factory itself.

© The Author(s), under exclusive license to Springer Nature Switzerland AG 2025
C. Cantagallo et al., *Digitization of Built Heritage*, Digital Innovations in Architecture, Engineering and Construction, https://doi.org/10.1007/978-3-031-96598-2_3

Therefore, there is a need to refine techniques of analysis and interpretation of historical artifacts through cognitive phases with different degrees of reliability, also in relation to their impact. The knowledge can be achieved with different levels of depth, depending on the accuracy of survey operations, historical research, and experimental investigations.

These operations will be a function of the objectives set and will affect all or part of the building, depending on the type of intervention planned. The study of the characteristics of the building aims to define an interpretative model that, in the different stages of its calibration, enables both a qualitative interpretation of structural functioning and structural analysis for a quantitative evaluation. The degree of reliability of the model will be closely related to the level of depth and the available data. From this point of view, different levels of knowledge are defined, at increasing depth, to which will be linked confidence factors to be used in the analysis aimed at both the assessment of the current state and following possible interventions.

The knowledge path consists of the following activities:

- **identification of the building**, its location in relation to particular areas at risk, and its relationship to the surrounding urban context; the analysis consists of an initial schematic survey of the building and the identification of any valuable elements (fixed decorative apparatus, movable artistic assets) that may condition the level of risk;
- **geometric survey** of the building in its current state, characterized by a complete stereometric description of the building, including any cracking and deformation phenomena;
- identification of the **building's evolution**, understood as the sequence of phases in its transformation, from the hypothetical original configuration to the present;
- identification of the **constituent elements** of the load-bearing structure, in both material and construction aspects, with special attention paid to **construction techniques**, construction **details**, and the connections between the elements;
- identification of **materials**, their state of degradation, and their mechanical properties;
- knowledge of the **subsoil and foundation structures**, with reference also to the variations occurring over time and related disruptions.

3.2 Identification of the Building

The first step of knowledge consists in the correct and complete identification of the building and its location on the territory, in order to identify the sensitivity of the factory with regard to the different risks and in particular the seismic one. This phase of the analysis also includes an initial schematic survey of the artifact, or a simple eidotype, capable of describing its general consistency and the location of any elements particularly sensitive to damage that, in the case of loss, would lead to irreparable damage to the cultural heritage, conditioning its level of risk.

As an example, a historic building located in San Martino sulla Marrucina (CH), also known as the ‘Episcopal Palace,’ is identified below. A schematic survey of the building is presented, aimed at describing its general condition and location.

The building is situated within the historic centre, not far from the parish church of San Cristinziano (Fig. 3.1a). Over the centuries, the building has undergone numerous additions and renovations and is currently in a state of evident decay. On the ground floor, there is a workshop for aluminium frame production, while the upper floor, which constitutes the attic, serves as a storage space. The main facade (Fig. 3.1b) is characterized by mixed masonry of stone and brick, with significant portions covered by plaster. On the left side, there is a buttress, likely added for structural reasons. Higher up, there is a portal (Fig. 3.1c) consisting of an arch supported by pilasters. The keystone of the arch has been replaced with bricks, as have parts of the pilasters. A visual interpretation of the facade suggests that there was once an external staircase leading to this portal, now evidenced by a brick protrusion at the base.

To the right of this portal, there is a window (Fig. 3.2a) surrounded on three sides by three bands, above which there is a Latin inscription (Fig. 3.2b) that was likely intended to be framed.

The side facades are extremely simple: one is entirely plastered (Fig. 3.3a), while the other is half plastered and half composed of relatively recent bricks, dominated by a large buttress (Fig. 3.3b).

The rear facade (Fig. 3.4) sits at a lower level than the main facade and features a combination of stone blocks, brick sections, and areas covered in plaster.

The interior of the ground floor features rooms with exposed masonry. The rooms are covered by barrel vaults with lunettes (Fig. 3.5a), cross vaults (Fig. 3.5b), or barrel vaults.

The interior of the upper floor, like the lower floor, consists of four rooms, but the recent interventions are even more evident here (Fig. 3.6). Additionally, numerous cracks and several newly created openings can be observed. In three rooms, traces of an ancient pavilion vault are present.

Fig. 3.1 Location and identification of the Episcopal Palace by means of the 2003 General Regulatory Plan (**a**, top) and 1996 Cadastral Map (**a**, bottom) and the images of main façade (**b**) and the detail of its portal (**c**)

(a) (b)

Fig. 3.2 Windows on the main façade of the Episcopal Palace, with the parastas and Latin inscription

(a) (b)

Fig. 3.3 Side facades of the Episcopal Palace

(a) (b)

Fig. 3.4 Rear façade of the Episcopal Palace

Before undertaking the delicate phase of recording dimensional information (geometric survey), it was necessary to dedicate sufficient time to drafting eidotypes or survey sketches of the Episcopal Palace in San Martino sulla Marrucina (CH, Italy), which would later be used to systematically record and organize direct

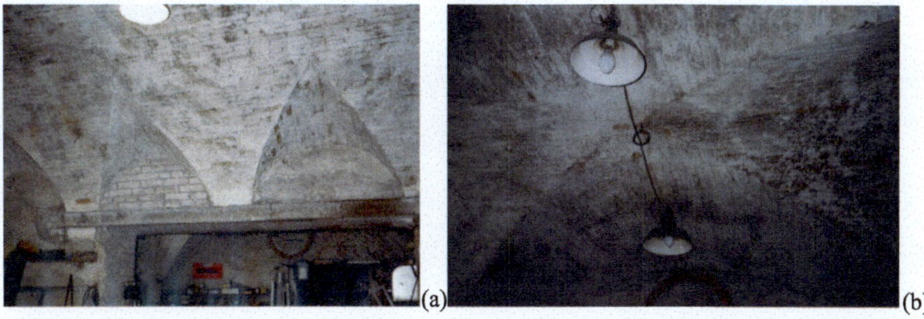

Fig. 3.5 Rooms of the ground floor covered by (a) barrel vaults with lunettes (b) and cross vaults

Fig. 3.6 The interior rooms on the first floor of the Episcopal Palace

measurements. The purpose of the survey eidotype is to create a preliminary graphical representation of the object to be surveyed, accurately depicting its dimensions and shape. This initial synthesis helps the surveyor approach the final graphical outcome.

The eidotypes of the Episcopal Palace in San Martino sulla Marrucina (CH) were drafted in both plan (Fig. 3.7) and elevation (Fig. 3.8). Following the initial site inspections, the key elements were identified, and simplified drawings were created, capturing the essential aspects without distorting proportions or geometry. The eidotypes were partially produced on-site and later in the studio, using photographs and preliminary dimensional data to appropriately record the necessary information. This approach facilitated a more informed and comprehensive subsequent direct survey, allowing for a rapid and critical understanding of the structure to be surveyed. By the end of this phase, a complete graphical overview of the objects to be surveyed had been outlined, with an evaluation of the number of measurements required in relation to the final graphical objectives.

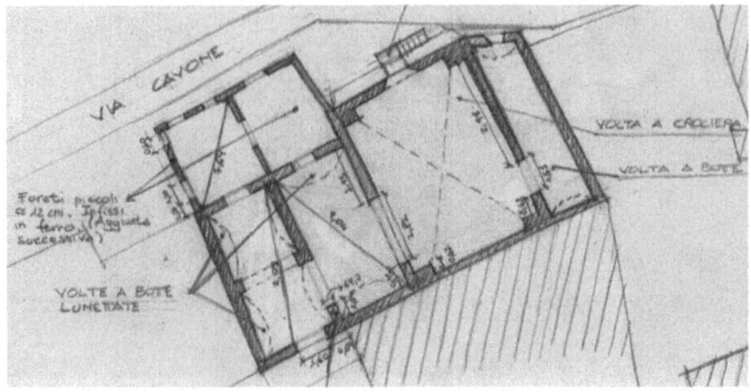

Fig. 3.7 Eidotype of the ground floor plan

Fig. 3.8 Eidotype of the main facade

Digitization Notes

The establishment of a digital database of built heritage not only facilitates the allocation of resources and the planning of interventions but also enhances collaboration across disciplines. For public administrations, access to this data enables more informed decision-making in terms of budgeting, scheduling maintenance, and prioritizing interventions. For professionals, it provides an essential reference point for ongoing research, restoration, and conservation activities, ultimately leading to more sustainable and effective management of cultural heritage assets. As for the "Identification of the building" step, this process can be carried out as follows:

Digitization and Cataloguing of Sketches and Eidotypes. All sketches and eidotypes created during the initial identification phase of the building must be systematically digitized and catalogued in standardized formats such as TIFF or PNG for raster images, and SVG or DXF for vector files. These formats ensure high resolution and scalability, preserving the quality of the original material for future retrieval and comparison, especially when updates or revisions are required over time. This process ensures that all graphical representations of the building are preserved in a structured database for future generations.

Photographic Survey for Data Collection and Digitalization. Alongside the sketches and eidotypes, the photographic survey plays a critical role in the comprehensive documentation of the building. High-resolution images of the façade, structural details, and interior spaces should be archived in lossless formats such as TIFF or RAW to maintain maximum image quality. Additionally, JPEG format may be used for more practical storage and faster retrieval in day-to-day operations. A well-organized digital archive of photographs, categorized by date, location, and content, can significantly improve the accuracy of future surveys and interventions, while also providing future generations with an invaluable record of the building's current condition.

3.3 Geometric Survey

3.3.1 General Aspects

Accurately capturing the geometry of a built heritage structure is fundamental to understanding its current condition, planning conservation interventions, and managing ongoing maintenance. A comprehensive geometric survey provides the precise spatial data required to create detailed models that serve as the basis for analysis, documentation, and restoration efforts.

A stereometric description of the building involves identifying the planimetric and altimetric characteristics of its constituent elements. Therefore, at each level, the geometry of all masonry elements, vaults (including thickness and profile), floors and roof structures (type and framework), and staircases (structural type) must be documented. The location of any niches, cavities, sealed openings (along with the methods used), chimneys, and foreign elements should also be recorded.

The survey results are represented through plans, elevations, and sections, along with detailed construction drawings. Three-dimensional representation of the structure may be complex, but it is certainly valuable for a more comprehensive survey. Once this process is completed, the structural framework is identified within the overall geometric survey. Since the geometric survey can also be used to define the geometry of the model for structural analysis, including constraints and acting loads, it is essential to ensure that all necessary information has been captured. Specifically, key points for a computational model must be identified, such as the position of floor supports and vaulted systems, and the extent of their bearing on the surrounding walls. Additionally, the masses of the elements and the loads applied to each wall segment must be thoroughly determined.

Challenges in conducting geometric surveys often arise due to the inaccessibility of certain spaces, such as attics, voids between false ceilings or vaults and roofs, or due to the excessive height of certain elements, such as bell towers, towers, or vaults in a nave. However, tools are available that allow for rapid surveying and accurate documentation, even for complex elements, as well as direct (endoscopy) or indirect (thermography, ground-penetrating radar, etc.) investigation techniques for inaccessible spaces.

Surveying and documenting any existing crack patterns aids in identifying the causes and potential developments of structural problems within the building. One or more damage mechanisms can be associated with each crack, or set of cracks, compatibly with the building's geometry, the alterations it has undergone, the materials present, and the events it has experienced.

3.3.2 *Traditional Technologies*

Traditional geometric surveying, also known as direct surveying, is a hands-on approach where the operator uses simple measuring and alignment tools (such as measuring tapes, meter sticks, plumb lines, levels, etc.) to take precise measurements in direct contact with the structure being documented. This method is particularly valuable for its simplicity and the tactile connection it provides between the surveyor and the building.

During the survey, measurements are carefully recorded on a simple sheet of paper, which also includes a rough sketch of the building, known as an eidotype. This sketch serves as a preliminary representation of the structure, capturing essential details and proportions. The eidotype is a crucial tool in the early stages of documentation, as it allows the surveyor to visualize the building's geometry in a straightforward manner.

Later, the information gathered from the direct survey and the eidotype can be transformed into a more accurate and detailed representation. This process involves refining the rough sketch into precise plans, elevations, and sections that fully reflect the building's dimensions and structural characteristics. Despite the advances in digital surveying technologies, traditional geometric surveying remains a fundamental practice, especially in cases where direct engagement with the structure is necessary for understanding its unique features and historical context.

As an example, to illustrate the knowledge acquisition process for a single building, the figures shown below present some graphic renderings of the geometric survey conducted on the same building analyzed in the previous section, the Episcopal Palace of San Martino sulla Marrucina (CH). Specifically, Figs. 3.9 and 3.10 display the planimetric measurements obtained from trilateration for the ground floor and first floor, respectively.

Trilateration is a surveying method used to determine the exact positions of points by measuring the distances between those points and known reference locations. Unlike triangulation, which involves measuring angles, trilateration relies solely on distance measurements. This method proves especially useful in complex environments or in cases where direct access to parts of the structure is challenging, as it allows for precise measurements even in hard-to-reach areas. The process typically begins with the selection of fixed reference points, also known as "stations," which are strategically placed in known positions within or around the structure. From these fixed points, measurements are taken to determine the distances to the unknown points of interest. By measuring the length of these lines (distances) from the known reference points to the unknown points, a geometric network of lines is formed. The intersection of these lines, when calculated accurately, reveals the exact position of the unknown points, creating a highly precise geometric representation of the surveyed area.

In Fig. 3.9 (Ground Floor Survey), the lines represent the trilateration measurements, showing the distances from key reference points within the building to the walls and structural elements. The network of trilateration lines creates an accurate representation of the layout of the ground floor, including the geometry of the rooms and their relative positioning. Similarly, Fig. 3.10 (First Floor Survey) employs trilateration to capture the precise dimensions of the first floor. The lines

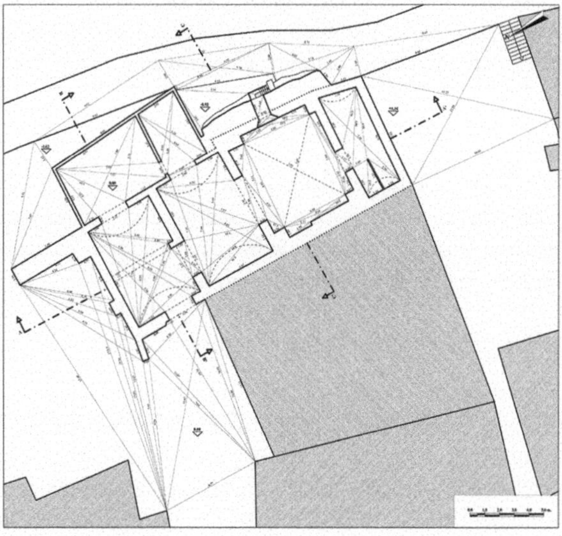

Fig. 3.9 Geometric survey of the Episcopal Palace of San Martino sulla Marrucina (CH)—Ground Floor Survey with the indications of the measurements obtained by trilateration

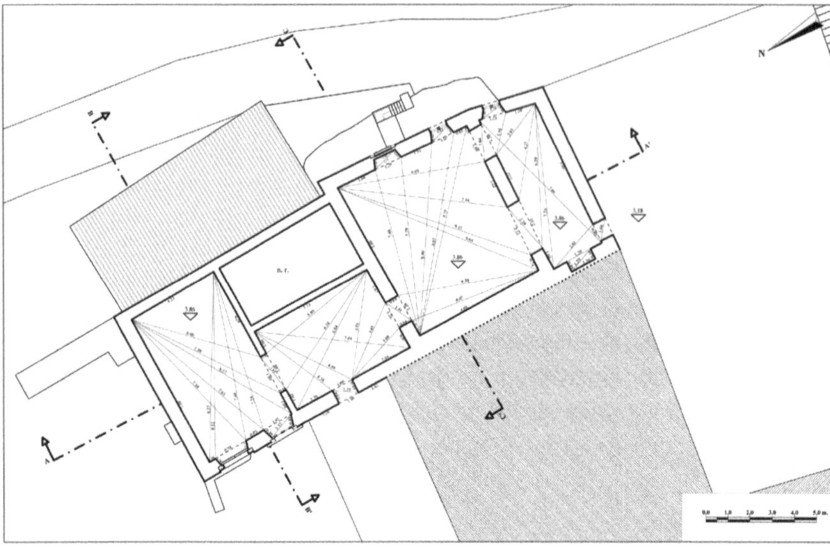

Fig. 3.10 Geometric survey of the Episcopal Palace of San Martino sulla Marrucina (CH)—First Floor Survey with the indications of the measurements obtained by trilateration

drawn between the reference points and various parts of the structure ensure that the more complex spatial relationships are also captured with high accuracy. This method allows for detailed geometric documentation of the first floor's layout, ensuring that even inaccessible areas or irregular angles are accurately represented.

Figures 3.11 and 3.12 show the longitudinal and transverse sections, respectively, while Figs. 3.13 and 3.14 depict the main and rear facades of the building, offering a comprehensive view of the Episcopal Palace's geometry.

Digitization Notes

The digitization of data obtained through traditional geometric surveys is essential for integrating manually collected information into modern conservation practices. Although traditional methods involve hands-on measurement and sketching, digitization enhances precision and facilitates the use of this data in computational models. The following steps outline the key actions required to digitize the geometric survey data obtained through traditional techniques:

- **Digital Conversion of Traditional Measurements**. Manual measurements taken during the survey should be digitized and converted into a precise, scalable format. This can be done using CAD software, which allows for the recreation of detailed plans, elevations, and sections. Digitizing these measurements preserves the precision of the traditional method while making the data usable for computational analysis and future updates.

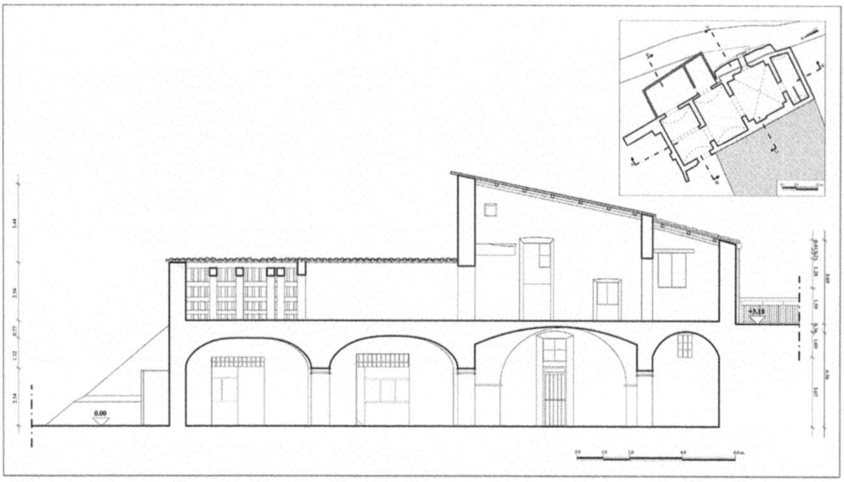

Fig. 3.11 Geometric survey of the Episcopal Palace of San Martino sulla Marrucina (CH)—Longitudinal section (A-A')

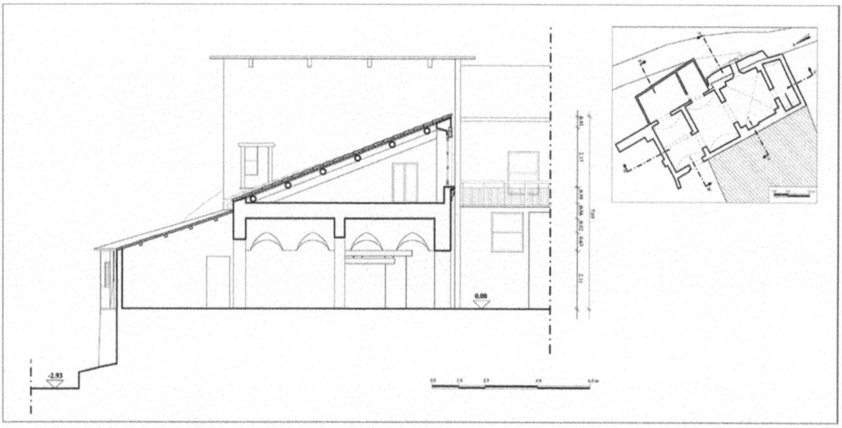

Fig. 3.12 Geometric survey of the Episcopal Palace of San Martino sulla Marrucina (CH)—Transversal section (B-B')

Common file formats for this step include **.dwg** (for AutoCAD), **.dxf**, and **.skp** (for SketchUp).

- **Integration of Geometric Data into 2D and 3D Digital Models**. The manually collected geometric data can be used to create both 2D and 3D digital models of the building. These models, built from traditional measurements, allow for more comprehensive analyses, including structural assessments. A precise digital model helps in visualizing the

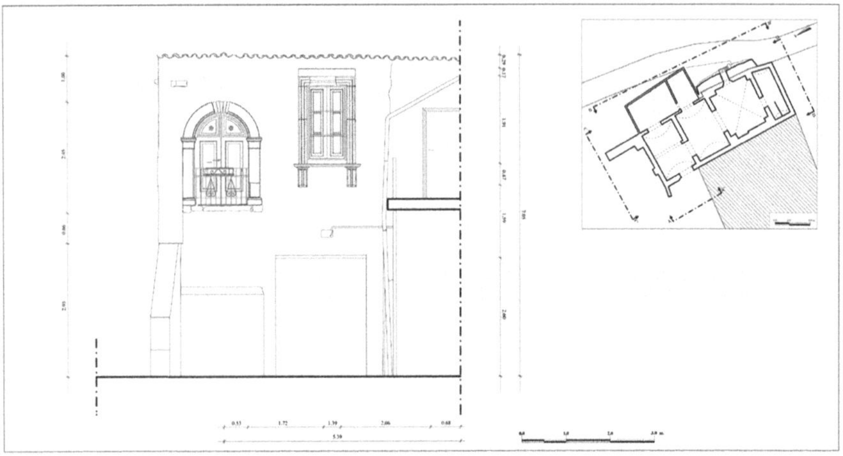

Fig. 3.13 Geometric survey of the Episcopal Palace of San Martino sulla Marrucina (CH)—Main Facade

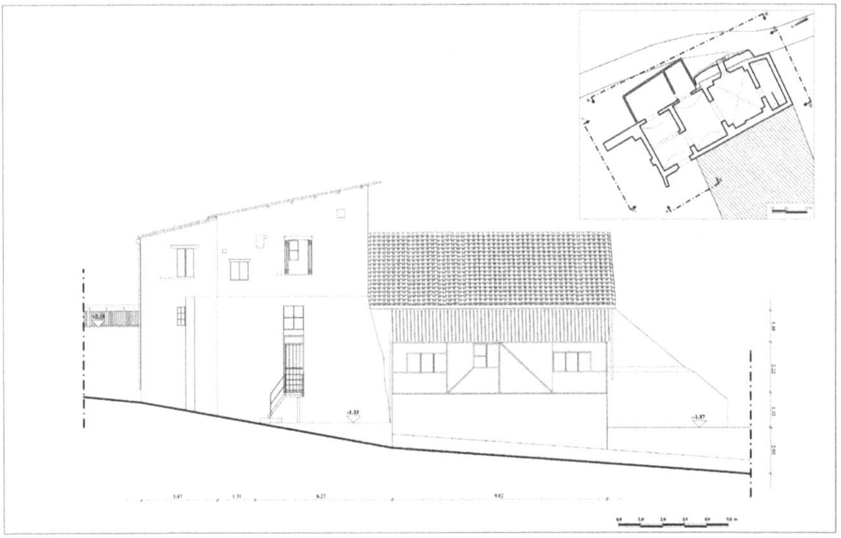

Fig. 3.14 Geometric survey of the Episcopal Palace of San Martino sulla Marrucina (CH)—Rear Facade

structural elements and in planning conservation interventions, ensuring that the unique features of the building are preserved in a digital format. Typical formats for 3D models include **.obj**, **.stl**, and **.3ds** for broader compatibility with various modelling software. The most commonly used free and commercial photogrammetry software are the following: OpenDroneMap, Regard3D, Meshroom, COLMAP, VisualSFM + MeshRecon, VisualSFM + PMVS, VisualSFM + OpenMVS, OpenMVG + MVS, OpenMVG + MVE, Regard3D + MVE, MVE + SMVS, MVE, MicMac, 3DF Zephyr, Recap Photo, Agisoft Metashape, and Reality Capture.

3.3.3 Reality Data Capture Technologies

In the modern context, traditional surveying methods have been significantly enhanced by the advent of advanced reality data capture technologies. These technologies offer unprecedented accuracy and efficiency in collecting geometric data, ensuring that even the most intricate details of a structure are captured and documented. As such, these tools are now considered essential for conducting thorough geometric surveys of heritage buildings.

In the context of historic building repair and maintenance, reality capture refers to the acquisition of precise and accurate data that reflect the current condition of a building (McGibbon 2020). This is a crucial first step in any restoration or maintenance project, as it provides the foundational "as-built" documentation needed for informed decision-making (Bosche et al. 2015). Reality Data Capture technologies, such as 3D laser scanning, high-resolution photogrammetry, and unmanned aerial vehicles (UAVs), offer innovative methods for documenting, mapping, and analyzing historic building information. These technologies bring numerous benefits to project management and on-site practices, including improved site logistics, enhanced quality control, and more effective progress monitoring.

Choosing the appropriate reality capture technology can be challenging, particularly as the selection process is influenced by project timelines and budget constraints. Different technologies may be more suitable depending on the specific requirements of the project (Valero et al. 2018). Therefore, it is essential to have not only a thorough understanding of industry practices but also a deep knowledge of the capabilities and limitations of each technology to ensure the most appropriate digital tools are applied (Forster et al. 2017).

However, it is important to recognize that no single technology can fulfil all the needs of a repair and maintenance project. The digitization of the surveying process involves a spectrum of complexity, ranging from basic data capture using structured documents to the use of advanced tools like 3D laser scanners and UAVs equipped with infrared cameras. These technologies enable the creation of accurate and intelligent 3D models that are integral to effective maintenance management (Arayici 2008).

Figure 3.15 illustrates the geometric survey process using reality capture technologies. It begins with the simplest methods, such as the documental survey and

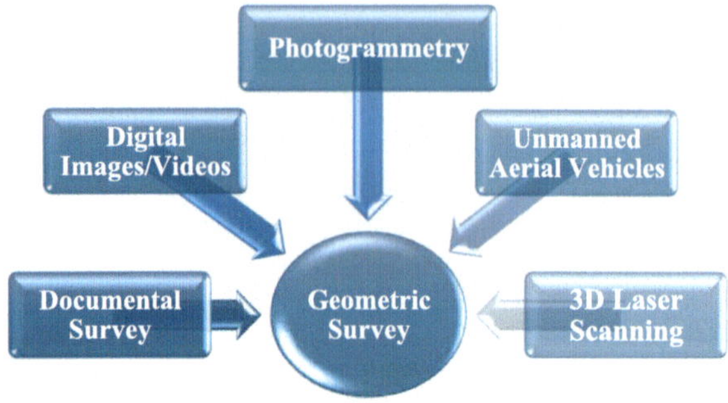

Fig. 3.15 The geometric survey process using the most used Reality Data Capture Technologies

the acquisition of photos and videos, and progresses through more sophisticated techniques, such as photogrammetry. Unmanned Aerial Vehicles (UAVs) extend the reach to areas difficult to access manually, and 3D laser scanning captures highly precise spatial information. The final step in this continuum is the integration of all these data sources into a comprehensive geometric survey, which supports the detailed documentation and management of the building's condition.

As highlighted by Bosche et al. (2015), the effectiveness of enhanced data capture relies heavily on the expertise of those utilizing the technology. Without the necessary knowledge and skills, the potential of these advanced tools may not be fully realized, underscoring the importance of ongoing professional development in this field.

3.3.3.1 Photogrammetry

Photogrammetry is an essential technique in the geometric survey of historic buildings, offering a versatile, cost-effective method for capturing detailed three-dimensional data. By capturing the geometry of a building's façade and architectural elements through a series of 2D digital images, photogrammetry enables the creation of a comprehensive 3D representation. This approach has been validated through numerous studies, which highlight its adaptability and precision in documenting buildings that require repair and maintenance (Stefani et al. 2014; Mitka and Pluta 2016). Photogrammetry's application extends beyond merely capturing the physical dimensions of a building. Studies such as those by Stefani et al. (2014) and Al-Omari et al. (2015) demonstrate its capability to integrate various types of data, such as material properties, degradation patterns, and historical information, into a single model. This integration is invaluable for creating a "**digital health record**" of a building, a tool that aids in the systematic documentation and monitoring of conservation efforts. For instance, in the conservation of Chambord Castle, a digital model was created that not only mapped the building's geometry but also included data on material degradation and historical repairs, enabling more informed and economically balanced decisions for future maintenance (Stefani et al. 2014).

As illustrated in Fig. 3.16, photogrammetry is applied to a historic statue where multiple digital cameras capture the statue from various angles. This network of visual data is processed to generate a detailed and precise 3D model, vital for advanced conservation analyses and the preservation of critical historical details.

Among the various methods employed in photogrammetry, the creation of **orthophotos** stands out for their precision and utility in surveying historic buildings. An orthophoto, or orthorectified photograph, is an **aerial photograph geometrically corrected** such that the scale is uniform: the photo has the same lack of distortion as a map. While UAVs equipped with advanced camera systems are instrumental in capturing aerial images, the photogrammetric processing transforms these images into accurate, top-down views of historic sites. For instance, the orthophotos of the "Throne Room" of the Melfi Castle (PZ, Italy) shown in Fig. 3.17 highlight how photogrammetry provides detailed visual documentation of heritage sites. These images are essential for conservation efforts, enabling precise measurements and condition assessments from perspectives previously difficult to achieve.

One of the significant advantages of photogrammetry is its accessibility and mobility, making it a practical choice for a wide range of scales and complexities. Techniques such as Structure from Motion (SfM), which construct 3D models from overlapping images taken from different viewpoints, have been particularly effective in this regard. These methods are not only low-cost but also require relatively simple equipment, making them ideal for use in various contexts, from detailed documentation of large architectural sites to smaller, intricate elements of historic buildings (Mitka and Pluta 2016).

Moreover, photogrammetry is increasingly being used in conjunction with other data capture technologies to enhance the accuracy and detail of the models produced. For example, combining photogrammetry with laser scanning allows for the creation of models that capture both the fine details and the broader structural form of a

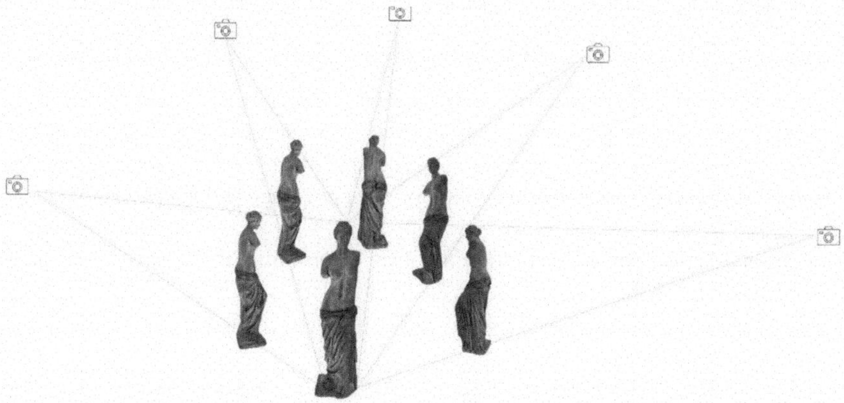

Fig. 3.16 Illustration of photogrammetry applied to a historic statue. Multiple digital cameras capture the statue from different angles, creating a network of visual inputs that are processed to generate a detailed and precise 3D model

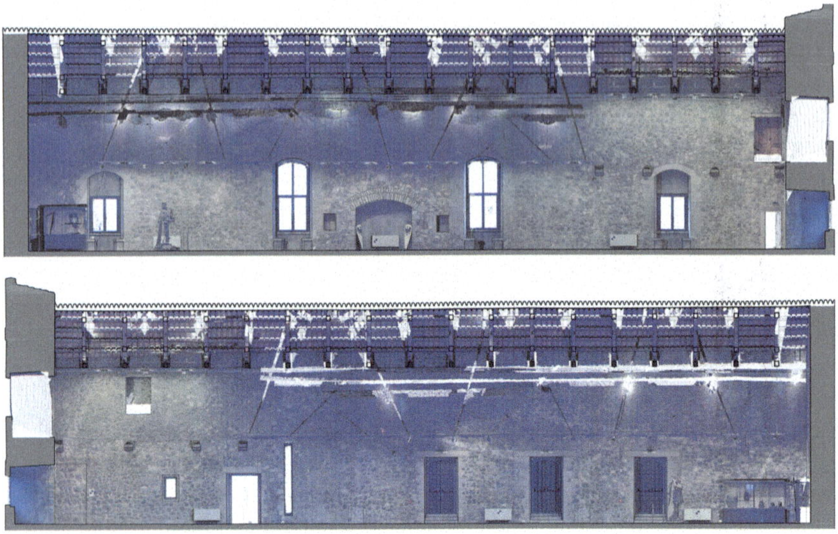

Fig. 3.17 Orthophotos of the "Throne Room" of the Melfi Castle (PZ, Italy). Longitudinal views. Images adapted from the Survey Documents of the "Final Project for Integrated Contract for the Museum Complex of Melfi-Venosa. Project for the Redevelopment and Functional Enhancement of the Castle of Melfi (PZ)" dated April 2012

building. This hybrid approach ensures that the 3D models produced are both geometrically accurate and rich in information, supporting a wide range of conservation and management activities (Brunetaud et al. 2012).

The process of creating a 3D model using photogrammetry involves several key stages (Fig. 3.18):

1. **Image Acquisition**: High-resolution photographs are taken from multiple angles around the subject, ensuring that all relevant surfaces are captured. The quality and number of images significantly impact the accuracy of the resulting model. For instance, in the case of cylindrical structures like towers, a convergent acquisition technique is often employed to optimize coverage and detail (Stefani et al. 2014).
2. **Camera Calibration**: Calibration involves determining the precise orientation and position of the camera during each shot. This step is crucial for aligning the images correctly in the 3D space, allowing for accurate reconstruction of the building's geometry.
3. **3D Modelling**: The core of photogrammetry is the generation of a 3D model by triangulating the position of points identified in multiple images. This model can be further refined and segmented into smaller architectural elements, allowing for detailed analysis and documentation.
4. **Texture Mapping**: Finally, textures from the original photographs are projected onto the 3D model, adding realistic surface details that enhance the visual quality

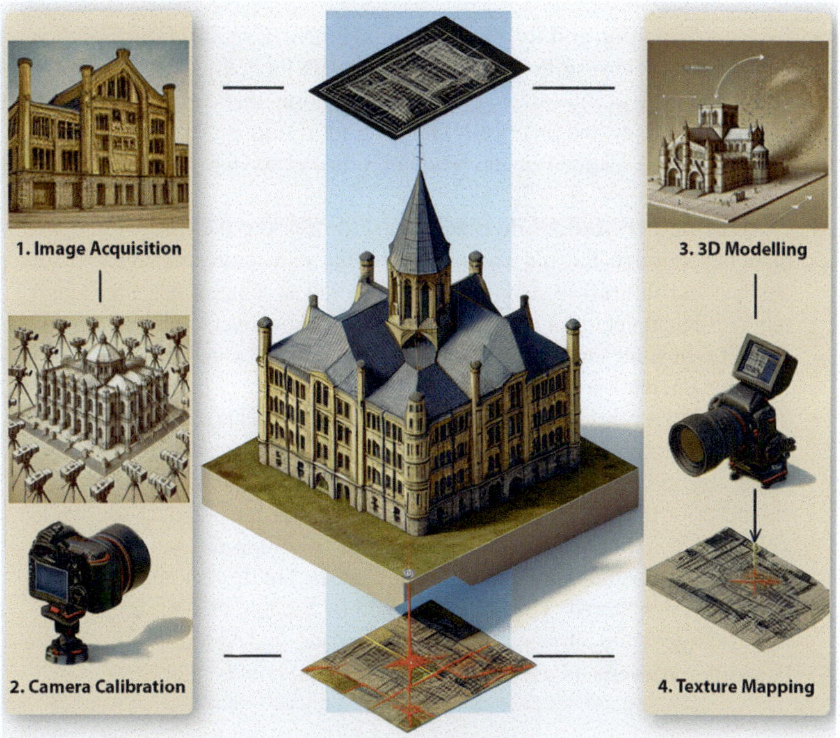

Fig. 3.18 Steps for creating a 3D model using photogrammetry

of the model and provide additional information about the material condition of the building.

As illustrated in Fig. 3.15, photogrammetry is part of a broader toolkit for reality data capture, which also includes technologies like 3D laser scanning and UAVs (drones). Each of these tools contributes unique strengths to the data capture process, and when used together, they provide a comprehensive, multi-faceted view of the building's condition. Photogrammetry, with its balance of accuracy, cost-effectiveness, and ease of use, is a cornerstone of modern heritage conservation practices, facilitating detailed, accurate documentation that informs all subsequent phases of building conservation and management.

Digitization Notes

Photogrammetry allows for the creation of detailed 3D models of heritage buildings, but managing file sizes and resolution is critical for efficient digital archiving. The following steps outline how to balance accuracy and storage efficiency in the digitalization process:

- **Image Acquisition and Resolution Management.** Capture high-resolution images that ensure sufficient detail for the 3D modelling process, while being mindful of file sizes. Select a resolution that strikes a balance between accuracy and practicality for long-term storage. Excessively large files should be avoided unless absolutely necessary for specific analysis purposes.
- **Optimizing 3D Model File Sizes.** When processing the images into 3D models, optimize the file sizes to make the data more manageable for storage and future use. It is important to find a compromise that maintains sufficient detail for conservation needs without overwhelming storage capacity. In some cases, a medium-resolution model may be sufficient for both documentation and analysis purposes.
- **Storing and Archiving.** Once the optimized 3D model is ready, it should be stored in a digital archive or IT platform with enough resolution to ensure future use in conservation, restoration, or research activities. The archive should prioritize a version that is detailed enough for practical use while keeping file sizes manageable. If necessary, very high-resolution files can be archived separately for specific future needs, but they should not be the default version stored. The archive should be capable of supporting long-term storage while allowing for periodic updates and additions and should be easily retrieved and used by professionals and public authorities.

3.3.3.2 Terrestrial Laser Scanning

Terrestrial Laser Scanning (TLS) is one of the most advanced technologies available for digital surveying, particularly in the field of historic building documentation and analysis. TLS is a form of remote sensing that utilizes laser technology to measure and capture the spatial geometry of a structure with high precision. This technology operates by emitting infrared laser light from a scanning device (Fig. 3.19), which then calculates and images the measurements of a surface's spatial relationships. The result is a collection of millions of spatially referenced points, known as "point clouds," which are used to create a detailed 3D digital representation of the scanned environment, building, or elevation (Kottke 2009; Smits 2011). The precision of TLS lies in its ability to capture the intricate details of surfaces, which is particularly crucial for historic buildings where even minor architectural details can be of significant cultural and historical importance. By providing a highly accurate representation of a structure, TLS enables comprehensive analysis and documentation, which are essential for effective conservation and restoration efforts.

The operation of TLS begins with the emission of a laser beam from the scanning instrument. This laser beam is projected onto the surface of the structure, and the time it takes for the beam to reflect back to the scanner is measured. This measurement allows the device to calculate the exact distance between the scanner and each point

Fig. 3.19 Example of a Terrestrial Laser Scanner used for surveying a cultural heritage buildings

on the surface of the structure. By repeating this process for millions of points across the surface, TLS generates a detailed map of the structure in the form of a 3D point cloud. The resolution of the point cloud is a key aspect of TLS, determined by the **density of the points** captured. A higher resolution point cloud, where points are closer together, provides more detailed surface information (Kottke 2009). A high level of detail is particularly beneficial when surveying historic buildings, as it allows for the accurate capture of complex surfaces and fine architectural details. Modern 3D laser scanning technology can capture hundreds of millions of points with single-point accuracy within millimetres in just a few minutes (Valero et al. 2018). This capability is crucial for preserving the exact details of historic structures, which can be invaluable for future restoration and maintenance efforts.

One of the primary advantages of TLS is its ability to generate **complete 3D models** that can be used for a variety of purposes, including structural analysis, building condition monitoring, and long-term documentation. These models provide a precise record of the building's current condition, which can be used to assess the need for repairs, plan restoration projects, and monitor changes over time (Smits 2011). In the context of stone conservation, for example, TLS has been instrumental in documenting and reconstructing historic stonework. This technology allows for the precise recording of the position and condition of individual stones, which is critical when stones need to be removed, restored, and replaced in their exact original locations (Xi et al. 2015). Moreover, the high level of detail provided by TLS makes it possible to create accurate profiles and sections of individual stones, which can be used to guide the carving and fitting of replacement stones, without the need for invasive procedures.

The applications of TLS in the conservation of historic buildings are numerous and have been extensively documented in both academic research and practical case studies. TLS is particularly effective in capturing the complex geometries of historic structures, including intricate architectural decorations and sculptural elements. For example, TLS was successfully employed in the digital reconstruction of the **Church of SS. Trinità** (Popoli Terme, PE), as shown in Figs. 3.20 and 3.21. This work was conducted as part of the GENESIS project, titled "*Seismic risk management for tourist valorisation of the historical centres of Southern Italy,*" funded by PON MIUR—Avviso MIUR, D.D. n. 1735 del 13/07/2017 (Cantagallo et al. 2024). The project's objective was to develop an integrated platform for managing seismic risk at various levels, ranging from the urban scale, which includes entire historical centres, to individual structural aggregates, and down to specific monumental buildings and the artistic goods contained within them. An innovative aspect of this platform is the integration of benefit–cost analysis, designed to assist in cultural heritage management and the planning of seismic mitigation interventions. This tool provides decision-makers with the ability to target investments more effectively, taking into account not only vulnerability assessments and **seismic risk evaluations** but also the **tourism potential** of the historical centre. The combined approach enhances both the preservation of cultural assets and the sustainable development of these areas, while maximizing the touristic potential of historical centres.

As demonstrated by TLS application within the GENESIS project, the TLS is able to create a **digital record** for the historic building, that can be used for future reference, allowing conservators to compare the current state of the building with its past condition. This capability is invaluable for monitoring the long-term effects of environmental or structural factors, such as crack progression or material deterioration due to weathering and pollution. Additionally, digital archives created using TLS can be shared with a wider audience, including scholars, architects, and the general public, thereby enhancing the accessibility and understanding of cultural heritage.

Another significant application of TLS is in the field of Building Information Modelling (BIM). When integrated with BIM workflows, TLS data can be used to create highly accurate as-built models that serve as the foundation for planning and managing conservation projects. The integration of TLS with BIM facilitates the coordination of different disciplines involved in a conservation project, ensuring that all stakeholders have access to the same accurate and up-to-date information (Bosche et al. 2015). Furthermore, the models obtained with the TLS techniques provide a reliable basis for structural analysis, allowing engineers to simulate the effects of various interventions and assess the potential impact on the building's integrity.

Fig. 3.20 Point cloud obtained from the TLS survey of the Church of SS. Trinità (Popoli Terme, PE) and its surroundings. Front view

Fig. 3.21 Point cloud obtained from the TLS survey of the Church of SS. Trinità (Popoli Terme, PE) and its surroundings. Rear view

Despite its many advantages, TLS is not without its challenges and limitations. One of the primary challenges is the **complexity of generating and manipulating point clouds**. The vast amount of data captured by TLS can be difficult to manage, requiring specialized software and expertise to process and analyse. Moreover, the creation of meshes and models from point clouds can be time-consuming and computationally intensive, particularly for large-scale projects or highly detailed scans

(Kottke et al. 2011). Another limitation of TLS is the high cost associated with the technology. The initial investment in 3D laser scanning equipment can be substantial, and the need for skilled operators and data analysts adds to the overall expense. This cost factor can be a barrier to the widespread adoption of TLS, particularly for smaller projects or organizations with limited budgets (Scott et al. 2013). However, as the technology continues to evolve, it is becoming more affordable and user-friendly, making it increasingly accessible to a broader range of users. Material properties and surface conditions also affect the accuracy and reliability of TLS data. Reflective surfaces, for example, can cause errors in point cloud generation, while weathered or irregular surfaces may require multiple scans from different angles to capture accurately. These factors must be carefully considered when planning a TLS survey to ensure that the resulting data meets the required standards of accuracy and detail (Kottke et al. 2011).

Recent research in TLS has focused on addressing some of these challenges by developing automated processes for point cloud data analysis. For instance, studies have explored the use of algorithms to automatically extract information from point clouds, such as identifying and labelling individual architectural elements, quantifying material volumes, and assessing structural conditions (Bosche et al. 2015). These advancements have the potential to streamline the workflow and reduce the time and effort required to process TLS data, making the technology more efficient and accessible. Another promising direction for TLS is its integration with mobile and cloud-based technologies. The development of portable 3D laser scanning devices, such as those combining a smartphone with a line laser, offers the potential for more flexible and on-the-go data capture (Kolev et al. 2014). Additionally, cloud computing enables the storage and sharing of large point cloud datasets, facilitating collaboration among project stakeholders and providing access to TLS data from remote locations (Schops et al. 2015).

In conclusion, TLS is a transformative technology in the field of heritage conservation, offering unparalleled accuracy and detail in the documentation and analysis of historic buildings. Its applications are wide-ranging, from creating detailed digital archives to supporting structural analysis and restoration planning. Despite the challenges associated with cost, data management, and material properties, TLS continues to evolve, becoming more accessible and integrated with other digital technologies. As the field advances, TLS will undoubtedly remain a cornerstone of conservation efforts, enabling the preservation of our cultural heritage with greater precision and care.

Digitization Notes

The effective use of **Terrestrial Laser Scanning (TLS)** for heritage conservation requires a structured approach to data management and digitalization. Below are the key steps for digitizing and utilizing TLS data:

- **Optimizing Point Cloud Resolution for Long-Term Use**. While high-resolution point clouds are essential for capturing detailed geometries, it is important to balance resolution with file size for efficient storage and future accessibility. Consider creating optimized versions of the point cloud data that maintain critical details but reduce file size for easier sharing and long-term storage in digital archives.
- **Data Integration into Digital Platforms**. TLS-generated point clouds should be integrated into digital archives or Building Information Modelling (BIM) platforms to ensure that they are accessible to all stakeholders. These platforms should support the visualization and manipulation of the data, enabling detailed analysis and enhancing collaboration between architects, engineers, and conservators. When integrating TLS data, it is important to ensure compatibility with the platform, considering common file formats such as **.e57**, **.las**, **.pts**, or **.ply**. These formats allow for efficient management of large datasets and ensure that the data can be used across different software environments for detailed heritage analysis.
- **Digital Archiving and Metadata Management**. Proper organization of TLS data in a digital archive is essential for ensuring that it can be easily retrieved and updated. Accompanying metadata, including scanning conditions, resolution details, and key project information, should be added to each dataset to facilitate future use and comparison with subsequent surveys.
- **Long-Term Monitoring and Data Comparison**. TLS provides a valuable record of the building's condition at a given point in time. To maximize its utility, the data should be structured in a way that allows for easy comparison with future surveys. This is particularly useful for tracking structural changes, such as the evolution of cracks or weathering effects on the building materials.
- **Cloud Storage and Data Sharing**. Given the large size of point cloud datasets, cloud-based storage solutions should be considered to facilitate sharing and collaboration across multiple locations. Cloud platforms allow stakeholders to access the data remotely, improving project coordination and ensuring that all relevant parties can work with the most up-to-date information.

3.3.3.3 Unmanned Aerial Vehicles (UAVs)

Unmanned Aerial Vehicles (UAVs), commonly referred to as drones, have become increasingly valuable in the construction industry due to their ability to capture high-resolution data in real time from locations that are difficult or impossible to access by traditional methods. This technology has evolved significantly, as evidenced by early systematic reviews in construction automation (Lu et al. 2010a, b; Kim et al. 2013; Sun et al. 2013; Vähä et al. 2013; Valero et al. 2015). These reviews have provided a comprehensive introduction to the role of UAVs in various aspects of construction, such as surveying, project management, and safety analysis.

UAVs are generally equipped with advanced imaging technologies, including High Dynamic Range (HDR) digital cameras and video devices, which facilitate precise data acquisition (Colomina and Molina 2014). These devices are capable of capturing detailed images that can be used for various applications, such as creating photogrammetric models, assessing structural integrity, and monitoring environmental conditions. One of the key advantages of UAVs is their ability to integrate with other surveying technologies. For example, combining UAVs with HDR cameras, Infrared Thermography (IRT), and photogrammetry allows for the creation of highly detailed 3D models that include both geometric data and thermal profiles (Hallermann et al. 2015). This integration is particularly valuable for identifying structural deformations, material degradation, and other critical issues that may not be visible to the naked eye.

The use of UAVs offers numerous benefits across different phases of construction projects. In the surveying phase, UAVs can quickly and accurately capture data over large areas, reducing the time and labour required compared to traditional methods (Fig. 3.22). Despite the evident advantages of UAVs, their application in the field of historic building repair and maintenance has been relatively underexplored. However, studies by Colomina and Molina (2014), Eschmann et al. (2012) and (Hallermann et al. 2015) have demonstrated that UAVs can be highly effective in this context. These studies show that UAVs are particularly useful for inspecting and monitoring hard-to-reach areas of historic buildings, such as roofs and facades, where traditional access methods might be unsafe or impractical. By capturing high-quality data, UAVs can help identify various issues, including structural cracks, stone decay, and other forms of deterioration that could compromise the integrity of historic structures (Ellenberg et al. 2014).

Fig. 3.22 Unmanned Aircraft System (UAS) or drone used to survey the South Obenchain Fire, 9/28/2020 *Source* https://www.flickr.com/. Licence: Public domain

In project management, UAVs can be used to monitor progress, manage material inventories, and ensure compliance with safety regulations. This capability is especially important in complex projects where real-time data is essential for making informed decisions (Chabot 2018). Moreover, UAVs offer several advantages over traditional methods of inspection and monitoring. One of the most significant is the reduction in logistical complexity. Traditional inspection methods often require scaffolding, cranes, or other heavy equipment, which can be costly and time-consuming to set up. UAVs, on the other hand, can be deployed quickly and with minimal infrastructure, significantly reducing costs and project timelines (Hallermann et al. 2015). Additionally, UAVs enhance safety by minimizing the need for workers to operate in hazardous conditions, such as at height or in confined spaces.

Recent advancements in UAV technology have further expanded their potential applications in construction and historic building maintenance. For instance, the integration of artificial intelligence (AI) and machine learning algorithms with UAV systems allows for the automated analysis of captured data. These technologies can identify patterns of wear and tear, predict potential failure points, and even suggest maintenance interventions based on historical data. Additionally, the development of more sophisticated sensors, such as LiDAR (Light Detection and Ranging), has enhanced the ability of UAVs to capture detailed 3D models of structures with unprecedented accuracy (Eschmann et al. 2012).

Furthermore, UAVs are increasingly being used in conjunction with Building Information Modelling (BIM) systems to create dynamic, real-time models of construction projects. By integrating UAV data with BIM, project managers can visualize and analyze the current state of a building in relation to its original design, allowing for more precise planning and execution of construction tasks. This integration also facilitates improved communication and collaboration among project stakeholders, as the UAV-captured data can be easily shared and accessed through cloud-based platforms.

In conclusion, UAVs represent a transformative technology in the construction and heritage conservation sectors. Their ability to capture high-resolution data quickly and safely, combined with their versatility in integrating with other technologies, makes them an invaluable tool for modern construction practices. As UAV technology continues to evolve, its applications are likely to expand, offering even greater potential for improving the efficiency, accuracy, and safety of construction projects, particularly in the maintenance and preservation of historic buildings.

Digitization Notes

To maximize the benefits of UAV technology in heritage conservation, it is crucial to implement an efficient digitalization strategy that ensures proper data management, integration, and long-term usability. Below are the key considerations for effectively digitizing UAV-generated data:

- **High-Resolution Data Collection for Digital Archives.** UAVs provide the capability to capture high-resolution images and videos of historic sites, particularly in hard-to-reach areas. These data should be systematically archived and organized within a digital platform, ensuring that images and 3D models created through photogrammetry or LiDAR are easily retrievable for future analysis, maintenance planning, and comparison with subsequent surveys.
- **Integration with Existing Digital Models.** UAV data should be integrated with existing digital models, such as BIM or GIS platforms, to provide a comprehensive view of the structure's condition. This integration allows for the combination of aerial data with other types of survey information, ensuring that a full digital representation of the site is available for stakeholders, facilitating conservation and intervention decisions.
- **Cloud-Based Data Management.** Given the large volume of high-resolution data generated by UAVs, storing and managing this data in cloud-based systems is essential for facilitating collaboration and sharing among project teams. Cloud storage allows stakeholders to access the most up-to-date information remotely, enabling more efficient coordination and decision-making in conservation projects.
- **Regular Monitoring and Updates.** UAVs enable regular and repeatable surveys of the same site over time. These repeat surveys should be systematically archived and compared with earlier datasets to monitor the progression of any deterioration or structural changes. Such an approach helps to track issues like crack propagation or material degradation, allowing for timely interventions based on real-time data.
- **Safety and Logistical Benefits in Digitization.** The use of UAVs minimizes the need for physical scaffolding or cranes, reducing both safety risks and the logistical complexity of conducting surveys. This ensures that even large-scale or complex heritage sites can be documented without invasive techniques, preserving the integrity of the site while efficiently gathering necessary data.

3.4 Critical-Historical Analysis

3.4.1 Documentary Research and Interpretation

The historical and critical analysis of a historic building serves as the foundation for understanding its origins and evolution. This process should begin with a visual examination of the building, followed by an in-depth documentary analysis that sheds light on its development over time. For this purpose, documents available at municipal and regional offices, heritage authorities, land registry agencies, state archives, and district notarial archives can be invaluable. These sources may provide surveys, architectural plans, cadastral records, and property transaction documents related to the building in question.

As an example, Table 3.1 provides a summarized chronology of the Episcopal Palace in San Martino sulla Marrucina (CH), derived from a thorough and well-considered documentary and archival analysis conducted at the Land Registry Agency and the State Archives of Chieti, and included both cadastral records and notarial deeds, which allowed the reconstruction of the history of the building starting from 1811.

Digitization Notes

The digitalization of historical documents is crucial for ensuring the long-term preservation and accessibility of valuable information related to historic buildings. In this context, the following steps should be considered:

- **Digital Scanning and Archiving of Paper Documents**. All documents, such as architectural plans, cadastral records, and notarial acts, currently stored in physical archives should be scanned at high resolution and archived digitally. This process ensures the preservation of these fragile records and facilitates their accessibility for future studies. Additionally, metadata should be attached to each document to enable efficient retrieval and classification.
- **Organizing Data in a Digital Database**. Once digitized, the documents should be organized into a structured digital database. This database should allow for easy cross-referencing between related documents, such as land registry records, architectural surveys, and property transaction documents. The ability to search for specific buildings or time periods can greatly enhance research efficiency and allow users to track the historical evolution of a building.
- **Preservation of Original Documents**. While digital versions offer easier access, the original documents must be preserved in a controlled environment to protect them from degradation. Digital copies ensure that the information remains accessible even if the physical records are damaged over time.
- **Digital Annotation and Interpretation**. Digital tools can be used to annotate scanned documents with interpretative notes, cross-referencing them with other sources of information. This allows for a more in-depth analysis of the evolution of the building and provides researchers with a comprehensive understanding of its historical context.

Table 3.1 Chronology of the Episcopal Palace in San Martino sulla Marrucina (CH)

Date	Transformation
Century XVI	After the sixteenth century, the noble floor was constructed, which displays evident Renaissance features, particularly in the window
1811	Angelo Pompilio is recorded as the owner of four houses located on the same street and at the same civic number (Via Porta di Mezzo 13), which were taken from Giuseppe Di Paolo, Giacomo Ricci, Giuseppe De Luca, and Saverio De Luca (*State Archives of Chieti, cadastral records, rural cadastre, general prospectus, file no. 191*)
1878	Angelo Pompilio is still the owner of the four houses, of which the first is two stories with twelve rooms, the second is two stories with two rooms, the third similarly has two stories and two rooms, and the fourth is two stories with five rooms. The taxable income for these houses was established at 2,250 lire for the first house, 600 lire for the second, 750 lire for the third, and 1,500 lire for the fourth. Unfortunately, only three of these four houses remain today and are the subject of our studies. The largest, which had a taxable income of 2,250 lire for its two stories and twelve rooms, was demolished following extensive damage caused by an earthquake, and today, we only know its perimeter (*State Archives of Chieti, cadastral records, rural cadastre, file no. 161*)
15/07/1895	Angelo Pompilio bequeaths to his sons, Rocco Pompilio and Giuseppe Nicola Pompilio, his numerous assets, including the house in the Porta di Mezzo district with several rooms, divided into two portions by the external staircase, which must be shared by both. From this information, we understand that the two northern portions of the house, one of which is the demolished section, passed to Giuseppe Nicola Pompilio, and the two southern portions to Rocco Pompilio. It also indicates that the main facade had a staircase that allowed access to the upper floor from the outside (*State Archives of Chieti, notarial records, not. Cellucci, mod. I, vol. 34, c. 182 v. and 185r.*)
1895	Rocco Pompilio becomes the owner of one house with two stories and two rooms and another with two stories and five rooms, both located at Via Porta di Mezzo 13, with a taxable income of 10 lire and 30 lire, respectively (likely the same house divided into two parts) (*State Archives of Chieti, cadastral records, rural cadastre, file no. 410*)
1906	In this year, Rocco Pompilio is no longer listed as the owner of two houses with two stories, one with two rooms and the other with five rooms, but rather of one house with two stories and four rooms (this is an internal reconfiguration that will be confirmed in the building analysis) (*State Archives of Chieti, cadastral records, urban cadastre, file no. 394*)
18/07/1926	Alberto, Attilio, and Iva Pompilio, children of Rocco Pompilio, inherit ownership of two portions of a house, both located at Via Porta di Mezzo 13, with two stories and two rooms and two stories and four rooms, respectively, while Maria Di Luzio, their mother, retains usufruct rights (*State Archives of Chieti, cadastral records, urban cadastre, file no. 801*)
22/08/1940	Giovanni, Rocco, Maria, and Lina Di Giuseppe, children of Giuseppe Di Giuseppe, become the owners of the two portions of the house (which still have the same number of stories and rooms). Giovanni and Rocco own 6/8 of the property, while Maria and Lina own 2/8. Maria Concetta Di Cola retains partial usufruct rights (*State Archives of Chieti, cadastral records, urban cadastre, file no. 888*)

(continued)

Table 3.1 (continued)

Date	Transformation
15/02/1945	The assessment and classification conducted on this date certify that Giuseppe Di Giuseppe, son of Giovanni, is the owner of the entire house on this plot. The building is designated for residential rental use, with four usable rooms on the ground floor and two storage spaces (under-stair areas). The cadastral size is recorded as 489, with a total of five usable rooms (*Territory Agency of Chieti, map sheet 4, map no. 482, subalt. 2*)
11/06/1963	The assessment and classification conducted on this date confirm that the property is still owned by Giuseppe Di Giuseppe. The ground floor is designated for residential rental use, with three usable rooms (*Territory Agency of Chieti, map sheet 4, map no. 482, subalt. 2*)
2003	The palace is owned by Rocco Serpellini. The ground floor contains four rooms, and the upper floor is divided into an equal number of rooms. There are numerous accretions, such as the addition of two rooms at the rear of the palace, which almost constitute a separate building. The ground floor is used as a workshop for aluminum frame production, while the upper floor is a storage space

3.4.2 Construction Phases

A complete historical–critical analysis, accompanied by a competent interpretation, generates a reconstruction of the construction phases, which are fundamental to understanding not only the evolution of the building, but also its structural behavior, facilitating a more coherent vulnerability assessment.

As for the other phases of the knowledge path, the process of identifying the construction phases is also carried out by referring, as an example, to the Episcopal Palace of San Martino sulla Marrucina (CH). The Episcopal Palace is a historic structure whose evolution spans several centuries. The origin of this work is uncertain. Local sources perceive it as the refectory of an ancient medieval abbey, but documentary and archival research has not confirmed this hypothesis. The building's history is divided into distinct construction phases, each reflecting the architectural styles and functional needs of its time. The earliest phase of the building dates back to the pre-sixteenth century. It is characterized by a core structure that is deeply integrated into the fabric of the town (Fig. 3.23). This original core may have been more than just a simple building. Its position and characteristics suggest that it may have been part of the town's defensive wall, possibly serving as the base of a tower. At this stage, the building was relatively modest in size. The ground floor, which provided direct access from the outside, featured a large opening on the southern elevation—an opening that has since been sealed (Fig. 3.24a). The rear elevation shows signs of a historic junction (Fig. 3.24b), indicating that the upper floor was once smaller than it is today. The layout during this phase reveals a structure built primarily for defense, with thick walls and limited access points, which were typical of the period's construction practices. The sixteenth century marked a significant period of growth and transformation for the Episcopal Palace. During this time, the building was expanded to accommodate changing needs. The upper floor was

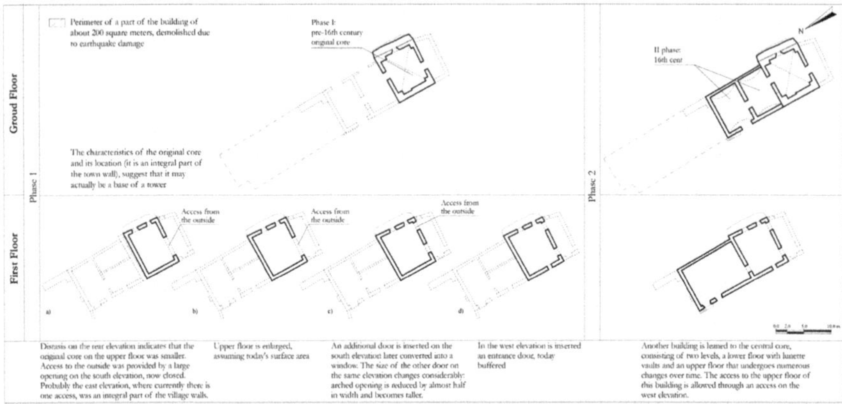

Fig. 3.23 Construction phases of the Episcopal Palace in San Martino sulla Marrucina (CH)—Phase 1 and Phase 2

enlarged, bringing it closer to the size we see today. One of the notable changes was the insertion of a new door on the south elevation, which was later converted into a window (Fig. 3.24c). The alteration of this door, which was once arched, involved reducing its width while increasing its height, reflecting the architectural trends of the time. Additionally, the west elevation saw the introduction of a new entrance door, which has now been sealed, suggesting changes in how the building was accessed and used (Fig. 3.24d). Throughout this period, the ground floor retained its traditional vaults with lunettes, a feature that highlights the structure's historic architectural character. The expansions and modifications during this phase indicate a shift from a purely defensive structure to one that accommodated residential or administrative functions, aligning with the evolving role of the building within the community.

As the centuries progressed, the Episcopal Palace saw continued expansion and significant structural changes. During a third phase, a small building was added, connecting the two main parts of the palace that had formed during the earlier phases. This new addition not only expanded the footprint of the building but also altered its internal layout (Fig. 3.25). On the ground floor, the two existing access doors on the south elevation were transformed into internal passage doors, effectively integrating the new structure with the old. This reconfiguration allowed for improved circulation within the building and facilitated access through a newly established entry point on the west side, enhancing the building's functionality. The upper floor underwent further significant changes. The various rooms were subdivided into smaller spaces and covered with vaulted ceilings, a modification that reflects the evolving needs and aesthetic preferences of the time (as shown by the base of the vault in Figs. 3.26a, b). These modifications mark a period in which the palace transitioned from its earlier defensive and residential roles to a more complex structure, accommodating a variety of functions and reflecting the architectural trends of the post-sixteenth century. When the walls lost their function of defending the village, an access was created on the east side, likely in the form of a proffered staircase that led directly to

Fig. 3.24 Transformations observed on the Episcopal Palace in San Martino sulla Marrucina (CH) in Phases 1 and 2—opening on the south elevation (**b**); additional door on the south elevation (**c**), later converted into a window; entrance door on the west elevation, today buffered (**d**)

the upper floor. Over time, the door leading to this staircase was transformed into a window, and the staircase itself was removed. The parastas of this window, however, were added later than the original construction of the walls, indicating a subsequent alteration in the building's design (Fig. 3.26c, d).

The most recent phase of the Episcopal Palace's evolution occurred during the twentieth and twenty-first centuries, as the building underwent further modifications to adapt to modern needs and to address the damage it sustained over time. In the twentieth century, significant changes were made to the ground floor and the building's overall structure. The side wall on the west façade was reconstructed in a new position, and a buttress was added to provide additional support. The openings between the various rooms were altered, with some being enlarged and others reduced in size. Notably, the staircase that once connected the outside to the upper floor was removed, and the building that housed 12 rooms on two floors was demolished following severe earthquake damage, leaving only its perimeter visible today. The

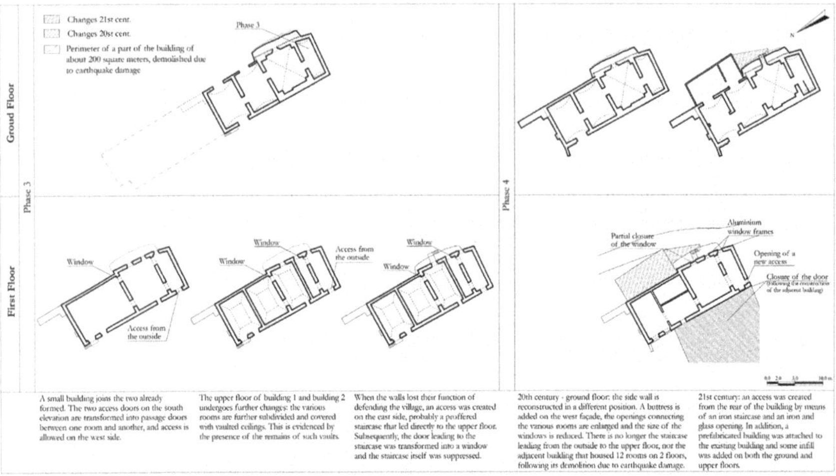

Fig. 3.25 Construction phases of the Episcopal Palace in San Martino sulla Marrucina (CH)—Phase 3 and Phase 4

twenty-first century brought additional alterations, driven by contemporary requirements and further preservation efforts. An access point was created at the rear of the building, with a modern iron staircase leading to a new iron and glass opening. A prefabricated structure was also attached to the existing building, and some infill was added to both the ground and upper floors, marking the latest chapter in the building's long history.

Digitization Notes

Documenting the construction phases of historic buildings is crucial for understanding their structural evolution and guiding preservation efforts. The following steps help ensure efficient digitalization and archiving of this information:

- **Digital Documentation of Construction Phases**. The detailed identification of construction phases should be digitally recorded, including visual evidence (photographs, drawings, and 3D models) of each phase. This documentation should be organized chronologically within a digital archive, allowing easy reference to each construction phase as new discoveries or restoration efforts occur.
- **Integration with 3D Models**. Information about the various construction phases should be integrated into 3D models of the building, allowing stakeholders to visualize how the structure evolved over time. These models can include layered representations of the building at different points in history, highlighting structural changes, modifications, and extensions.

Fig. 3.26 Transformations observed on the Episcopal Palace in San Martino sulla Marrucina (CH) in Phases 3 and 4—opening on the south elevation (**b**); additional door on the south elevation (**c**), later converted into a window; entrance door on the west elevation, today buffered (**d**)

- **Cross-Referencing Archival Research with Construction Phases**. Archival documents, such as cadastral records and historical plans, should be cross-referenced with the identified construction phases to provide a comprehensive view of the building's development. This data should be linked within a digital database to facilitate connections between documentary evidence and physical transformations.
- **Digital Preservation of Structural Alterations**. Any significant structural modifications, such as the removal or addition of elements, should be digitally documented. This includes the digitization of plans, drawings, and any existing physical evidence, ensuring that these changes are preserved for future reference.
- **Data Sharing and Accessibility**. The digital archive documenting construction phases should be easily accessible to researchers, conservators, and

stakeholders. Cloud-based platforms can facilitate the sharing of this information, enabling collaboration and ensuring that all parties have access to the most up-to-date data on the building's historical evolution.

3.5 Building Techniques and Construction Details

3.5.1 General Objectives

The material and construction survey is essential for thoroughly identifying the structural system. It takes into account both the quality and conservation state of materials and structural components. This process often requires uncovering hidden information—such as elements concealed beneath plaster or behind false ceilings—which can be accessed through non-destructive, indirect investigation techniques (such as thermography, ground-penetrating radar, sonic tomography) or minimally invasive direct inspections (such as endoscopies, plaster removal, sampling, and small excavations).

A critical aspect of this survey is the careful selection of the number, type, and location of tests to be conducted. Ideally, a comprehensive understanding would involve widespread testing; however, due to potential impacts and economic considerations, tests should be carried out judiciously—only when they provide significant value to the evaluation and planning of interventions. To minimize the impact of these investigations, it is crucial to have not only an understanding of the building's construction history but also a deep knowledge of construction practices in the region and during the relevant historical periods.

Particular attention must be given to evaluating the masonry quality, considering the local construction techniques and identifying the geometric and material characteristics of individual components as well as their assembly methods. Of particular importance are:

- The presence of transverse elements (known as diatons) that connect the masonry layers;
- The shape, type, and size of the materials used;
- Recognition of a regular, almost horizontal arrangement of courses (or alternatively, the presence of evenly spaced bonding courses);
- The quality of the masonry bond, achieved through the interlocking of elements (including the number and extent of contacts and the presence of chips) and the regular staggering of joints;
- The nature of the mortar and its state of conservation.

Understanding the structural behaviour of the building requires detailed knowledge of construction techniques and the connections between different structural elements, including:

- The type of masonry (whether brick, stone—cut, rough-hewn, split, cobble—or mixed; single-wall, double-wall, or multi-layered) and its construction characteristics (regular or irregular bonding; with or without transverse connections, etc.);
- The quality of connections between vertical walls (interlocking at corners and intersections, the presence of tie rods, etc.);
- The quality of connections between floors (including floors, vaults, and roofs) and walls, noting the presence of horizontal ties or other connecting devices (such as tie rods);
- Discontinuities caused by shafts, chimneys, and similar features;
- The type of floors, vaults, and roof structures, with particular reference to their in-plane stiffness;
- The type and effectiveness of lintels above openings;
- The presence of structurally efficient elements designed to counteract lateral forces;
- The presence of non-structural elements that exhibit high vulnerability.

3.5.2 Material Survey

The analysis of materials in a historic building begins with an initial observation to establish whether the building is plastered or the masonry is partially or fully exposed. In the latter case, the identification of the type of masonry and its construction characteristics becomes easier, and a complete architectural survey that includes the masonry texture already provides important insights into the type of masonry and its construction details. A detailed architectural survey makes it possible to determine whether the masonry consists of of bricks, stone, dressed stone, pebbles, or a mixed type, and whether the texture is regular or irregular, including any bonding stones.

When the masonry is plastered, it becomes necessary to conduct targeted local investigations, such as thermographic surveys or partial removal of the plaster, provided that such actions do not compromise the architectural and cultural value of the building. These investigations are crucial for revealing hidden construction details that are not visible on the surface, thus providing a more accurate understanding of the building's material composition and structural integrity.

For instance, the **architectural survey** of the main facade of the Episcopal Palace in San Martino sulla Marrucina (CH), shown in Fig. 3.27, reveals the intricate composition and detailing of the masonry. The facade is characterized by a mixed masonry texture, predominantly composed of irregularly shaped stones and pebbles and bricks. The larger stones are roughly shaped and arranged with varying degrees of precision, while smaller stones fill the gaps between them. This type of construction indicates a traditional building technique, where local materials were utilized in a way that reflects the available resources and the craftsmanship of the time. The facade also features dressed stone around the openings, such as windows and doors, adding a

Fig. 3.27 Architectonic Survey of the main façade of the Episcopal Palace in San Martino sulla Marrucina (CH)

level of refinement to the otherwise rustic texture. The presence of a cornice and arch elements above the windows further the structure's historical and architectural significance.

The analysis of the four different masonry panels, each with a surface area of 1 m × 1 m, as shown in Figs. 3.28, 3.29, 3.30, and 3.31 confirms the primary material characterization observed during the visual inspection and architectural survey. This investigation further revealed several significant details about the construction techniques and material choices used in the building.

Figures 3.28 and 3.29 depict masonry panels from the main façade of the building. These panels demonstrate a combination of medium-sized bricks, approximately 4.5 × 12 × 27 cm, and roughly hewn stones, arranged with a regular pattern in some areas and a more irregular layout in others. The brickwork in Fig. 3.28 is characterized by thin, consistent mortar joints, indicating a construction approach focused on uniformity and stability. The use of lime and sand mortar, with some recent integrations of cement, ensures the durability of these sections. In contrast, the stones in Fig. 3.29 vary significantly in size and shape, reflecting a traditional method where larger stones form the base structure, with smaller pieces used to fill in the gaps. Mortar application in this section is more generous to accommodate the uneven surfaces, suggesting a construction technique focused on robustness.

Figure 3.30 shows a masonry panel from the rear façade composed entirely of bricks. These bricks, measuring approximately 4.3 × 9.5 × 21 cm, are laid in a

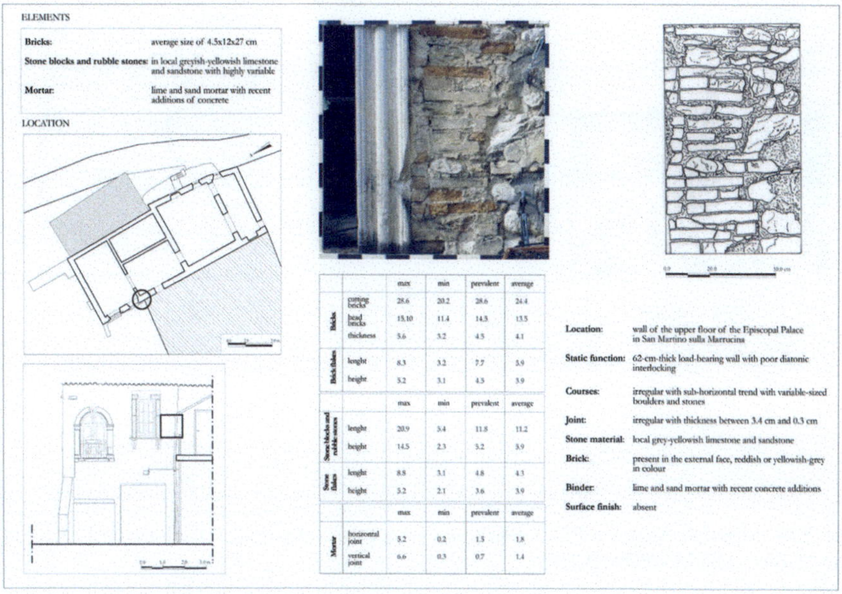

Fig. 3.28 Materials Analysis. Part 1. Rough-hewn stone and brick masonry of the main façade of the Episcopal Palace in San Martino sulla Marrucina (CH)

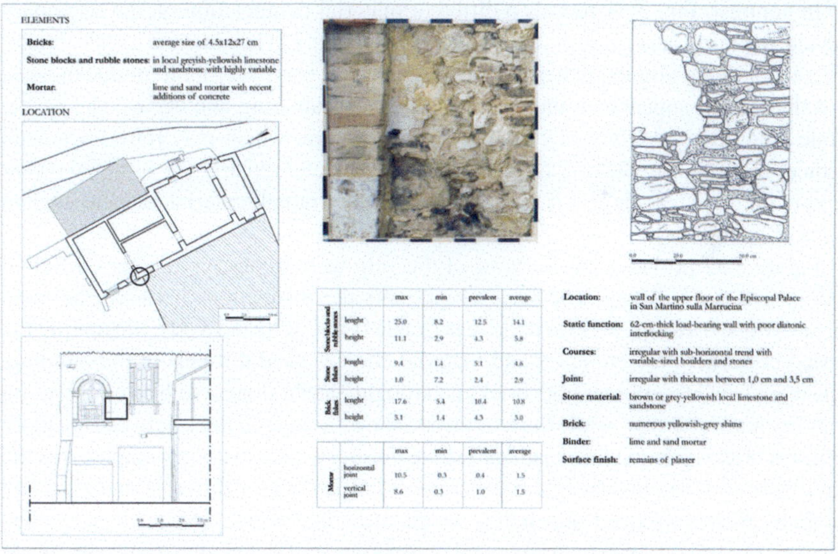

Fig. 3.29 Materials Analysis. Part 2. Rough-hewn stone and brick masonry of the main façade of the Episcopal Palace in San Martino sulla Marrucina (CH)

Fig. 3.30 Materials Analysis. Part 3. Brick masonry of the rear façade of the Episcopal Palace in San Martino sulla Marrucina (CH)

regular pattern with uniform, thin mortar joints. The wall appears well-preserved, with minimal signs of deterioration.

In contrast, Fig. 3.31 reveals a different material composition on the secondary façade, featuring dark greyish stones, likely sedimentary in origin, such as calcarenite. These stones are aligned in a sub-horizontal pattern, with varying joint thicknesses, and the overall appearance indicates significant weathering and decay. The darker stone coloration adds visual contrast to the structure, which may reflect aesthetic choices made during the original construction or in later repairs. The minimal recent repairs on this area suggest it retains much of its original material, despite evident signs of aging.

Figure 3.32 provides an overview of the various masonry types across different sections of the building. The image highlights the distinctions between the main façade (Figs. 3.28 and 3.29) and the rear façade (Figs. 3.30 and 3.31), illustrating the varied construction techniques and material choices used throughout the building. The main façade combines both bricks and roughly hewn stones, with an emphasis on stability and durability, while the rear façade shows a stark contrast between the more uniform brickwork and the weathered stone masonry. This overview emphasizes the building's complex history of construction and repair, where different materials and methods were employed depending on the structural needs and aesthetic preferences of the time. It also underscores the importance of detailed material surveys in understanding the building's integrity and guiding conservation efforts.

Overall, these detailed observations from the masonry panels and the comprehensive overview provided in Fig. 3.32 offer valuable insights into the material composition and construction techniques of the Episcopal Palace. This analysis not only confirms the findings from the visual inspection but also enriches our understanding of the building's historical development and structural condition, which are crucial for any ongoing restoration and conservation efforts.

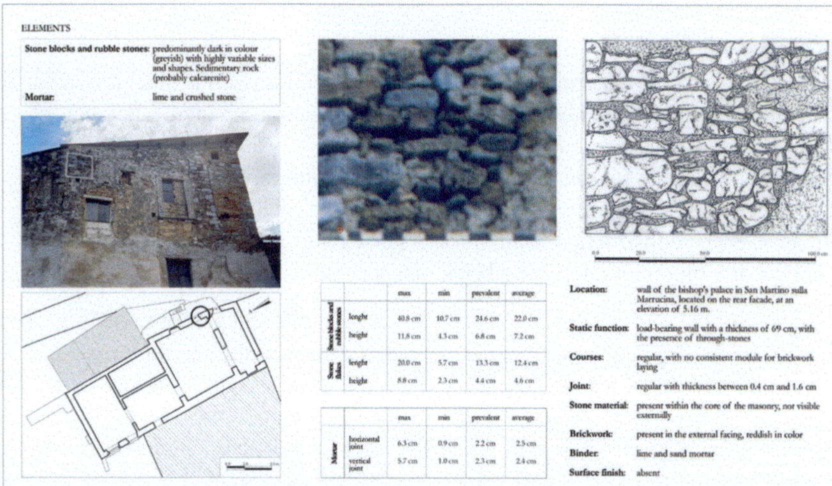

		max	min	prevalent	average
Stone blocks and rubble stones	lenght	40.8 cm	10.7 cm	24.6 cm	22.0 cm
	height	11.8 cm	4.3 cm	6.8 cm	7.2 cm
Stone flakes	lenght	20.0 cm	5.7 cm	13.3 cm	12.4 cm
	height	8.8 cm	2.3 cm	4.4 cm	4.6 cm

		max	min	prevalent	average
Mortar	horizontal joint	6.3 cm	0.9 cm	2.2 cm	2.5 cm
	vertical joint	5.7 cm	1.0 cm	2.3 cm	2.4 cm

Fig. 3.31 Materials Analysis. Part 4. Rough-hewn stone masonry of the rear façade of the Episcopal Palace in San Martino sulla Marrucina (CH)

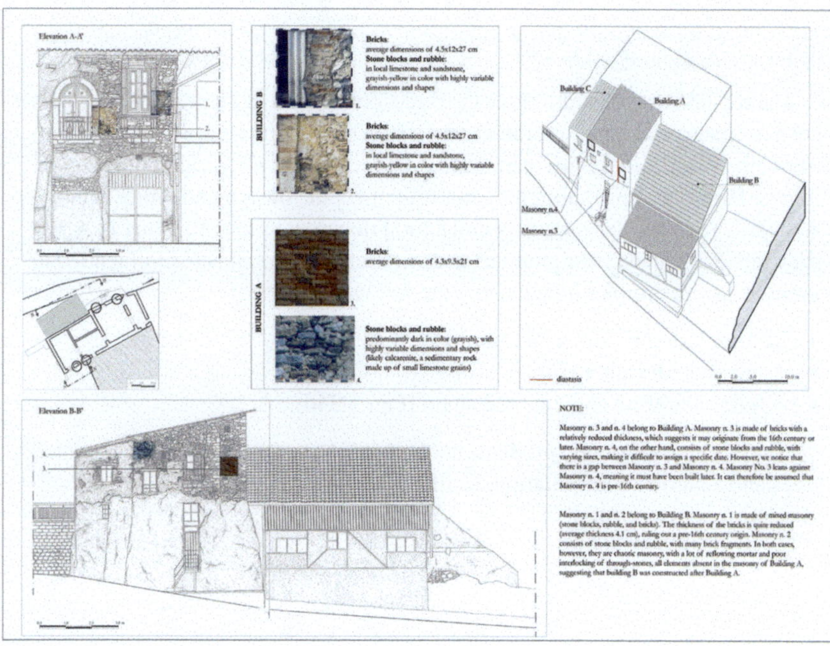

Fig. 3.32 Materials analysis. Overview

Digitization Notes

In the context of material surveys for historic buildings, the integration of advanced digital tools significantly enhances the accuracy and efficiency of architectural documentation. Currently, several software solutions and free online tools allow for the rapid conversion of photographic data into detailed CAD models, facilitating a more comprehensive analysis of the building's materials and construction techniques. As an example, Fig. 3.33 illustrates the conversion of a photo depicting brick masonry into a **.dxf** format.

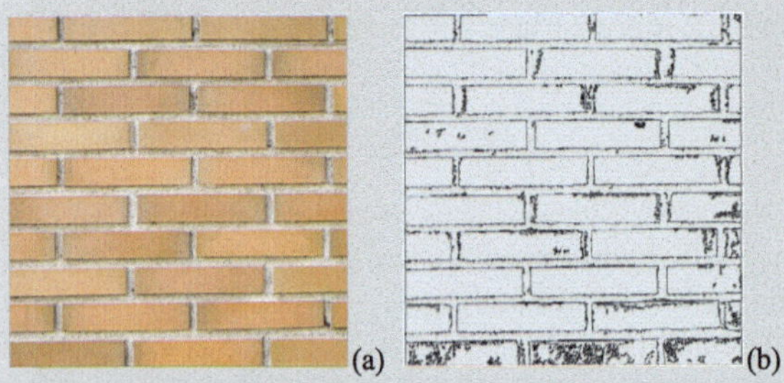

Fig. 3.33 Automatic conversion of an image in a .dxf file carried out with an online free tool (https://imagetostl.com/)

Figure 3.33 demonstrates how the direct conversion of an image (in formats such as .jpeg, .tiff, .png, or others) into a CAD model produces a .dxf file with segmented and dotted lines, capturing even the shadows present in the original image. However, this output may not fully meet the design and representation standards required for precise architectural documentation. To generate CAD drawings with clean, straight lines and without shadows, it is recommended to preprocess the original image using the following steps:

- Increase contrast and saturation (Fig. 3.34b)
- Apply a colour filter (Fig. 3.34c)
- Adjust saturation and contrast again (Fig. 3.34d).

These steps ensure the production of clear CAD models, free of segmented lines and shadow representation, as illustrated in Fig. 3.34.

Fig. 3.34 Conversion .jpeg to .dxf file after the image processing—**a** original image; **b** increasing contrast and saturation, **c** application of a colour filter; **d** adjusting saturation and contrast again; **e** conversion from .jpeg to .dxf (https://imagetostl.com/)

Automated photogrammetric software can extract and digitize the geometrical and textural details of building facades directly from photographs. This capability is particularly useful in documenting the various masonry types, providing a dimensional and typological analysis of facade elements, which is essential for both historical accuracy and the planning of restoration interventions.

The ability to digitize building materials and textures directly from photographs not only saves time but also preserves the integrity of the structures by minimizing physical contact and intervention. The resulting digital models serve as a valuable repository of information that can be used for detailed studies, structural assessments, and the creation of "digital health records" for the buildings.

These technological advancements underscore the importance of incorporating digital tools into the heritage conservation workflow, ensuring that the rich details and historical significance of architectural elements are preserved and understood in their entirety.

3.5.3 Non-destructive Survey Techniques

3.5.3.1 Overview

Non-destructive testing (NDT) methods are essential in the conservation of historic buildings, allowing for the detailed analysis and monitoring of structures without causing damage. These techniques provide crucial insights into the condition and integrity of architectural elements, facilitating informed conservation decisions. Here are some of the key NDT methods commonly used in the field of heritage conservation:

1. **Infrared Thermography (IRT)**: Infrared thermography involves capturing thermal images of a structure to detect heat variations that can indicate issues such as moisture ingress, voids, and thermal anomalies. This technique is highly effective for identifying hidden problems and monitoring the uniformity of materials under varying environmental conditions.
2. **Endoscopies**: Endoscopic inspection allows for the visual examination of internal structures through small access points. This method is instrumental in assessing hidden structural components, such as wall cavities and behind façades, providing a direct view of areas typically concealed from the naked eye.
3. **Ultrasonic Testing**: Ultrasonic testing utilizes high-frequency sound waves to detect internal defects within materials. This technique is invaluable for assessing the integrity of structural elements like stone, metal, and wood, identifying flaws such as cracks, delaminations, and other discontinuities that can compromise a building's stability.

Each of these techniques offers unique advantages and can be employed singly or in combination to achieve a comprehensive understanding of a structure's condition. For instance, while infrared thermography is excellent for preliminary assessments of moisture and insulation, ultrasonic testing provides a deeper insight into material properties and internal defects. Endoscopies are particularly useful when visual access to concealed areas is necessary without causing significant disruption to the historic fabric.

Digitization Notes

The integration of digital technologies with non-destructive testing (NDT) methods offers substantial benefits for the conservation of historic buildings by enhancing the effectiveness and accuracy of inspections and analyses. The following notes highlight how specific digital tools and software can be utilized in conjunction with NDT methods to optimize data collection, analysis, and documentation:

- **Advanced Imaging Software**: Utilizing advanced software for processing thermal images and ultrasonic scans significantly enhances the resolution and interpretation of collected data, allowing for more precise diagnostics of structural conditions.

- **3D Modelling**: The combination of data from various NDT techniques can be used to create detailed 3D models of internal structures and anomalies. Notably, these technologies enable the development of comprehensive Historic Building Information Modeling (HBIM) models. HBIM models integrate detailed architectural and engineering data, providing a digital representation that is invaluable for both analysis and restoration planning. This capability is essential for visualizing the complexities of historical structures and facilitates a deeper understanding of their construction and condition.
- **Digital Archiving**: Digitally archiving data obtained through NDT investigations allows for easy access and longitudinal comparison, which is crucial for monitoring the structural health of historic buildings over time.
- **AI and Machine Learning**: Applying artificial intelligence and machine learning to data collected via NDT techniques can help predict the progression of structural pathologies, optimizing maintenance and intervention plans.

3.5.3.2 Infrared Thermography

Infrared thermography (IRT) is a **non-destructive testing method** that has gained widespread application in the field of historic building conservation over the past two decades. It is particularly valued for its ability to provide rapid, remote, and accurate imaging of masonry structures without the need for direct contact or intrusive sampling. IRT operates by detecting infrared radiation (heat) emitted from objects, which is then converted into an image or thermogram. This method allows for the visualization of temperature variations on the surface of a structure, which can be indicative of underlying issues such as material degradation, moisture accumulation, or structural weaknesses (Kylili et al. 2014).

The IRT process typically involves the use of an infrared camera, which can capture the thermal emissions from a building's surface. Modern infrared cameras have become increasingly sophisticated, with improved accuracy, resolution, and cost-effectiveness, making IRT a more reliable tool for the diagnosis of historic buildings (Lim and Cao 2013). The cameras are often complemented by software that enables the detailed analysis of thermal data, allowing for both qualitative and quantitative assessments of a structure's condition.

Infrared thermography can be classified into two main types: passive and active. In passive IRT, the camera simply records the natural thermal radiation emitted by the building, which is particularly useful for detecting temperature anomalies due to factors like poor insulation or moisture. Active IRT, on the other hand, involves artificially heating the building's surface with external sources (such as halogen lamps) and then observing the cooling patterns. This method is particularly effective for identifying subsurface defects such as voids or delaminations (Maierhofer et al. 2011).

IRT is extensively used for monitoring the performance of masonry interventions and for the preventive diagnostics of historic (Fig. 3.35) and recent (Fig. 3.36) buildings. It can detect crack configurations, structural failures, states of decay, and residual moisture that may not be visible through traditional visual inspection (Paoletti et al. 2013; Bisegna et al. 2014). The ability to conduct these surveys without physically disturbing the structure is particularly advantageous for historic buildings, where preservation of the original materials is paramount. Several studies highlight the effectiveness of IRT in the **evaluation of post-seismic damage**. Through the application of pre- and post-earthquake thermographic surveys, these studies revealed that many of the thermal anomalies detected prior to the earthquake corresponded to areas of damage observed afterward. This predictive capability underscores the value of IRT as a tool for assessing the vulnerability of historic structures in seismic regions (Paoletti et al. 2013; Bisegna et al. 2014).

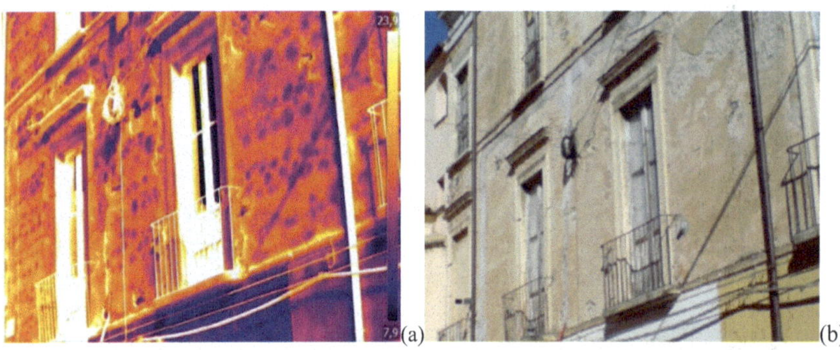

Fig. 3.35 Thermographic survey on the facade of a historic building. Thermographic image (**a**) and comparison photo (**b**)

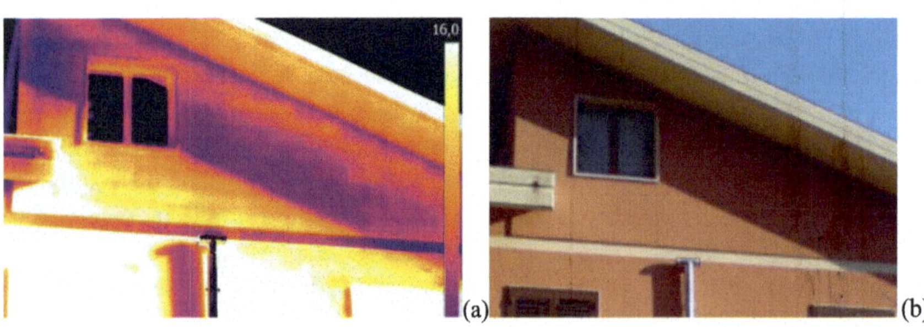

Fig. 3.36 Thermographic survey on the facade of a masonry building. Thermographic image (**a**) and comparison photo (**b**)

While IRT offers numerous benefits, its effectiveness depends heavily on the accurate assessment of thermal parameters and a thorough understanding of the materials and construction techniques of historic buildings. For example, the thermal emissivity of different building materials can affect the accuracy of the temperature readings, necessitating careful calibration of the equipment (Kirimtat and Krejcar 2018). Additionally, environmental factors such as weather conditions can influence the thermal images, potentially leading to false positives or negatives. To enhance the reliability of IRT, it is often integrated with other non-destructive techniques, such as 3D laser scanning, which can provide precise geometrical data that complement the thermal images (Maierhofer et al. 2011). This combination allows for a more comprehensive analysis, enabling the discrimination between surface effects due to material properties and actual internal defects. For instance, Lim and Cao (2013) discuss the integration of IRT with Ground Penetrating Radar (GPR) and Ultrasonic Pulse Velocity (UPV) in the evaluation of concrete and masonry structures. Such combinations can significantly improve the accuracy of defect detection and provide a more holistic understanding of the building's condition.

In conclusion, infrared thermography represents a critical tool in the arsenal of techniques available for the preservation and maintenance of historic buildings. Its ability to provide non-intrusive, accurate assessments of structural integrity, moisture presence, and material conditions makes it invaluable for ongoing monitoring and preventive maintenance. As IRT technology continues to advance, its role in historic building conservation is likely to expand, offering even greater precision and utility in preserving our architectural heritage.

3.5.3.3 Endoscopies

Endoscopies represent a crucial non-invasive diagnostic tool for the investigation and conservation of built heritage. This technique allows for the exploration of cavities, voids, and other hard-to-reach areas within a building's structure, offering a detailed visual assessment of its internal conditions without causing significant damage to the material fabric.

The endoscopic examination involves inserting a flexible tube, equipped with a high-resolution camera and a light source, into pre-existing openings or small drilled holes in the structure (Fig. 3.37). This equipment transmits real-time images to a monitor, enabling experts to inspect and analyze the inner composition of walls, ceilings, and other structural elements. This method provides valuable insights into the building's construction techniques, the condition of internal materials, and the presence of voids, cracks, or other anomalies that could compromise structural integrity.

In the context of heritage conservation, endoscopies are particularly useful for:

1. **Internal Wall Examination**: This technique is highly effective for analyzing the internal composition of masonry, identifying voids, cracks, or areas where materials may have deteriorated. For example, endoscopic surveys can reveal

Fig. 3.37 Fibre-optic video endoscope. Model MIGS 200

whether a wall contains hidden cavities, inconsistent construction methods, or internal decay that would not be visible from the exterior.

2. **Vault Integrity Assessment**: Endoscopies are instrumental in evaluating the internal conditions of vaults, especially when signs of deformation or cracking are present. By exploring these areas, experts can diagnose structural issues such as weakening due to load stresses or material fatigue, which might not be evident from external observation alone.
3. **Inspection of Narrow Cavities**: The ability to access and examine tight spaces such as interstitial voids between walls or within complex joinery elements is another significant advantage of endoscopies. These inspections can often reveal problems like moisture infiltration, hidden structural faults, or evidence of past repairs that could affect the building's stability.

The primary advantage of endoscopy is its ability to provide a clear view of a building's internal conditions without requiring destructive interventions. This non-invasive approach is particularly valued in heritage conservation, where preserving the original fabric is of utmost importance. However, the technique has limitations. The visual range is restricted to the length and flexibility of the endoscope, and the quality of the imagery can be affected by lighting conditions, the presence of dust, or debris obstructing the view. Additionally, the requirement to drill small access holes, although minimally invasive, still necessitates careful planning to avoid damage to significant features.

Figures 3.38 and 3.39 report an example of endoscopy of a vaulted system and the corresponding results in a specific report, respectively. Such visual documentation reveals the interior conditions of the masonry, showing areas of material loss and structural cracks and is crucial in developing targeted conservation strategies.

A practical application of endoscopic techniques can be observed in the analysis performed on the Royal Palace in Caserta, as documented by Guadagnuolo et al. (2020). In this case, endoscopies were used in conjunction with other non-destructive methods, such as infrared thermography and surface penetrating radar, to assess the

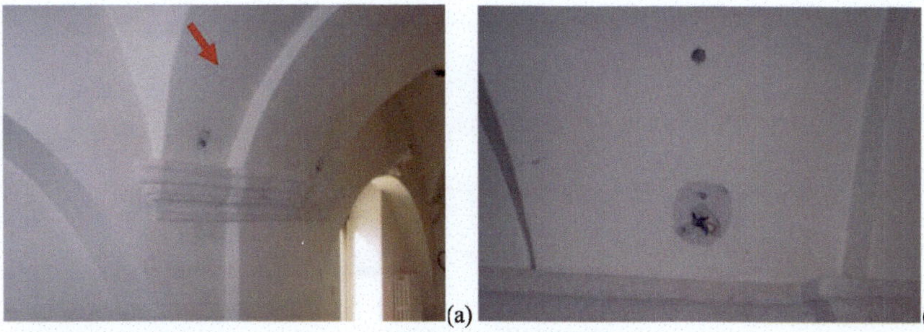
(a) (b)

Fig. 3.38 Hole for the endoscopy carried out on a vaulted system

condition of the façades. The endoscopic inspections provided detailed views of the internal structure, revealing the extent of metal cramp oxidization, which had caused several large stone slabs to detach and fall. This information was critical in planning the necessary conservation measures to prevent further deterioration and ensure the safety of the façade.

Another example of the use of the endoscopies in combination with other diagnostic techniques is provided by Fatiguso et al. (2017), that use of endoscopies alongside radar surveys and thermographic analysis in the study of Italian school heritage buildings. This integrative approach allowed for a more detailed understanding of the construction methods and the extent of decay, particularly in cases where different materials had been used in combination, leading to complex patterns of degradation.

In conclusion, endoscopies are an invaluable tool in the conservation of historic structures, offering detailed insights into the internal state of buildings that are otherwise inaccessible. When combined with other non-invasive techniques, endoscopies contribute to a holistic understanding of a building's health, allowing for informed decision-making in its preservation. As demonstrated in the case studies of Italian heritage sites, this method not only aids in diagnosing current issues but also helps prevent future structural failures by providing early detection of potential problems.

3.5.3.4 Ultrasound

Ultrasonic testing is a non-destructive technique that employs high-frequency sound waves to inspect the internal integrity of materials and structures without causing any damage. This method is invaluable for identifying internal anomalies such as cracks, voids, and other imperfections that are not visible through visual inspection.

The primary objectives of ultrasonic testing include:

- **Assessing Material Quality**: Determining the density and uniformity of materials to identify the presence of flaws or deterioration.
- **Locating Defects**: Pinpointing the exact location of internal defects or inconsistencies within the material, critical for assessing structural integrity.

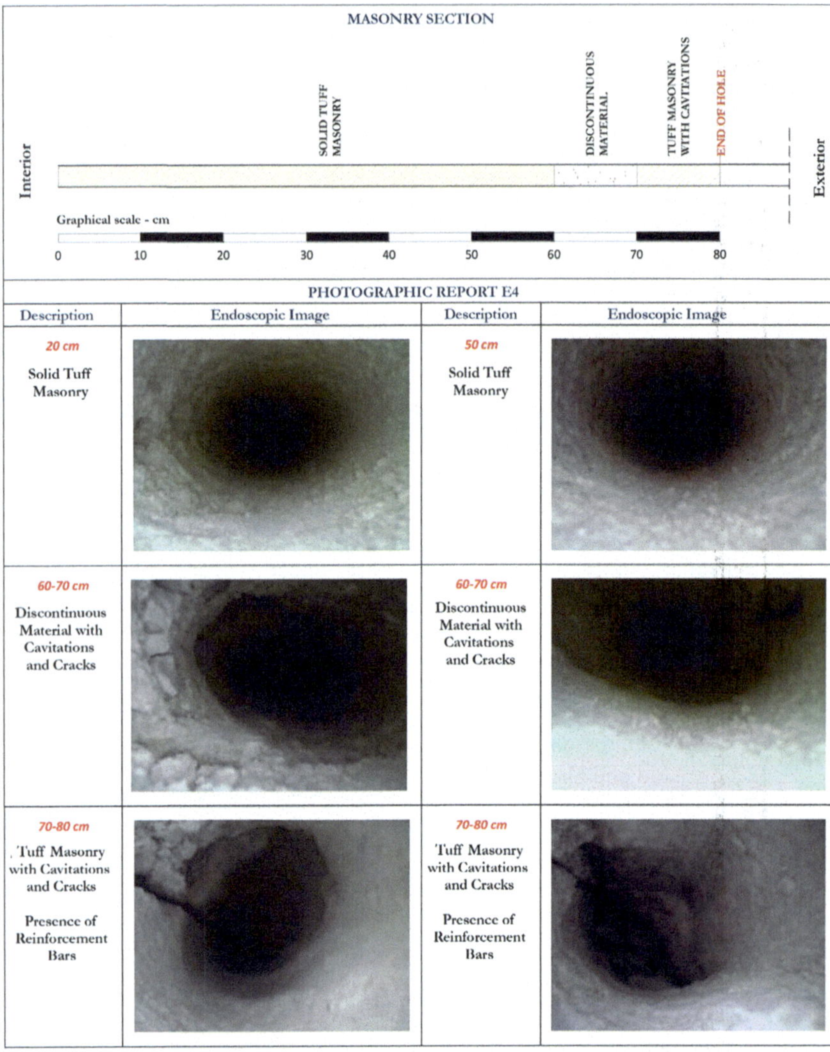

Fig. 3.39 Example of the report obtained from one endoscopy

- **Monitoring Degradation**: Tracking changes in material properties over time to predict potential failure points and assess the overall longevity of the structure.
- **Evaluating Repairs**: Assessing the effectiveness of restoration efforts by comparing the condition of materials before and after treatment.
- Ultrasonic testing involves several key components:
- **Transducer**: This device sends and receives high-frequency ultrasonic waves. When placed against the material, it emits sound waves that travel through the material and reflect back upon encountering any discontinuities.

- **Flaw Detector**: The main unit that displays the data. It interprets the reflected sound waves, with the time and amplitude of echoes analysed to determine the presence and severity of defects.

The testing process involves placing the transducer against the material. The device emits ultrasonic waves that penetrate the material and reflect off any internal features or boundaries. These echoes are converted into electrical signals and displayed as waveforms on the flaw detector's screen, allowing for the identification and characterization of potential issues within the material.

Several studies have demonstrated the utility of ultrasonic testing in evaluating the integrity of structural elements in historical structures, revealing significant differences in material quality that could impact structural stability. As an example, Babacan and Gelisli (2015) conducted extensive research on the use of ultrasonic testing to analyse the integrity of marble columns in historical structures. Their study provided a comprehensive assessment of the variability in material properties across different columns, which directly influences the structural stability and longevity of historical architectures. The research illustrated how ultrasonic velocity measurements could differentiate between varying qualities of marble, thereby aiding in targeted conservation strategies.

Moreover, research by Işık et al. (2020) highlights the technique's effectiveness in monitoring conditions and interventions in historical city walls. Ultrasonic tests were used in combination with other non-destructive methods to evaluate and monitor the condition of the historic Diyarbakır City Walls. The research emphasized the role of ultrasonic testing in identifying areas of the wall that were subject to degradation due to environmental factors and previous restoration efforts. The study underscored the importance of integrating ultrasonic testing in regular maintenance schedules to pre-emptively address potential structural failures.

Moropoulou et al. (2013) discussed the integration of ultrasonic testing with other non-destructive techniques, emphasizing its utility in characterizing materials and detecting degradation, as well as assessing the effectiveness of interventions. Their research validated the use of ultrasonic testing across various scenarios, from evaluating the preservation state of materials to guiding the planning of conservation interventions with precision.

These collective insights illustrate the critical role of ultrasonic testing in the conservation of historical buildings. By facilitating a deep understanding of structural conditions and material characteristics, ultrasonic testing enables conservators to make informed decisions that enhance the durability and historical authenticity of heritage structures.

3.5.4 Decay and Crack Survey

3.5.4.1 Decay Analysis

The preservation of historical monuments requires a thorough understanding of the causes of their aging and decay. This understanding is essential not only for implementing preventive measures but also for halting or at least slowing down the processes of alteration and degradation. Therefore, the study of decay is a critical phase in any restoration project. This scientific and methodological process aims to identify the causes, mechanisms, and final effects of decay on the structure. A detailed analysis of degradation pathologies is necessary to select the most appropriate type and method of intervention.

The degradation of a structure can be caused by various factors, which are generally classified into intrinsic and extrinsic causes. Intrinsic causes refer to flaws that are almost congenital, stemming from the initial construction process, and are inherent to the materials or techniques used by the architect. In contrast, extrinsic causes are external factors that affect the building over time, often related to environmental exposure or historical events that have impacted the building's structural integrity.

When analyzing decay, the Nor.Ma.L. Recommendations (CNR-ICR 1990) provide a critical framework. According to these guidelines, alteration refers to a modification of the material that does not necessarily worsen its conservation state, while degradation implies a change that always results in a deterioration of the material's properties. The surface decay can affect various materials, including plasters, paintings, stuccoes, stone elements, wood, and iron components. The analysis typically identifies different forms of decay, such as cracks, superficial deposits, losses, alveolar erosion, and water infiltration, each of which requires specific interventions.

The Episcopal Palace in San Martino sulla Marrucina (CH), an architectural gem of significant historical value, has stood the test of time, bearing witness to the region's rich cultural heritage. However, like many historic structures, the palace's main façade has not escaped the inevitable passage of time, revealing a variety of decay patterns that now threaten its structural integrity and aesthetic charm. Figure 3.40 shows the decay analysis of the main façade, providing a comprehensive overview of the various forms of decay present across the building's surface. This graphic restitution highlights the different areas affected by deterioration, offering a map of the Palace's current condition. The visual representation serves as a prelude to the more focused observations that follow, where each form of decay tells a story of the building's interactions with its environment over the centuries.

Moving from the overview to the specific, we observe the façade up close through the lens of the photographs in Figs. 3.41 and 3.42. These images capture the intricate details of the decay, each frame a testament to the diverse challenges the structure faces. In Fig. 3.41a, the viewer's eye is drawn to a section of the stonework suffering from severe alveolarization and disintegration. The once-solid stone has been eroded by the relentless forces of wind and the crystallization of salts, leaving behind a honeycomb-like texture where the stone's surface has disintegrated into

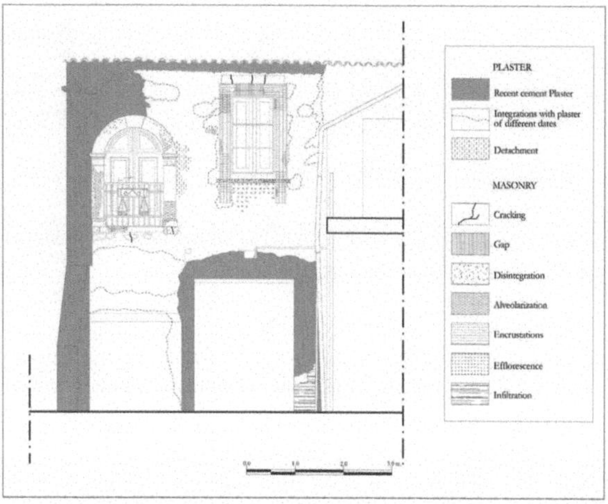

Fig. 3.40 Decay Survey of the main façade of the Episcopal Palace in San Martino sulla Marrucina (CH)

small, crumbling particles. The surface encrustation, likely a result of prolonged exposure to pollutants, adds another layer of complexity to the decay, suggesting that the façade has been under assault from both natural and man-made forces. Adjacent to this, Fig. 3.41b reveals a different, yet equally troubling, form of decay. Here, the stonework shows pronounced gaps where material has been lost, likely due to structural settling or ongoing erosion. These voids are not merely superficial; they signify a deeper vulnerability within the masonry. The encrustation observed within these gaps further exacerbates the decay, as foreign materials fill the voids, trapping moisture and accelerating the degradation process. As we shift our focus to Fig. 3.42a, the theme of gaps and encrustation continues. The encrustation here suggests a prolonged period of exposure, where pollutants and environmental debris have been allowed to accumulate unchecked. Figure 3.42b presents another striking visual—a thick, uneven layer of encrustation clinging stubbornly to the stone surface. This buildup, likely from a combination of soot, dust, and other environmental pollutants, speaks to the changing atmosphere surrounding the palace over the years. Each layer of encrustation represents a different period in the building's history, a stratification of time that now poses a threat to the integrity of the underlying stone. Finally, in Fig. 3.42c, we encounter a different kind of decay—cracking. The vertical fissures cutting through the stone are more than just a visual blemish; they are indicators of deeper structural issues. These cracks may have originated from the natural settling of the building or from thermal expansion and contraction over the years. Regardless of their origin, they now act as conduits for water infiltration, which, if left unaddressed, could lead to further deterioration of both the façade's interior and exterior.

Together, these images paint a vivid picture of the challenges faced by the Episcopal Palace. The narrative they tell is one of a structure that has weathered the forces of nature and the passage of time yet now stands at a crossroads where intervention is necessary to preserve its legacy. The detailed analysis provided by this decay survey offers invaluable insights that will guide the restoration efforts, ensuring that the

Fig. 3.41 Decay observed on the main façade of the Episcopal Palace in San Martino sulla Marrucina (CH). Alveolarization, disintegration and encrustation (**a**); gap and encrustation (**b**)

Fig. 3.42 Decay and cracking pattern observed on the main façade of the Episcopal Palace in San Martino sulla Marrucina (CH). Gap and encrustation (**a**); encrustation (**b**); cracking (**c**)

palace can continue to be a part of the cultural fabric of San Martino sulla Marrucina for generations to come.

Digitization Notes

The study of decay in historical buildings emphasizes the critical role of accurately understanding and documenting various forms of decay to effectively plan conservation efforts. The integration of artificial intelligence (AI) significantly enhances the accuracy and efficiency of detecting and classifying decay, leading to more informed conservation decisions.

1. **AI-Enhanced Imaging and Analysis**: Advances in deep learning, particularly the use of convolutional neural networks (CNNs), have revolutionized the **automatic detection and classification of decay patterns from images**. Bruno et al. (2023) demonstrated the effectiveness of using image-based deep learning for identifying decay in historic buildings, which supports the need for automated, non-invasive inspection techniques crucial for preliminary conservation assessments.

2. **Integration with HBIM**: The integration of AI-driven analysis into Historic Building Information Modelling (HBIM) systems enhances these models with accurate decay data. Giovannini et al. (2021) introduced DECAI - DEcay Classification using Artificial Intelligence, which uses machine learning to analyse decay and integrates the results directly into HBIM, demonstrating the practical application of AI in enriching HBIM environments with detailed decay information.
3. **Automated Monitoring and Reporting**: AI technologies facilitate **continuous monitoring and real-time reporting of decay**, enabling early detection of potential issues requiring intervention. This automated process, highlighted by the work of Karimi et al. (2024), who explored the integration of various non-destructive testing methods with AI to enhance decay monitoring in historical buildings, shows the potential for significant improvements in maintenance strategies. The ongoing research into AI applications for decay analysis is also highlighted by Mishra and Lourenço (2024), who assessed different machine learning models for their effectiveness in predicting **decay progression**, which is crucial for pre-emptive conservation strategies.

By leveraging AI for decay analysis, conservation efforts are significantly optimized, ensuring interventions are timely, targeted, and effective in preserving the integrity and longevity of historic structures.

3.5.4.2 Crack Pattern Analysis

Crack pattern analysis is a fundamental aspect of the structural assessment of historic buildings. It serves as a diagnostic tool that provides insights into the issues affecting the stability and integrity of the structure. Cracks in masonry buildings can result from various causes, including foundation settlement, material degradation, thermal movements, and structural overloading. Each type of crack, whether it is a surface crack, a through crack, or a more complex fissure, can indicate different levels of concern, ranging from superficial cosmetic damage to serious structural distress that may compromise the building's safety.

Understanding the crack pattern involves not only mapping the location and orientation of cracks but also interpreting their causes and potential impacts. This analysis can help conservators, engineers, and architects determine the severity of the issues and prioritize interventions accordingly. The pattern, size, and development of cracks over time can also provide valuable information about the building's structural history, revealing past modifications, load redistributions, and the effects of environmental factors such as moisture ingress or seismic activity.

Moreover, crack pattern analysis is essential for planning restoration and reinforcement strategies. By identifying the origins of cracks and understanding the mechanisms behind their formation, professionals can design targeted interventions that address the root causes rather than merely treating the symptoms. This approach ensures that conservation efforts are both effective and sustainable, preserving the historic integrity of the building while enhancing its structural resilience.

The Episcopal Palace in San Martino sulla Marrucina presents a compelling case study for the application of crack pattern analysis. The building complex, composed of three distinct sections, each exhibits varying degrees of cracking, reflecting the differences in construction methods, materials, and historical modifications.

The crack pattern observed in the Episcopal Palace (Fig. 3.43) reveals a range of structural behaviors across its different sections. The most notable cracks are found in building "A," where through cracks and significant material displacements suggest a history of structural strain, particularly in the upper levels. This contrasts with buildings "B" and "C," which show little to no cracking, indicating that these parts of the complex have maintained better structural integrity over time.

The building identified as "B" shows no visible cracks, while building "C" presents primarily superficial cracks, which are likely due to surface-level issues rather than deep structural concerns. On the other hand, building "A" exhibits more severe

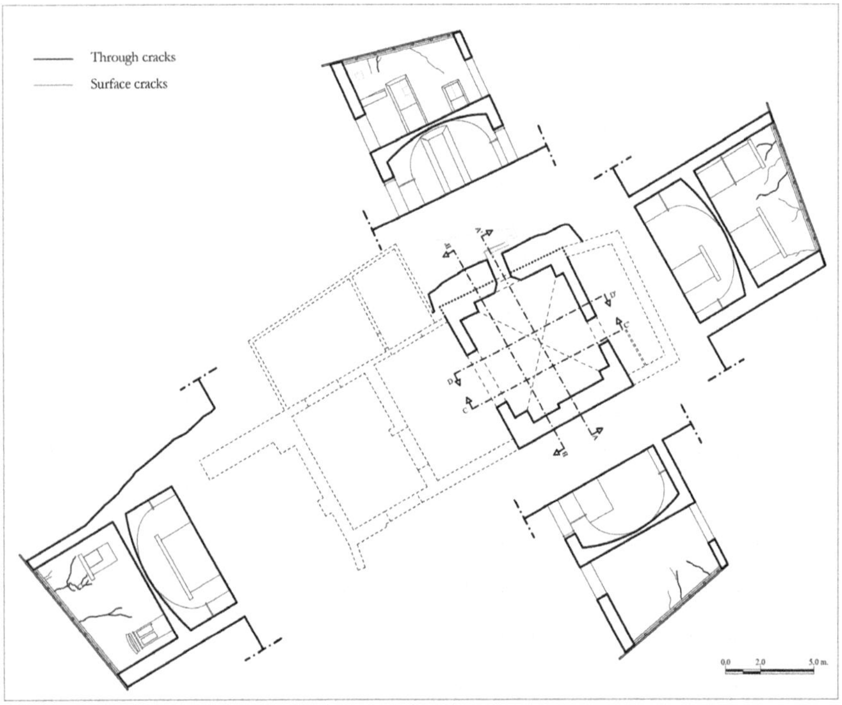

Fig. 3.43 Cracking patten of the building A of the Episcopal Palace in San Martino sulla Marrucina (CH)

cracking, including through cracks that can be seen in several locations, particularly in the interior elevations. These cracks are most prominent on the upper part of the second level, while the ground floor remains free of such damage. This suggests that the stress causing the cracks is concentrated in the upper levels, possibly due to structural changes or load redistributions over time. The detailed crack pattern analysis reveals that the cracking has led to the expulsion of a considerable amount of material, indicating severe structural distress. The hypothesis is that these cracks result from localized crushing caused by concentrated loads, likely originating from the beams of a previous roof structure.

This analysis underscores the importance of understanding the building's history and the structural interventions it has undergone. The differential cracking across the three sections of the building complex highlights the varying levels of structural integrity and the impact of historical modifications, such as the replacement of the roof. This type of detailed investigation is crucial for planning any necessary restoration or reinforcement work, ensuring that interventions address the root causes of structural distress and are sensitive to the building's historical significance.

Digitization Notes

While the decay of materials represents a general concern in heritage conservation, crack patterns provide crucial insights into the **structural integrity** of historic buildings. The challenge lies in accurately identifying, documenting, and predicting the progression of cracks over time, which is essential for preventing structural failures. Digital technologies have significantly enhanced our ability to tackle these challenges, particularly through the integration of **AI**, **automated crack detection**, and **predictive analysis**.

1. **Automated Crack Detection**: Advances in image processing and machine learning have enabled the automatic detection and classification of crack patterns in high-resolution images. Unlike more general forms of decay, cracks can reveal deep structural issues. By using **Convolutional Neural Networks (CNNs)** trained specifically for identifying different types of cracks (surface, through, or fissures), the system can continuously analyze the **growth** and **severity** of cracks, thus supporting proactive interventions (Fig. 3.44).

Fig. 3.44 Automated crack detection on a heritage building façade using AI-powered image analysis. The red lines indicate detected cracks of varying severity and progression, providing crucial insights into the structural health of the building

2. **Crack Mapping and HBIM Integration**: Integrating **crack pattern data** into **Historic Building Information Modelling (HBIM)** provides a comprehensive approach to managing structural risks. As cracks evolve, their location, size, and type are mapped in real-time and linked to the HBIM model. This allows engineers and conservators to visualize the progression of damage within the digital twin of the building, enabling more accurate structural assessments. The **real-time synchronization** of crack data within HBIM also enhances long-term maintenance plans by keeping an up-to-date record of all structural changes.
3. **AI-Driven Predictive Models**: Beyond simple detection, **machine learning models** can predict the future progression of cracks. By analysing historical data on crack development, **AI algorithms** assess patterns and provide forecasts on the risk of further deterioration. This predictive capability is crucial in planning pre-emptive interventions, reducing the likelihood of sudden structural failure. Research by Karimi et al. (2024) has shown that integrating non-destructive testing (NDT) methods with AI models can significantly improve the ability to track crack propagation, providing early warnings of potential structural problems before they become critical.

By focusing on the **specific challenges of crack pattern analysis**, these digital tools and AI solutions not only improve the accuracy of assessments but also ensure that conservation efforts are **targeted** and **timely**, ultimately preserving the structural stability of historic buildings for future generations.

3.5.5 Construction Techniques

Understanding construction techniques is fundamental in the study and conservation of built heritage. These techniques reveal the methods used by builders in the past and provide insight into the historical, cultural, and material contexts that influenced a building's creation. Knowledge of these construction methods is crucial for accurate preservation, restoration, and structural analysis. By understanding how a building was constructed, conservators can make informed decisions that respect the original structure and intent. Construction techniques can often be discerned through a thorough historical–critical analysis, which involves examining the construction details present in the various interventions that have shaped the building over time. The Episcopal Palace in San Martino sulla Marrucina (CH) serves as an exemplary case study for analyzing historic construction techniques. This building showcases a variety of methods used in the construction of arches, walls, and decorative elements, providing a tangible link to the construction practices of the past. The following analysis of specific construction elements, illustrated by Figs. 3.45, 3.46, 3.47 and 3.48, offers a deeper understanding of the building's construction techniques. Figure 3.45 specifically illustrates the construction technique used for the window with a wooden frame present on the façade of the Episcopal Palace. The window features a molded reveal made of limestone blocks, topped by a monolithic architrave and a frieze with a Latin inscription. The Latin characters are difficult to interpret, even for experts in the field.

Figure 3.46 presents an exploded axonometric drawing that highlights the construction characteristics of the window and the assembly of its stone elements, as well as the masonry texture surrounding it. The drawing also includes detailed views

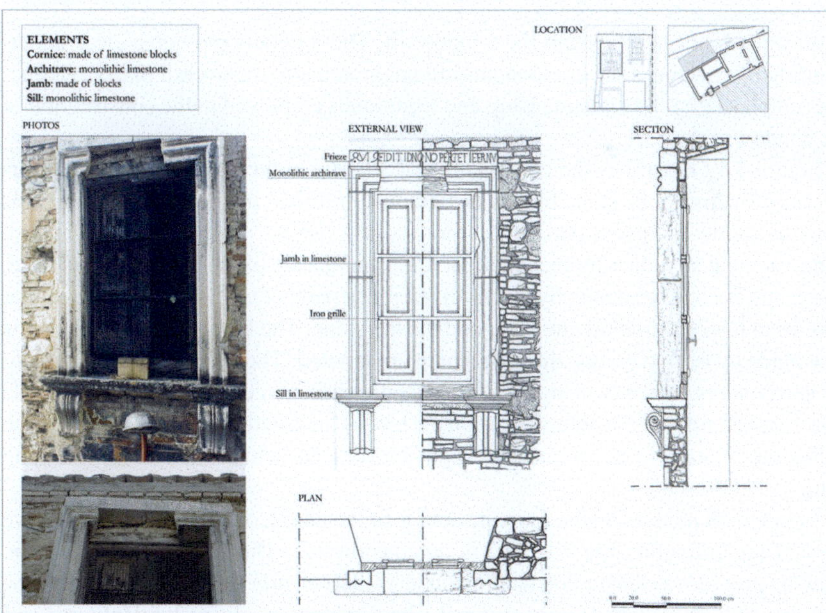

Fig. 3.45 Analysis of construction techniques. Window with wooden frame

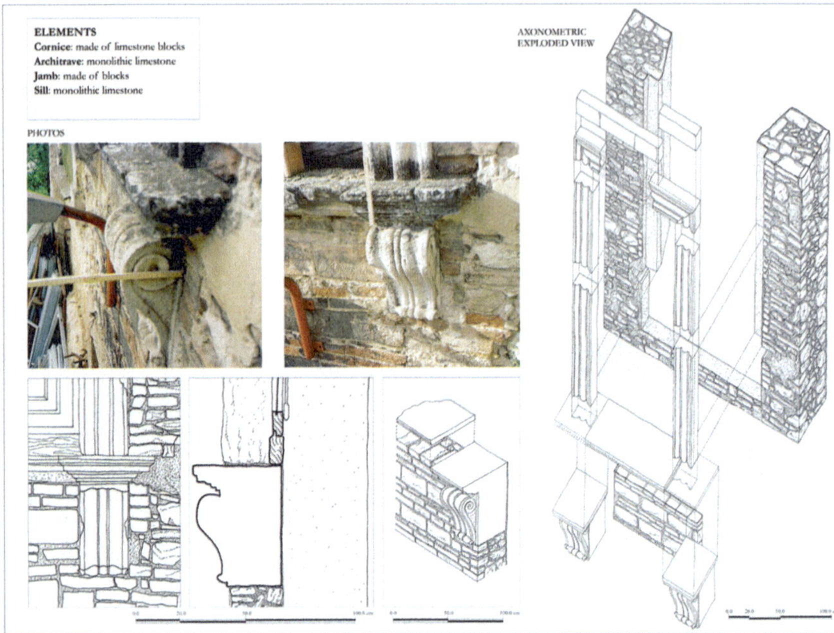

Fig. 3.46 Analysis of construction techniques. Details of the window with wooden frame

of the corbels located beneath the windowsill. These details provide a clear understanding of how the stone components are integrated into the overall masonry structure, emphasizing the craftsmanship and techniques employed in the construction of the window.

Figure 3.47 illustrates the construction of a portal in the Episcopal Palace, characterized by limestone stone blocks meticulously finished on the exposed face. This construction technique is particularly noteworthy for its robustness and the visual impact it creates, reflecting the grandeur and significance of the building's entrance. The portal is complemented by bricks that replace parts of the original structure that have been lost, specifically the jamb and the keystone. The keystone, in particular, is now made up of five bricks, three of which are tapered. The portal is integrated into the surrounding masonry using traditional stone working techniques, such as precise chiseling and fitting. The joints between the blocks are filled with lime mortar, which is flexible enough to accommodate slight movements while maintaining a strong bond.

Figure 3.48 focuses on the intricate details of the portal, offering a comprehensive view of the craftsmanship involved in its construction. The exploded axonometric drawing provides a clear understanding of the wall's construction, demonstrating how each stone block is shaped and positioned. The combination of structural integrity and aesthetic detail in this portal exemplifies the mastery of stone construction techniques during the period. The details in Fig. 3.48 also include ornamental elements, such

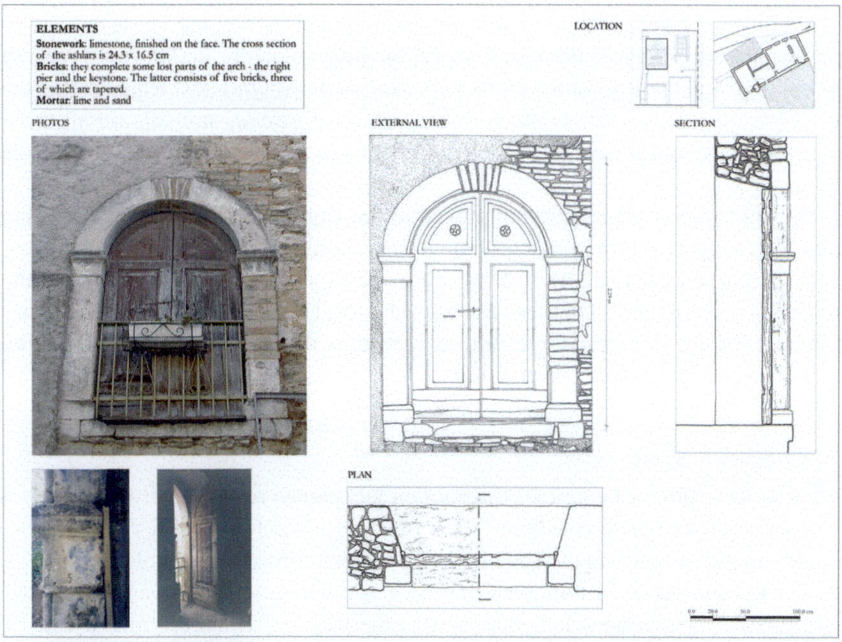

Fig. 3.47 Analysis of construction techniques. Portal with a stone block reveal

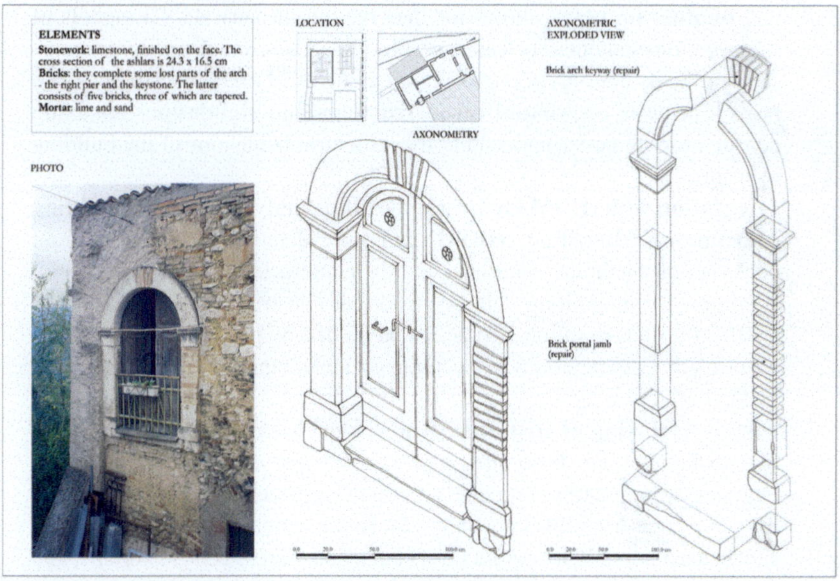

Fig. 3.48 Analysis of construction techniques. Details of the portal with a stone block reveal

as moldings or carvings, that are integrated into the stone blocks. These decorative features not only enhance the visual appeal but also reflect the architectural style of the building and the skill of the craftsmen. The accompanying text explains that the portal's components, such as the cornice, architrave, reveal, and sill, are all made from finely worked limestone, which contributes to the decorative richness of the entrance.

The construction techniques employed in the Episcopal Palace of San Martino sulla Marrucina (CH) stand as a testament to the skill and ingenuity of the builders. Through the detailed analysis of specific elements, valuable insights into the building's construction techniques can be evaluated. Understanding these techniques is crucial for any conservation effort, as it ensures that restoration and renovation works respect the original construction.

Digitization Notes

While the study of historical construction techniques relies heavily on direct observation and analysis, digital tools offer significant advantages for documenting and preserving these methods. By integrating digital technologies such as **3D scanning** and **photogrammetry**, it is possible to capture and archive intricate construction details that would otherwise be difficult to record with traditional methods.

1. **3D Modelling of Construction Techniques**: The use of **laser scanning** and **photogrammetry** allows for the creation of accurate 3D models of construction elements, such as stone blocks, arches, and decorative details. These models not only serve as digital archives of craftsmanship but also provide a basis for **virtual reconstructions** and **simulations**, enabling conservators to better understand the structural behaviour of the building over time.
2. **Integration with HBIM for Craftsmanship Analysis**: **Historic Building Information Modelling (HBIM)** can store detailed information about construction methods, materials, and their interactions within the overall structure. By integrating data on historical construction techniques into HBIM, future restoration efforts can be guided by precise digital records, ensuring that interventions respect the original methods and materials used by the builders.
3. **Digital Archiving of Historical Knowledge**: Many traditional construction techniques are becoming rare or even lost due to modernization. Digital documentation, especially through **high-resolution imagery** and **3D scans**, preserves this knowledge for future generations. These archives can be used not only by conservators but also by architects, historians, and educators to study and teach historical construction practices.

3.6 Material Properties

Masonry is a composite material consisting of the assembly of elements that can be either natural (erratic stones, split stones, roughly shaped or squared stones) or artificial (bricks). The variable characteristics include:

- the material constituting the elements (stone, brick, raw earth, etc., sometimes used in combination);
- the dimensions and shape of the elements;
- the assembly technique (dry-laid or with mortar joints);
- the texture, meaning the geometric arrangement of the elements in the masonry surface; and
- additional details such as layering or the use of chips.

The mechanical behavior of this composite material depends on all these variables.

Typically, the elements exhibit an elastic-brittle behavior, with tensile strength lower than compressive strength but nonetheless significant. Mortar, on the other hand, exhibits elastic-brittle behavior in tension, with much lower strength compared to the elements and, in absolute terms, very low strength. In compression and shear, it dislays ductile and strongly non-linear behavior. The mechanical response of the joints is strongly influenced by friction and shows significant non-linearities.

In masonry, the elements are typically arranged in successive layers, usually horizontally. This configuration leads to the formation of continuous main joints and discontinuous secondary joints, the latter due to the staggered arrangement of the elements (interlocking). The orientation of the main joints is generally orthogonal to the prevailing compressive stresses, optimizing the behavior of the masonry under service loads. However, the main joints become potential planes of discontinuity, reducing the strength of the masonry under tensile and shear stresses induced by seismic actions. In such cases, the friction generated along the joints by the normal compressive stresses associated with inertial forces contributes to both strength and energy dissipation.

The interlocking within the masonry plane affects shear behavior; in general, it is a function of the average ratio between the base and height of the elements and the criteria for staggering the secondary joints. The composition of the masonry in section influences compressive strength and out-of-plane behavior; in the case of two or more wythes, the presence of through-stones that create a connection between the two external wythes (diatones) is significant.

Visual surveys and several investigations (Sect. 3.5.2) can lead to a good understanding and assessment of the quality of the materials and their degradation. However, in some cases, modelling the structural behavior requires precise knowledge of the mechanical parameters governing deformability and strength of the materials, particularly masonry.

Non-destructive diagnostic techniques, such as sonic and ultrasonic tests, can evaluate the homogeneity of mechanical parameters in different parts of the construction but do not provide reliable quantitative estimates of their values, as they are

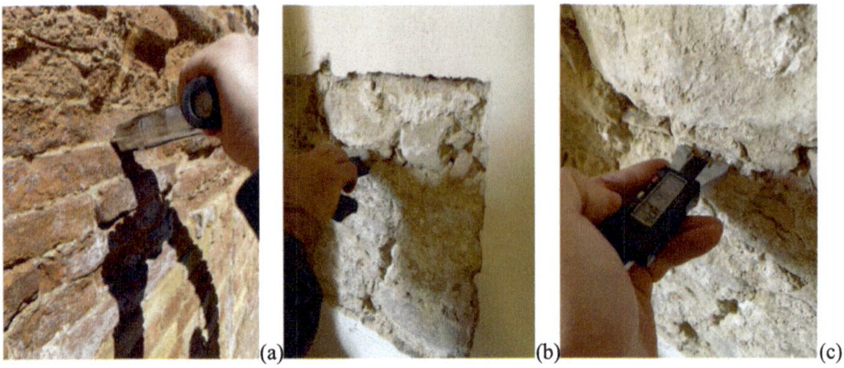

Fig. 3.49 Penetrometric tests carried out on **a** a brick masonry walls and **b**, **c** an irregular stone masonry wall

inferred from the measurement of other quantities (e.g., the propagation speed of volume waves). Direct measurement of the mechanical parameters of masonry, particularly strength, can only be carried out through **slightly destructive or destructive tests**, even if on limited portions. Characterization of the constituent elements (mortar; bricks or stone elements) can be performed on-site or on small samples, taken and subsequently analyzed in a laboratory (DPR 380/2001). Regarding mortars, among other tests, the following can be performed: (i) sclerometric and penetrometric tests (Fig. 3.49); (ii) chemical analyses on samples taken at depth, so as not to be subject to surface degradation, for the characterization of the mortar. For bricks, in addition to determining physical characteristics, it is possible to evaluate the elastic modulus and tensile and compressive strengths through laboratory mechanical tests, including compression and bending tests. For stone elements, lithological characterizations can be performed.

If the **mechanical characteristics of masonry** (understood as a homogeneous assembly of mortar and bricks or stone elements) need to be determined, the following destructive or semi-destructive tests should be referenced:

- For determining the normal elastic modulus and compressive strength: (a) **Double flat-jack test** (a slightly destructive technique, as it is performed on a limited portion of a masonry wall subjected to maximum stress corresponding to the initiation of cracking, carried out by making small cuts, preferably in the mortar joints and therefore easily repairable, as indicated in Fig. 3.50); (b) **Compression test on a masonry panel** (very invasive, as it involves a significant portion of masonry, on the order of a meter, and requires large cuts to accommodate the jacks and often costly contrast structures). The type (b) test should be limited to cases where other investigation methods do not provide sufficiently reliable assessments or when sacrificial masonry is available.
- For determining shear strength and modulus, tests on panels can generally be used, with the same considerations as in the previous paragraph, in two modes:

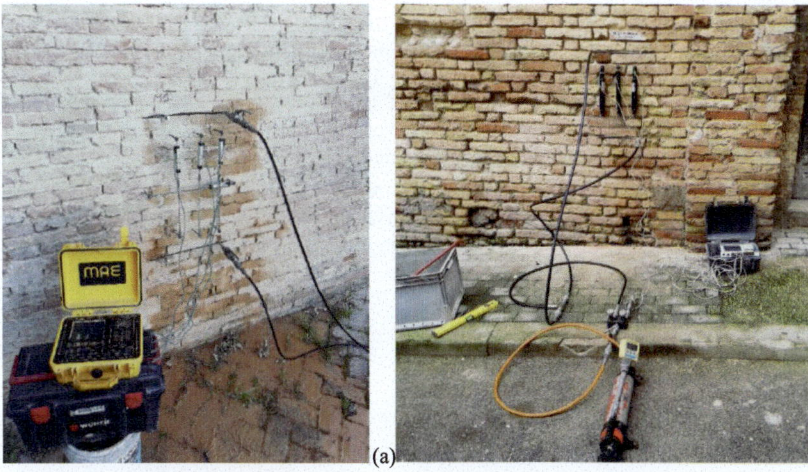

(a) (b)

Fig. 3.50 Double flat-jack tests carried out on two different masonry walls

(a) Diagonal compression test on a square panel;
(b) Compression and shear test on a rectangular panel with a height twice the width. Both of these tests are highly invasive.

Given the destructive nature of the aforementioned tests, they should **only be employed if well-motivated and justified** not only by the use of the results in structural modelling but also by the fact that they are discriminative in evaluating or choosing the intervention. In identifying possible sacrificial areas for potential destructive analyses, consideration should be given to the outcomes of historical research, the state of material conservation, and the survey of valuable surfaces. The number of tests that can be performed on homogeneous material will generally be very limited and will not allow for a statistically significant treatment of the results in relation to formal safety verification procedures based on probabilistic or semi-probabilistic methods. Therefore, the planning of investigations and the interpretation of results should be framed within broader procedures, where the use of even a single experimental datum can be meaningful.

The identification of mechanical characteristics can also be obtained by **analogy with similar masonry**, considering, as much as possible, the effects of degradation. For this purpose, it is desirable that territorial protection and control authorities establish permanent archives containing: (a) almanacs of the various masonry types present over time in the area under examination; (b) tables with reference values of mechanical properties, derived from experiments organized by the same authorities and/or using campaigns conducted for individual interventions and studies. It is also possible to refer, for each masonry type, to the **values of mechanical parameters defined by current regulatory codes**, as for example those indicated in Tables C8.5.I and C8.5.II of the Italian National Code (NTC2018 2018) and its explication document (Circolare 7/2019 2019), as reported below (Table 3.2).

Table 3.2 Mechanical parameters suggested by the Italian Regulatory Code in Table C8.5.I of Circolare 7/2019 (2019). Reference values for the mechanical parameters of masonry and the average specific weight for different types of masonry. The values refer to: **f** = average compressive strength, $\boldsymbol{\tau_0}$ = average shear strength in the absence of normal stresses, $\mathbf{f_{v0}}$ = average shear strength in the absence of normal stresses, **E** = average value of the normal elastic modulus, **G** = average value of the shear elastic modulus, **w** = average specific weight

Types of masonry	**f** (N/mm^2)	$\boldsymbol{\tau_0}$ (N/mm^2)	$\mathbf{f_{v0}}$ (N/mm^2)	**E** (N/mm^2)	**G** (N/mm^2)	**w** (kN/m^2)
	min–max	min–max		min–max	min–max	
Irregular stone masonry (cobbles, erratic and irregular stones)	1.0–2.0	0.018–0.032	–	690–1050	230–350	19
Roughly hewn stone masonry with non-homogeneous wall thickness	2.0	0.035–0.051	–	1020–1440	340–480	20
Split stone masonry with good bonding	2.6–3.8	0.056–0.074	–	1500–1980	500–660	21
Irregular masonry of soft stone (tuff, calcarenite, etc.)	1.4–2.2	0.028–0.042	–	900–1260	300–420	13–16 (**)
Regular ashlar masonry of soft stone (tuff, calcarenite, etc.)	2.0–3.2	0.04–0.08	0.10–0.19	1200–1620	400–500	
Squared stone block masonry	5.8–8.2	0.09–0.12	0.18–0.28	2400–3300	800–1100	22
Solid brick masonry with lime mortar	2.6–4.3	0.05–0.13	0.13–0.27	1200–1800	400–600	18
Hollow brick masonry with cement mortar (e.g., double UNI with 40% perforation)	5.0–8.0	0.08–0.17	0.20–0.36	3500–5600	875–1400	15

(*) For roughly hewn stone masonry, the reference strength values can be increased if there is systematic evidence of deep wedges of stone improving the contacts and damping; in the absence of more precise assessments, a coefficient equal to 1.2 is used
(**) Due to the lithological variability of soft stone, the average specific weight can be estimated with direct tests. In the case of ashlar masonry of soft stone, if there is clear evidence of compressive strength of the units, the compressive strength can be assessed in accordance with §11.10 of NTC2018
(***) For solid brick masonry, if joints exceed 13 mm in thickness, the strength values are reduced using a coefficient of 0.7, and the elastic moduli are reduced using a coefficient of 0.8 in the absence of more precise evaluations

Table C8.5.I of Circolare 7/2019 (2019) (here reported in Table 3.2) provides guidelines on the possible values of mechanical parameters for the most common types of masonry, identified through the survey of construction aspects. These values are given under specific conditions: lime mortar with modest characteristics (average compressive strength $\mathbf{f_m}$ estimated between 0.7 and 1.5 N/mm^2), absence of string courses (horizontal layers of bricks or stones), simply adjacent or poorly connected wall faces, well-executed bonding (in the case of regular elements), and non-consolidated masonry. For seismic verification purposes only, if the mortar exhibits particularly poor characteristics (average compressive strength $\mathbf{f_m}$ estimated below 0.7 N/mm^2), a reduction coefficient of 0.7 should be applied to the strength values and 0.8 to the elastic modulus values (Circolare 7/2019 2019).

In the presence of consolidated masonry or when designing a reinforcement intervention, the explication document of the Italian regulatory code indicates that the values of Table 3.2 can be increased by applying the additional coefficients indicated in Table C8.5.II. These adjustments are made based on the consolidation techniques used, following the methods illustrated below.

- **Consolidation with Injection of Binding Mixtures**. The coefficient indicated in Table 3.3, differentiated for various masonry types, can be applied to both the resistance parameters ($\mathbf{f}$, $\boldsymbol{\tau_0}$ and $\mathbf{f_{v0}}$) and the elastic moduli (**E** and **G**). The achievable benefits are significantly influenced by the original quality of the mortar, with greater improvements observed when the original mortar is of poor quality. It is important to note that the actual benefits of injections depend on the real possibility of the injected mortars filling existing gaps in the masonry structure and adhering to the existing materials. In all cases, it is advisable to carry out preliminary and verification tests to assess the results actually achieved.
- **Consolidation with Reinforced Plaster**. In the absence of specific evaluations, it is possible to adopt the coefficient indicated in Table 3.3, applicable to both the resistance parameters ($\mathbf{f}$, $\boldsymbol{\tau_0}$ and $\mathbf{f_{v0}}$) and the elastic moduli (**E** and **G**). In this case, the coefficients related to the transverse connection and reinforced repointing should not applied. It is noted that consolidation with reinforced plaster is ineffective in the absence of systematic transverse connections and its effectiveness is reduced when applied to only one face of the wall. When adopting any improvement coefficients, the characteristics of the mortars used (cement-based or lime-based) and the reinforcements (metallic or fiber) should be considered. Finally, it is necessary to verify in advance that the wall does not show excessive disintegration or the presence of voids, which would render the coupling with reinforced plaster ineffective; in such cases, it is advisable to combine the intervention with injections.
- **Consolidation with Artificial Diatones or Anti-Expulsion Ties**. In the case of the insertion of artificial diatones with significant shear stiffness and sufficiently widespread, the coefficient indicated for originally well-connected masonry can be applied to all resistance parameters. The tensile connection elements (anti-expulsion ties) have a significant effect only on the compression resistance (**f**).

Table 3.3 Maximum corrective coefficients suggested by the Italian Regulatory Code in Table C8.5.II of Circolare 7/2019 (2019) for consolidated and reinforced masonry

Types of masonry	Current condition			Consolidation interventions			
	Good mortar	Courses or string courses	Transverse connection	Injection of binding mixtures (*)	Reinforced plaster (**)	Reinforced repointing with connection of faces (**)	Maximum overall coefficient
Irregular Stone Masonry (cobbles, erratic and irregular stones)	1.5	1.3	1.5	2	2.5	1.6	3.5
Roughly Hewn Stone Masonry with Non-Homogeneous Wall Thickness	1.4	1.2	1.5	1.7	2.0	1.5	3.0
Split Stone Masonry with Good Bonding	1.3	1.1	1.3	1.5	1.5	1.4	2.4
Irregular Masonry of Soft Stone (tuff, calcarenite, etc.)	1.5	1.2	1.3	1.4	1.7	1.1	2.0
Regular Ashlar Masonry of Soft Stone (tuff, calcarenite, etc.)	1.6	–	1.2	1.2	1.5	1.2	1.8
Squared Stone Block Masonry	1.2	–	1.2	1.2	1.2	–	1.4
Solid Brick Masonry with Lime Mortar	(***)	–	1.3 (****)	1.2	1.5	1.2	1.8
Hollow Brick Masonry with Cement Mortar (e.g., double UNI with 40% perforation)	1.2	–	–	–	1.3	–	1.3

(*) The corrective coefficients related to the injection of bonding mixtures must be adjusted according to the actual benefit provided to the masonry. This benefit should be verified both during the execution phase (injectability) and afterward (experimental verification through sonic tests or similar methods)
(**) Values should be appropriately reduced in the case of walls with considerable thickness (e.g., greater than 70 cm)
(***) In the case of brick masonry, "good mortar" is defined as mortar with an average compressive strength f_m greater than 2 N/mm^2. In such instances, the corrective coefficient can be set equal to $f_m{}^{0.35}$ (f_m in N/mm^2)
(****) In the case of brick masonry, "transversely connected masonry" refers to masonry that is constructed according to traditional standards of workmanship

- **Consolidation with Reinforced Repointing and Connection of Faces**. The coefficient indicated in the table, differentiated for the various masonry types, can be applied to both the resistance parameters (**f,** $\boldsymbol{\tau_0}$ and $\mathbf{f_{v0}}$) and the elastic moduli (**E** and G), with the latter being reduced by 50%.

In case of combined use of different consolidation techniques, the coefficients can be applied multiplicatively; however, the overall coefficient value must not exceed the maximum coefficient indicated in the last column of Table 3.3.

In the mechanical characterization of materials, three levels of testing can be distinguished based on the depth of the analysis:

1. **Limited in-situ inspection**: These are preliminary investigations that are not detailed or extensive, primarily based on **visual examinations** of surfaces and limited checks of the components of the masonry. Localized removal of plaster is conducted to identify the materials constituting the building. With the aid of **historical and critical analysis**, it is possible to divide the masonry walls into areas considered homogeneous. The purpose of these investigations is to identify the types of masonry to be referenced for determining mechanical properties, which includes surveying the masonry texture of the facades and estimating the wall section.
2. **Extended in-situ inspection**: These investigations involve **widespread and systematic visual inspections**, accompanied by **local in-depth studies**. Extensive tests are performed, both on the surface and within the thickness of the masonry (including endoscopies), aimed at understanding the materials and internal morphology of the masonry, identifying homogeneous areas in terms of materials and masonry texture, and detecting transverse connection devices and degradation phenomena. Additionally, the execution of mortar analysis, and if significant, analysis of the constituent elements, is planned. These are accompanied by non-destructive diagnostic techniques (such as penetrometric, sclerometric, sonic, thermographic, and radar methods), which may be integrated with moderately destructive techniques (such as flat-jack testing), aimed at more accurately classifying the masonry type and its quality.
3. **Comprehensive in-situ inspection**: In addition to the requirements of the previous category, direct testing of materials is planned to determine their mechanical parameters. The type and quantity of tests are established by the designer based on the knowledge needs of the structure. The tests must be carried out either in situ or in a laboratory on undisturbed samples taken from the site. These may include, if significant: compression tests (for example, on panels or using double flat-jack tests); shear tests (for example, compression and shear, diagonal compression, direct shear on the joint), selected according to the masonry type and the resistance criterion adopted for the analysis. The tests must be conducted on all masonry types or, at a minimum, on those related to elements that, from the sensitivity analysis based on preliminary data, are deemed significant for the safety assessment. Verification values will be obtained from the average values listed in Table 3.2, using direct experimental measurements on the building, considering the reliability of the testing method. Alternatively, results from tests conducted on other structures in the same area may be

considered, provided there is clear and proven typological correspondence in materials and morphology.

Following the investigations, it is necessary to evaluate, for each test, the representativeness of both the typological class attributed to the material and the average values of the mechanical characteristics of the building to be used in modelling. To this end, **methods that provide estimates of these characteristics through quality indicators of the masonry**, obtained by visually reading the facades and the section, may be useful.

Digitization Notes

The **digital storage of material properties** is crucial for ensuring that data derived from tests and diagnostic analyses are preserved and easily accessible for future interventions. This allows for rapid access to information and comparison between different investigations over time. The use of cloud-based platforms or HBIM systems enables the integration of results from non-destructive, semi-destructive, and destructive tests into a **permanent archive**. This ensures the traceability of material conditions over time, facilitating the consultation of historical data for structural verification and modelling, while preventing the loss of critical information. Moreover, by standardizing the archived data according to regulatory codes such as NTC2018 (2018), results can be interpreted consistently even years later, aiding comparisons between similar materials and different geographic regions.

Storing material data in digital form creates a searchable and updatable **database** where material properties can be correlated with other factors, such as degradation or environmental conditions. This system enables conservation professionals to employ the information for more accurate predictive modelling, accounting for changes in material properties over time and future structural loads.

3.7 Soil and Foundations

The identification and evaluation of the type and integrity of the foundation system, along with the geotechnical characterization of the subsurface soil within the relevant volume, are crucial elements for assessing seismic actions and their effects on the structure.

All investigations and assessments should be preceded by a thorough review of the available documentation concerning the building, covering both its historical and recent background. Based on these preliminary studies, it is possible to develop an **investigation plan** aimed at determining the shape, dimensions, and materials that constitute the foundation structures. Among the various possible investigations, **non-destructive methods**—such as geophysical and tomographic tests—are generally preferred. These tests should be conducted even after the excavation of inspection pits and trenches to expose the foundations and reveal the foundation levels. If necessary, continuous core drilling may be performed, with various orientations, to reach the foundation levels and adequately penetrate the underlying soil layers. Boreholes can also be effectively used for video inspections and geophysical testing (MiBACT 2011).

Special attention must be given to the presence or absence of an archaeological substratum, as it can significantly influence the seismic response of the structure and limit the potential types of intervention on foundations and soils.

Geotechnical investigations should enable the **physical–mechanical characterization** of the foundation soils through in-situ and laboratory testing, aimed at developing geotechnical models suitable for local seismic response analyses and soil-structure dynamic interaction. These investigations must also focus on evaluating the stability of the site where the building is located, with particular emphasis on slope instability and soil liquefaction phenomena. As indicated in the *Guidelines for the Seismic Risk Evaluation and Reduction of the Cultural Heritage* (MiBACT 2011), the investigation methods and geotechnical characterization should comply with the general principles of geotechnical indicated in the current technical standards (EN 1998-1 2023; NTC2018 2018; Circolare 7/2019 2019).

3.8 Knowledge Levels and Confidence Factors

Once the construction has been identified, and depending on the depth of the geometric survey, material-construction investigations, mechanical analysis, and studies of the soil and foundations, the designer assigns a confidence factor (F_C) ranging between 1 and 1.35 EN 1998-1; FEMA 273; FEMA 356).

F_C is a numerical value used to adjust the mechanical properties of the structural materials. For the built heritage, this factor can be determined using the formulation provided in the Italian *Guidelines for the seismic risk evaluation and reduction of the cultural heritage* (MiBACT 2011):

$$F_C = 1 + \sum_{k=1}^{4} F_{Ck} \tag{3.1}$$

The formula considers four different partial Confidence Factors, F_{Ck} (k = 1 to 4), which are assessed based on the level of knowledge attained in each of the following areas: (1) Geometric survey, (2) Historical and construction specifics, (3) Material properties and (4) Soil and foundations. The *Guidelines for the Assessment and Reduction of Seismic Risk in Cultural Heritage* (MiBACT 2011) associate different levels of detail to each of these four types of investigation, and for each level of detail, a corresponding partial confidence factor is defined, as shown in Table 3.4.

The confidence factor is applied differently depending on the models used for the evaluation of seismic safety, which can be classified as follows:

- Models that consider the deformability and strength of materials and structural elements;
- Models that consider the ultimate equilibrium of the various elements of the construction, treating the masonry material as rigid and non-resistant to tension (creating a kinematic mechanism of rigid blocks through the introduction of appropriate disconnections).

Table 3.4 Definition of the levels of detail in investigations for various aspects of knowledge and identification of the corresponding partial confidence factors according to the *Guidelines for the Assessment and Reduction of Seismic Risk in Cultural Heritage* (MiBACT 2011)

Geometric survey	Complete geometric survey	$F_{C1} = 0.05$
	Complete geometric survey, with graphic restitution of the crack and deformation pattern	$F_{C1} = 0$
Identification of the historical and constructive specificities of the building	Hypothetical reconstruction of the construction phases based on a limited material and construction elements survey, combined with an understanding of the transformation events (documentary and thematic investigations)	$F_{C2} = 0.12$
	Partial reconstruction of the construction phases and interpretation of the structural behaviour based on: (a) limited material construction elements survey, combined with an understanding and verification of transformation events (documentary and thematic investigations, diagnostic verification of historiographical hypotheses); (b) extensive material survey and analysis of construction elements, associated with the understanding of transformation events (documentary and thematic investigations)	$F_{C2} = 0.06$
	Complete reconstruction of the construction phases and interpretation of the structural behaviour based on a comprehensive material and construction elements survey, combined with a thorough understanding of the transformation events (documentary and thematic investigations, and any necessary diagnostic tests)	$F_{C2} = 0$
Mechanical properties of materials	Mechanical parameters derived from existing data	$F_{C3} = 0.12$
	Limited investigations on the mechanical parameters of materials	$F_{C3} = 0.06$
	Extensive investigations of the mechanical parameters of materials	$F_{C3} = 0$
Soil and foundations	Limited investigations on the soil and foundations, in the absence of geotechnical data and information on the foundations	$F_{C4} = 0.06$
	Availability of geotechnical data and information on foundational structures; limited investigations on the soil and foundations	$F_{C4} = 0.03$
	Extensive or comprehensive investigations on the soil and foundations	$F_{C4} = 0$

In the first case, the confidence factor is generally applied to the material properties, particularly by reducing the strengths. The initial values of the mechanical characteristics, to which the confidence factor may be applied, will be defined based on the level of knowledge related to the mechanical properties of the materials, using the ranges reported in Tables 3.2 and 3.3, and considering the level of knowledge achieved according to the current Italian Regulatory Code (NTC2018 2018) and its corresponding explanatory document (Circolare 7/2019 2019, Sect. C8.5.4.1). In the second case, namely for rigid body models where the material strength is not considered, the confidence factor is applied directly to the capacity of the structure, that is, by reducing the acceleration corresponding to the different limit states.

A thorough knowledge acquisition process results in a lower FC, which is advantageous in structural analysis because the mean values of the mechanical properties are divided by both the FC and the material's partial safety factor. Thus, the structural analysis of existing buildings is intrinsically linked to the extent of knowledge about the structure. Furthermore, this process incentivizes designers to achieve a comprehensive understanding of the building, enabling the use of higher mechanical properties for the structural materials.

3.9 Heritage Building Information Modelling

3.9.1 Introduction to BIM

The term BIM stands for "*Building Information Modelling*", which refers to the creation of a digital representation of a building. This digital model contains all the necessary information about the building throughout its lifecycle, including the design, construction, management, maintenance, and eventual demolition. BIM has revolutionized the construction industry by enabling more efficient and coordinated workflows among all stakeholders involved in a project, from architects and engineers to contractors and facility managers.

The construction industry is traditionally fragmented, with various disciplines operating in silos, leading to disorganized information flows and inefficiencies. BIM addresses these challenges by providing a centralized, digital model that integrates all relevant data, ensuring transparency, collaboration, and accuracy throughout the project lifecycle.

A particularly intuitive and perhaps even compelling way to understand the scope of digitalization in construction is by referring to the "dimensions" of BIM. While the three spatial dimensions (3D) are sufficient for geometric descriptions, additional dimensions are introduced to incorporate further information such as time, cost, and more. The concept of BIM dimensions helps to categorize the different aspects of information that can be managed within a BIM model as each dimension corresponds to a specific type of model, data or analysis. The most common dimensions referenced in BIM include (Fig. 3.51):

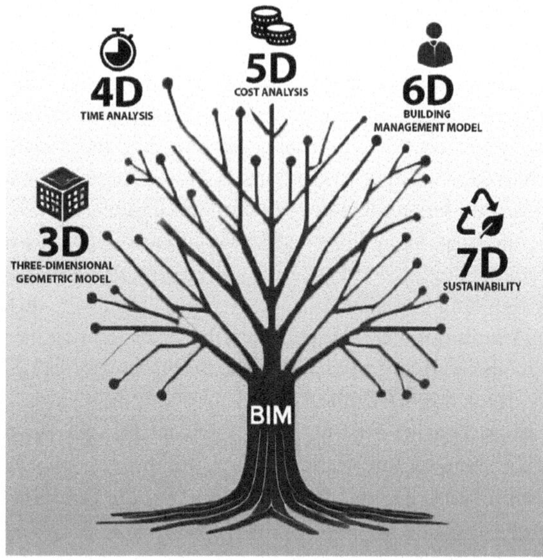

Fig. 3.51 The seven dimensions of BIM

- **3D: Geometric Modelling**. At its core, BIM starts with the creation of a three-dimensional geometric model of the building. This model serves as a digital representation of the building's physical characteristics, enabling detailed visualization and design accuracy. Advanced tools allow for the refinement of this model, ensuring that every architectural and structural element is accurately represented.
- **4D: Time Analysis**. The fourth dimension adds the element of time, allowing for the scheduling and management of construction sequences. This time analysis integrates project timelines, helping stakeholders to visualize the progression of construction activities and identify potential scheduling conflicts before they arise. Techniques like the Work Breakdown Structure (WBS) are employed to break down the project into manageable parts that can be analyzed and tracked over time.
- **5D: Cost Analysis**. The fifth dimension introduces cost analysis into the BIM model. This includes the estimation and management of project costs throughout the lifecycle of the building. By linking material quantities to cost data, BIM allows for accurate budgeting and cost forecasting, helping to minimize financial risks. The dynamic nature of BIM ensures that any changes to the design or construction schedule are automatically reflected in the cost estimates.
- **6D: Building Management**. Beyond construction, BIM extends into the operational phase of the building's lifecycle with the sixth dimension, which focuses on facility management. This dimension incorporates information about the building's operation, maintenance, and eventual decommissioning. An "as-built" model, which reflects the final constructed state of the building, is used as a reference for ongoing management and maintenance tasks, ensuring that the building is efficiently operated over its lifetime.

- **7D: Sustainability**. The seventh dimension of BIM deals with sustainability and environmental impact. This dimension focuses on the integration of sustainable practices into the design, construction, and operation of buildings. It includes analysis of energy efficiency, resource use, and the environmental impact of materials and processes. Sustainable BIM practices help ensure that buildings are designed and operated in ways that minimize their environmental footprint, aligning with broader goals of sustainable development.

These additional dimensions are not merely theoretical but are rooted in practical applications, as outlined in international standards like ISO 19650-1, BS 1192 and ASTM E2506-15, followed by national standards such as the Italian UNI 11,337. These standards specify the methods for managing time (4D), cost (5D), facility operations (6D), and sustainability (7D) in the context of a BIM-based project. Each dimension adds a layer of information that enhances the overall functionality of the BIM model, making it a comprehensive tool for managing every aspect of a building's lifecycle.

3.9.2 Evolution from BIM to HBIM

Building Information Modelling (BIM) has revolutionized the architecture, engineering, and construction (AEC) sector by managing large volumes of data and information throughout the entire building process. Initially designed to enhance new construction projects, BIM's application has now extended to existing buildings, particularly those of historical significance. This extension has led to the development of Historic Building Information Modelling (HBIM), a specialized branch of BIM that focuses on the preservation, restoration, and management of built heritage. HBIM involves creating detailed digital models of historic buildings, which integrate a wide range of data, from geometric and material properties to historical records and conservation plans.

The integration of Building Conditions Assessment (BCA) into HBIM models has been identified as a critical step in effectively managing and refurbishing historic buildings. BCA involves conducting detailed diagnostics of a building's structural, material, and environmental conditions, which are then incorporated into the HBIM model. This integration enhances the model's accuracy and provides a solid foundation for informed conservation strategies, ensuring that refurbishment efforts are based on a comprehensive understanding of the building's current state (Bruno and Fatiguso 2018).

HBIM serves as a comprehensive tool for understanding, documenting, and managing historic structures. By incorporating various data types—architectural, structural, and historical—HBIM provides a holistic view of a building's condition and history, facilitating informed decision-making in conservation efforts.

The concept of HBIM was first introduced by Murphy et al. (2009), who described it as a new method for modelling existing building entities through the generation of intelligent three-dimensional objects that can store and manage information. Their research outlined the use of remote data capture techniques, such as laser scanning and digital imaging, to create detailed engineering drawings and 3D models of historic buildings. These models, enriched with point cloud data and digital images, provided a foundation for creating detailed BIM objects used in both 2D and 3D architectural representations. The study demonstrated that HBIM significantly advances the digital documentation of historical architecture by combining the thoroughness of traditional surveys with the precision of modern 3D modelling techniques.

Over the past decade, the adoption of digital informational modelling in historic architecture has generated a growing body of scientific literature. However, the application of BIM to architectural heritage remains complex due to the specificity of Cultural Heritage and the need to manage various types of information, such as historical-documentary, metric-morphological, material-constructive, stratigraphic, and degradation data. These complexities require a multidisciplinary approach to handle both analytic-scientific investigations and critical-interpretative analyses related to understanding the material and immaterial values of the architectural heritage, such as its authenticity.

The scientific community has increasingly recognized HBIM as a process for creating models from survey data aimed at generating libraries of parametric objects combined with informational data. Key reviews by López et al. (2018), Saygi and Remondino (2013), and (Volk et al. 2014) provide comprehensive overviews of the main challenges associated with BIM for historic architectural heritage. Bianchini et al. (2016) and Osello et al. (2016) also contribute to understanding these challenges, particularly in relation to the irregularity and uniqueness of historic architecture, which necessitates specific reflections on modelling techniques.

Furthermore, studies by Cursi et al. (2015) and Simeone et al. (2014) have explored the integration of BIM modelling with knowledge systems based on ontologies, while Chiabrando et al. (2017) have examined the potential of parametrization and the creation of shared libraries for architectural elements.

Contemporary research has highlighted several areas within HBIM that require further exploration, including the precision and reliability of surveys, the fidelity of models to reality, and the management of temporal and historical data. Informational models, when viewed as tools for the digital visualization of cultural assets, are guided by the principles outlined in the London Charter (Beacham et al. 2006; Brusaporci and Trizio 2013). This charter establishes general methodological guidelines for digital visualization in Cultural Heritage research and communication, emphasizing intellectual integrity, reliability, documentation, sustainability, and accessibility.

When applied to built heritage, BIM significantly enhances its potential by going beyond simple digital and geometric replication. It transforms into a powerful data collection tool, characterized by parametric configuration and well-defined semantics. This approach enables deeper analysis of heritage sites, providing an essential tool for sustainable conservation. In HBIM systems, each process is guided by critical analysis, which informs the modelling and semantic structuring of the building's components. This method not only facilitates a better understanding of the building's history and composition but also supports the creation of models for maintenance, management, and planning future restorations. These models ensure that interventions are both informed and respectful of the building's historical and cultural integrity.

The development of fully interoperable 3D models, rich in informational content, represents a significant advancement in the management of Architectural Heritage (Quattrini et al. 2017). HBIM models, composed of intelligent objects, can be reused and adapted to other buildings with similar constructional and architectural styles. Additionally, the parametric libraries within HBIM models contain detailed information about graphic documentation, typology, and the primary construction features of the building. Ultimately, the HBIM paradigm creates robust data collectors that encompass both geometric and non-geometric information, including historical documents, monitoring data, structural details, and the state of conservation or maintenance.

3.9.3 Components of HBIM

3.9.3.1 Data Collection and Surveying Techniques

The foundation of any HBIM project is the accurate collection of data about the existing building. This typically involves a combination of traditional surveying methods and advanced digital technologies. Key techniques include:

- **Laser Scanning (LiDAR)**: Laser scanning, or Light Detection and Ranging (LiDAR), is a crucial tool in HBIM. It involves using laser beams to capture precise measurements of a building's geometry, producing a detailed point cloud that can be used to create a highly accurate 3D model. This technique is particularly useful for capturing complex or irregular architectural features that are common in historic buildings.
- **Photogrammetry**: Photogrammetry involves taking multiple photographs of a building from different angles and using software to create a 3D model. This method is often used in conjunction with laser scanning to enhance the detail and accuracy of the HBIM model.
- **Unmanned Aerial Vehicles (UAVs)**: Drones equipped with cameras and sensors can be used to capture high-resolution images and data from areas that are difficult to access, such as rooftops and façades. UAVs provide a cost-effective and efficient means of surveying large or complex historic sites.
- **Documentary Research**: Historical documents, such as architectural drawings, maps, and written records, are essential sources of information for HBIM. These documents provide insights into the building's original design, construction methods, and modifications over time. Integrating this historical data into the HBIM model enriches the understanding of the building's heritage.

3.9.3.2 Creation of the Digital Model

Once the data has been collected, the next step is to create the digital model. This involves several stages:

- **3D Modeling**: The point cloud generated from laser scanning and photogrammetry serves as the basis for creating a detailed 3D model of the building. This model captures the geometry and dimensions of the structure with high accuracy.
- **Parametric Object Creation**: In HBIM, buildings are modelled using parametric objects—intelligent 3D elements that represent real-world components such as walls, windows, and doors. These objects are not just geometric representations; they also contain metadata about the materials, construction methods, and historical context. For historic buildings, parametric objects may need to be custom-created to accurately reflect the unique features and materials of the structure.
- **Level of Development (LOD)**: A crucial concept in BIM, and by extension HBIM, is the Level of Development (LOD). LOD defines the level of detail and reliability of the information contained in the model at various stages of the project, as outlined in standards such as the American Institute of Architects (AIA) G202-2013 and the BIM Forum's LOD Specification. The LOD framework ranges from LOD 100 (conceptual) to LOD 500 (as-built). In the context of HBIM, LOD is particularly important because it guides the modelling process, ensuring that the model accurately reflects the building's current state and historical integrity.
    - **LOD 100 (Conceptual Design)**: At this stage, the model includes basic massing and volumetric information, representing the building's overall shape and size. This is useful for early feasibility studies and conceptual planning.
    - **LOD 200 (Design Development)**: The model becomes more detailed, with approximate geometry and basic information about key elements. This stage is typically used for schematic design and initial analysis.
    - **LOD 300 (Detailed Design)**: At this level, the model includes accurate geometry and detailed information about the building components, such as materials, dimensions, and connections. This stage is used for detailed design and construction documentation.
    - **LOD 350 (Detail and Coordination)**: LOD 350 adds more detail to the model, including information about interfaces between building components and systems. It includes precise geometry and spatial relationships between elements, making it suitable for coordination among different disciplines. This level is particularly valuable in HBIM, where the interaction between various historical elements and modern interventions must be carefully managed and documented.
    - **LOD 400 (Construction)**: The model is further refined to include detailed fabrication and installation information. This level is used during the construction phase to guide the assembly and installation of building components.

  - **LOD 500 (As-Built)**: The model represents the building as it was actually constructed, with all components accurately modelled and documented. This level is used for facility management and maintenance.

- **Integration of Historical Data**: Historical information collected during the research phase is integrated into the HBIM model. This includes data on construction techniques, material properties, and historical events that have impacted the building. This information is often linked to specific elements within the model, allowing for a detailed and contextual understanding of the building's history.

3.9.3.3 Interoperability and Data Management

One of the key advantages of HBIM is its ability to integrate and manage a wide range of data. However, this requires the model to be interoperable with other software and systems used in the conservation and management of historic buildings.

- **Industry Foundation Classes (IFC)**: IFC is an open data format that facilitates the exchange of BIM data between different software platforms. It is crucial for ensuring that HBIM models can be used by various stakeholders, including architects, engineers, conservators, and facility managers. The use of IFC also ensures that the model can be easily updated and maintained over time, preserving its value as a long-term asset.
- **Data Storage and Management**: HBIM models can generate large amounts of data, including 3D geometry, metadata, historical documents, and sensor data. Efficient data management is essential to ensure that this information is accessible and usable throughout the building's lifecycle. Cloud-based platforms are increasingly being used to store and manage HBIM data, providing secure access and facilitating collaboration among stakeholders.

Digitization Notes

Digitalizing complex heritage buildings through HBIM faces several challenges related to the vast amount of data generated during the survey process. Techniques like **laser scanning** and **photogrammetry** create highly detailed point clouds, but managing this volume of data requires advanced **data processing** strategies. A significant challenge is converting these raw data into usable **parametric models** while maintaining accuracy, especially when dealing with irregular geometries often found in historic structures. The key to efficient digitalization lies in **simplifying data** without losing critical details, a process that can strain computational resources and data storage capacity. For instance, Fig. 3.52 illustrates the downsizing of a 3D STL model, which consequently results in a reduction of the model's quality and detail, dependent on the number of vertices or triangles in the mesh. This reduction is achieved using Meshmixer™ software (Autodesk®, Inc., San Rafael, CA, USA), by applying the 'Reduce Mesh' command repeatedly (as shown in Fig. 3.53). Achieving an optimal reduction hinges on the necessity to maintain specific details for a particular case study, the computational power of the tools employed, and the requirements for data sharing among different users.

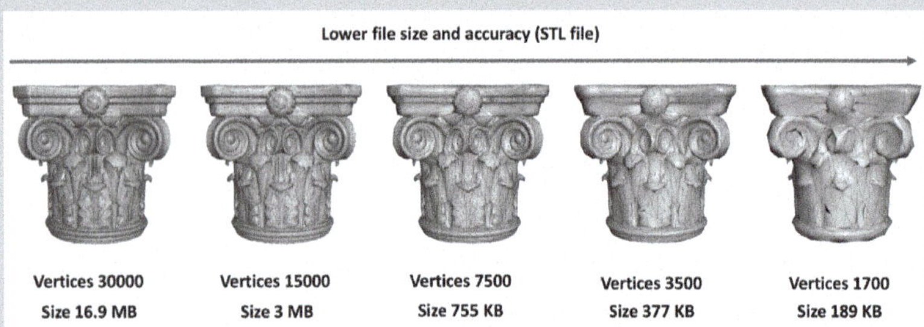

Fig. 3.52 Example of the reduction in the size of a 3D STL model

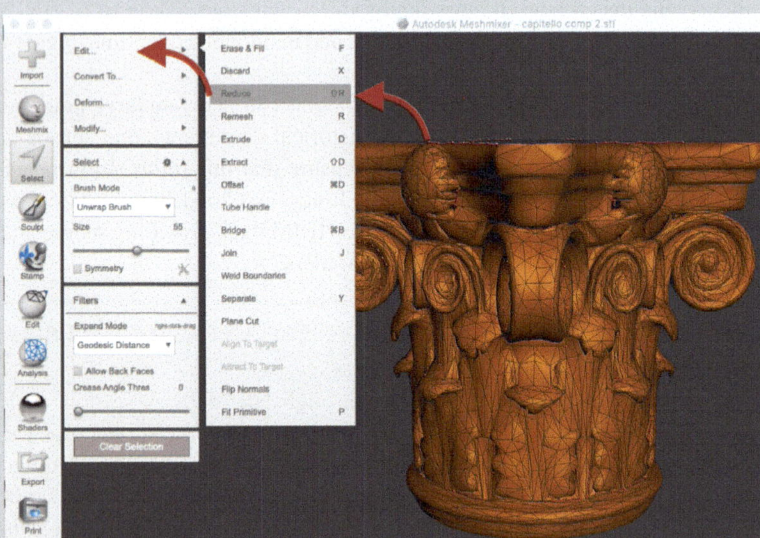

Fig. 3.53 How to reduce the size of a 3D STL model in the software Meshmixer

3.9.4 Applications of HBIM

HBIM has a wide range of applications in the conservation, restoration, and management of historic buildings. Some of the key areas where HBIM is being used include conservation and restoration, facility management and documentation and archiving.

3.9.4.1 Conservation and Restoration

HBIM is a powerful tool for planning and executing conservation and restoration projects. By providing a detailed and accurate digital model of the building, HBIM allows conservators to assess the condition of the structure, identify areas of concern, and plan interventions in a way that minimizes impact on the building's historical fabric. The main roles of the HBIM into the conservation and restoration process are the following:

- **Condition Assessment**: HBIM models can be used to conduct detailed assessments of the building's condition, including structural integrity, material degradation, and environmental impacts. This information is essential for planning conservation strategies that are both effective and sensitive to the building's historical significance.
- **Restoration Planning**: The detailed information contained in an HBIM model allows for precise planning of restoration work. This includes selecting appropriate materials and techniques, scheduling work to minimize disruption, and ensuring that all interventions are documented and reversible.
- **Simulation and Analysis**: HBIM models can be used to simulate various scenarios, such as the impact of environmental conditions, structural loading, or proposed alterations. This capability is invaluable for assessing the potential outcomes of different conservation approaches and ensuring that the chosen strategy is sustainable and effective.

Recent advances in Diagnosis-Aided Historic Building Information Modelling and Management (DA-HBIMM) suggest that incorporating diagnostic data with performance assessment features can significantly enhance the effectiveness of conservation strategies. By integrating cognitive automation and AI, DA-HBIMM enables automated evaluation of building conditions, early identification of potential issues, and optimization of restoration and maintenance activities. This approach not only improves the accuracy of the model but also streamlines decision-making processes, ensuring that conservation efforts are both efficient and effective (Beacham et al. 2006; Bruno et al. 2018).

3.9.4.2 Facility Management

HBIM is not only useful during the conservation and restoration phases but also plays a crucial role in the ongoing management of historic buildings.

- **Maintenance Planning**: The detailed information in an HBIM model provides a comprehensive record of the building's condition and maintenance history. This allows facility managers to plan and schedule maintenance activities more effectively, ensuring that the building remains in good condition over time.
- **Monitoring and Diagnostics**: HBIM models can be integrated with monitoring systems that use sensors to track various environmental and structural parameters, such as temperature, humidity, and movement. This real-time data can be used to identify potential issues early and take preventive action before significant damage occurs.
- **Space Management**: For historic buildings that are still in use, such as museums, churches, or government buildings, HBIM models can be used to manage space efficiently. This includes tracking occupancy, planning for events, and ensuring that the building is used in a way that preserves its historical integrity.

3.9.4.3 Documentation and Archiving

One of the key advantages of HBIM is its ability to create a detailed and accurate digital record of a historic building. This digital archive serves as a valuable resource for future research, conservation efforts, and educational purposes.

- **Digital Archiving**: HBIM models provide a comprehensive digital archive of the building, including its geometry, materials, and historical data. This archive can be used for future research, providing a detailed record of the building's condition at a specific point in time.
- **Educational Use**: HBIM models can be used as educational tools, allowing students and researchers to explore historic buildings in detail. Virtual tours, 3D visualizations, and interactive models provide an immersive experience that can enhance understanding and appreciation of the building's historical significance.
- **Public Engagement**: HBIM models can also be used to engage the public in the preservation of historic buildings. By providing virtual access to the building, HBIM allows people to explore and learn about the structure's history, architecture, and cultural significance, fostering greater awareness and support for conservation efforts.

Digitization Notes

The integration of HBIM into conservation efforts requires careful planning of digital workflows. Digital tools must ensure that **real-time data** from sensors or monitoring systems is interoperable with HBIM models for **predictive maintenance** and early detection of structural issues. Digitalization also introduces challenges in maintaining a balance between **preserving historical integrity** and integrating modern data management systems. For example, in **facility management**, the use of HBIM can streamline ongoing maintenance but requires constant updates and accurate synchronization between the physical building and the digital model.

3.9.5 Technical Challenges and Solutions in HBIM

While HBIM offers numerous benefits, it also presents several technical challenges that must be addressed to ensure its successful implementation.

3.9.5.1 Data Accuracy and Model Fidelity

One of the primary challenges in HBIM is ensuring the accuracy of the data collected during the survey phase. Historic buildings often have complex geometries and irregularities that can be difficult to capture accurately using traditional surveying methods.

- **High-Resolution Scanning**: Advances in laser scanning technology have significantly improved the accuracy of data collection, allowing for high-resolution scans that capture even the most intricate details of a building's geometry. However, the sheer volume of data generated by these scans can be challenging to manage and process.
- **Modelling Complex Geometries**: Creating a digital model that accurately reflects the complex geometries of historic buildings requires advanced modelling techniques. This includes the use of specialized software that can handle irregular shapes and non-standard elements, as well as the creation of custom parametric objects that accurately represent the building's unique features.

3.9.5.2 Interoperability and Data Exchange

Another significant challenge in HBIM is ensuring interoperability between different software platforms and stakeholders. Historic building projects often involve a wide range of professionals, including architects, engineers, conservators, and historians, each using different tools and systems.

- **Open Standards**: The use of open standards, such as IFC, is crucial for ensuring that HBIM models can be easily shared and used by all stakeholders. However, not all software platforms fully support these standards, leading to potential issues with data exchange and model integrity.

- **Collaboration Platforms**: Cloud-based collaboration platforms are increasingly being used to facilitate data exchange and collaboration among project teams. These platforms provide a centralized repository for HBIM data, allowing all stakeholders to access and update the model in real time.

3.9.5.3 Managing Heterogeneous Data

Historic buildings are characterized by the diversity of data types that must be managed, including geometric data, material properties, historical records, and conservation plans.

- **Data Integration**: Integrating these diverse data types into a single HBIM model requires a robust data management framework. This includes the use of relational databases, linked data technologies, and metadata standards to ensure that all information is accurately represented and easily accessible.
- **Semantic Modelling**: Semantic modelling techniques are increasingly being used in HBIM to manage the complexity of historic buildings. These techniques involve the use of ontologies and vocabularies to define the relationships between different elements of the model, allowing for more sophisticated querying and analysis of the data.

Digitization Notes

One of the central challenges in digitalizing heritage buildings through HBIM is managing the **accuracy** and **complexity** of the data collected during surveys. Historic structures often feature intricate, irregular geometries that are difficult to capture and model accurately. These buildings do not conform to modern construction standards, which adds to the difficulty of creating precise digital representations. The **scan-to-BIM** process, which involves transforming point clouds into parametric models, is a crucial step in the HBIM workflow. However, the high resolution of **laser scanning** and **photogrammetry** produces dense point clouds that can be difficult to process due to their size. This poses a significant problem: how to **simplify** the point cloud without losing important geometric details, especially when dealing with complex features such as curved surfaces, vaults, or decorative elements typical of historical architecture.

In their study, Pepe et al. (2024) propose a novel method for addressing this issue by using **geometric feature analysis** to identify the most critical points in a point cloud and simplify the dataset without compromising accuracy. Their method leverages tools like **ShrinkWrap**, an algorithm that creates a mesh around selected geometric features of the structure. This allows for the generation of **polygonal meshes** that preserve the essential shape and details of the building, while significantly reducing the data size and computational requirements. This approach is particularly beneficial for complex architectural elements that are challenging to model, as it reduces errors such as non-manifold surfaces or overlapping geometries.

The process reduces a dense point cloud to a manageable and accurate **mesh model**, ready for further processing in BIM environments like **Rhinoceros/Grasshopper** and **Revit**, ensuring that the final HBIM model remains **lightweight** and **accurate** (Fig. 3.54), even when managing large datasets typical of heritage buildings (Pepe et al. 2024).

Fig. 3.54 Output obtained from the point cloud simplification process using the ShrinkWrap algorithm for the Beach Patrol Headquarters located in Miami Beach

3.9.6 Future Directions in HBIM

The field of HBIM is rapidly evolving, with new technologies and methodologies continually being developed to address the challenges and expand the possibilities of digital heritage management.

3.9.6.1 Integration with Emerging Technologies

Several emerging technologies are poised to play a significant role in the future of HBIM, including artificial intelligence (AI), machine learning, and the Internet of Things (IoT).

- **Artificial Intelligence and Machine Learning**: AI and machine learning can be used to automate the creation of parametric objects, improve the accuracy of historical data interpretation, and enhance the overall efficiency of the HBIM process. For example, AI algorithms can be trained to recognize and classify architectural features from point cloud data, significantly reducing the time and effort required to create a detailed model.

- **Internet of Things (IoT)**: IoT devices, such as sensors and smart meters, can be integrated into HBIM models to provide real-time monitoring and data collection. This enables more proactive maintenance and management of historic buildings, as well as the ability to simulate and analyse the impact of environmental conditions on the building's structure.
- **Digital Twins**: The concept of digital twins, which involves creating a virtual replica of a physical asset, is increasingly being applied to historic buildings through HBIM. These digital twins can be used to simulate different scenarios, such as the impact of climate change or natural disasters, allowing for more informed decision-making in the preservation and management of built heritage.

3.9.6.2 Sustainable Conservation Practices

As the field of heritage conservation continues to evolve, there is a growing emphasis on sustainability and the need to balance conservation efforts with environmental considerations.

Energy Efficiency: HBIM models can be used to assess and improve the energy efficiency of historic buildings, identifying areas where improvements can be made without compromising the building's historical integrity. This includes the integration of energy modelling tools and the simulation of different retrofit scenarios to determine the most effective strategies for reducing energy consumption.

Lifecycle Management: Sustainable conservation also involves considering the entire lifecycle of a building, from construction to demolition. HBIM models provide a comprehensive view of the building's lifecycle, allowing for more informed decisions about the long-term management and preservation of the structure. This includes planning for future maintenance, assessing the environmental impact of different materials, and ensuring that conservation efforts are aligned with broader sustainability goals.

Historic Building Information Modelling represents a significant advancement in the digital management of built heritage. By integrating detailed and accurate digital models with a wealth of historical and material data, HBIM provides a comprehensive tool for the conservation, restoration, and management of historic buildings. As the technology continues to develop, its applications will expand, offering new possibilities for the sustainable preservation of cultural heritage.

The adoption of HBIM is not just about preserving the past but also about ensuring that our cultural heritage is accessible and appreciated in the future. By embracing this technology, we can bridge the gap between tradition and innovation, creating a digital record of our built heritage that will stand the test of time.

References

Al-Omari A, Brunetaud X, Beck K, Al-Mukhtar M (2015) Preliminary digital health record of limestone walls in Al-Ziggurat, Al-Nimrud city, Iraq. J Cult Herit 16:737–740. https://doi.org/10.1016/j.culher.2014.11.005

Arayici Y (2008) Towards building information modelling for existing structures. Struct Surv 26:210–222

ASTM E2506-15 (2020) Standard guide for developing a cost-effective risk mitigation plan for new and existing constructed facilities. American Society for Testing and Materials

Babacan AE, Gelisli K (2015) Ultrasonic investigations of marble columns of historical structures built in two different periods. Carbonates Evaporites 30:357–363. https://doi.org/10.1007/s13146-015-0233-x

Beacham R, Denard H, Niccolucci F (2006) An introduction to the London charter

Bianchini C, Inglese C, Ippolito A (2016) The role of BIM (Building Information Modeling) for representation and managing of built and historic artifacts. In: Disegnare con...

Bisegna F, Ambrosini D, Paoletti D, Sfarra S, Gugliermetti F (2014) A qualitative method for combining thermal imprints to emerging weak points of ancient wall structures by passive infrared thermography—a case study. J Cult Herit 15:199–202. https://doi.org/10.1016/j.culher.2013.03.006

Bosche FN, Forster AM, Valero E (2015) 3D surveying technologies and applications: point clouds and beyond. Heriot-Watt University

Brunetaud X, Stefani C, Badosa SJ, Beck K, Al-Mukhtar M (2012) Comparison between photo-modelling and laser scanning to create a 3D model for a digital health record. Euro J Environ Civil Eng

Bruno S, Fatiguso F (2018) Building conditions assessment of built heritage in historic building information modeling. Int J Sustain Dev Plan 13:36–48. https://doi.org/10.2495/SDP-V13-N1-36-48

Bruno S, De Fino M, Fatiguso F (2018) Historic building information modelling: performance assessment for diagnosis-aided information modelling and management. Autom Constr 86:256–276

Bruno S, Galantucci RA, Musicco A (2023) Decay detection in historic buildings through image-based deep learning. VITRUVIO—Int J Architectural Technol Sustain 8:6–17. https://doi.org/10.4995/vitruvio-ijats.2023.18662

Brusaporci S, Trizio I (2013) La "Carta di Londra" e il Patrimonio Architettonico: riflessioni circa una possibile implementazione. SCIRES-IT—Sci Res Inf Technol 3:55–68. https://doi.org/10.2423/I22394303V3N2P55

Cantagallo C, Cianchino G, Sangiorgio V, Masciotta MG, Pierantozzi M, Lops C, Di Loreto S, Brando G, Spacone E (2024) GENESIS: a web-based platform for managing the seismic risk of historic centres of Southern Italy. In: Boffill Y, Lombillo I, Blanco H (eds) REHABEND 2024. 10th Euro-American congress on construction pathology, rehabilitation technology and heritage management. University of Cantabria, Building Technology R&D Group, pp 2522–2530

Chabot D (2018) Trends in drone research and applications as the journal of unmanned vehicle systems turns five. J Unmanned Veh Syst 6:vi–xv

Chiabrando F, Lo Turco M, Santagati C (2017) Digital invasions: From point clouds to historical building object modeling (h-BOM) of a unesco WHL site. In: International archives of the photogrammetry, remote sensing and spatial information sciences—ISPRS archives. International Society for Photogrammetry and Remote Sensing, pp 171–178

Circolare 7/2019 (2019) Istruzioni per l'applicazione dell'«Aggiornamento delle "Norme tecniche per le costruzioni"» di cui al D.M. 17 gennaio 2018 (in Italian)

CNR-ICR (1990) Raccomandazioni Normal 1/88. Alterazioni Macroscopiche dei Materiali Lapidei: Lessico—Macroscopic Alteration of Stone Materials: Glossary

Colomina I, Molina P (2014) Unmanned aerial systems for photogrammetry and remote sensing: a review. ISPRS J Photogramm Remote Sens 92:79–97

Cursi S, Simeone D, Toldo I (2015) A semantic web approach for built heritage representation. In: Communications in computer and information science. Springer, pp 383–401
DPR 380/2001 Testo unico delle disposizioni legislative e regolamentari in materia edilizia. G.U. n. 245 del 20 ottobre
Ellenberg A, Kontsos A, Bartoli I, Pradhan A (2014) Masonry crack detection application of an unmanned aerial vehicle. In: Computing in civil and building engineering, pp 1788–1795
EN 1998-1 Eurocode 8: Design of structures for earthquake resistance, Part 1: general rules, seismic actions and rules for buildings, Brussels, CEN
Eschmann C, Kuo C-M, Kuo C-H, Boller C (2012) Unmanned aircraft systems for remote building inspection and monitoring. In: 6th European workshop on structural health monitoring (EWSHM 2012)
Fatiguso F, De Fino M, Scioti A (2017) Performance assessment of italian school heritage architecture by non-destructive investigation. Int J Heritage Architecture: Stud Repairs Maintence 1:564–579. https://doi.org/10.2495/ha-v1-n4-564-579
FEMA 273 (1997) NEHRP guidelines for the seismic rehabilitation of buildings
FEMA 356 (2000) Prestandard and commentary for the seismic rehabilitation of buildings
Forster AM, Bosché F, Valero E (2017) Digitally Surveying the Damage. RICS Build Conserv J
Giovannini EC, Tomalini A, Pristeri E, Bergamasco L, Lo Turco M (2021) Decay classification using artificial intelligence. In: International archives of the photogrammetry, remote sensing and spatial information sciences—ISPRS archives. International Society for Photogrammetry and Remote Sensing, pp 847–854
Guadagnuolo M, Titomanlio I, Faella G, Gambardella C (2020) The royal palace in caserta: nondestructive tests for safety and conservation. Int J Adv Res Eng Technol 11:537–546. https://doi.org/10.34218/IJARET.11.5.2020.055
Hallermann N, Morgenthal G, Rodehorst V (2015) Vision-based monitoring of heritage monuments: unmanned Aerial Systems (UAS) for detailed inspection and high-accuracy survey of structures. In: Structural studies, repairs and maintenance of heritage architecture XIV. WIT Press, pp 621–632
Işık N, Halifeoğlu FM, İpek S (2020) Nondestructive testing techniques to evaluate the structural damage of historical city walls. Constr Build Mater 253. https://doi.org/10.1016/j.conbuildmat.2020.119228
ISO 19650-1 (2018) Organization and digitization of information about buildings and civil engineering works, including building information modelling (BIM)—Information management using building information modelling. Part 1: concepts and principles
Karimi N, Mishra M, Lourenço PB (2024) Deep learning-based automated tile defect detection system for Portuguese cultural heritage buildings. J Cult Herit 68:86–98. https://doi.org/10.1016/j.culher.2024.05.009
Kim C, Park T, Lim H, Kim H (2013) On-site construction management using mobile computing technology. Autom Constr 35:415–423. https://doi.org/10.1016/j.autcon.2013.05.027
Kirimtat A, Krejcar O (2018) A review of infrared thermography for the investigation of building envelopes: advances and prospects. Energy Build 176:390–406
Kolev K, Tanskanen P, Speciale P, Pollefeys M (2014) Turning mobile phones into 3D scanners. In: Proceedings of the IEEE conference on computer vision and pattern recognition. pp 3946–3953
Kottke J, Matero F, Hinchman J (2011) Terrestrial laser scanning: imaging, quantifying, and monitoring microscale surface deterioration of stone at heritage sites. Change over Time 1:268–287
Kottke J (2009) An investigation of quantifying and monitoring stone surface deterioration using three dimensional laser scanning. Master of Science in Historic Preservation, University of Pennsylvania
Kylili A, Fokaides PA, Christou P, Kalogirou SA (2014) Infrared thermography (IRT) applications for building diagnostics: a review. Appl Energy 134:531–549
Lim MK, Cao H (2013) Combining multiple NDT methods to improve testing effectiveness. Constr Build Mater 38:1310–1315. https://doi.org/10.1016/j.conbuildmat.2011.01.011

López FJ, Lerones PM, Llamas J, Gómez-García-Bermejo J, Zalama E (2018) A review of heritage building information modeling (H-BIM). Multimodal Technol Interaction 2

Lu X, Clements-Croome D, Viljanen M (2010a) Integration of chaos theory and mathematical models in building simulation. Part I: literature review. Autom Constr 19:447–451

Lu X, Clements-Croome D, Viljanen M (2010b) Integration of chaos theory and mathematical models in building simulation. Part II: conceptual frameworks. Autom Constr 19:452–457. https://doi.org/10.1016/j.autcon.2010.01.003

Maierhofer C, Krankenhagen R, Röllig M, Schlichting J, Schiller M, Seidl T, Mecke R, Kalisch U, Hennen C, Meinhardt J (2011) Investigating historic masonry structures with a combination of active thermography and 3D laser scanner. Quant Infrared Thermogr J 8:115–118. https://doi.org/10.3166/qirt.8.115-118

McGibbon S (2020) A Common Structured Integrated Collaborative Digitised (CrOsS) Framework for the Historic Building Repair and Maintenance (R&M) Sector. PhD Thesis, Heriot-Watt University

MiBACT (2011) Guidelines for the seismic risk evaluation and reduction of the cultural heritage—alignment to the Technical Standards for Constructions (in Italian)

Mishra M, Lourenço PB (2024) Artificial intelligence-assisted visual inspection for cultural heritage: state-of-the-art review. J Cult Herit 66:536–550

Mitka B, Pluta M (2016) Comparative analysis of the process of creating a 3Dmodel od architecture object with using laser scanning and structure from motion technologies. Surv Geol Mining Ecol Manag 2:847–854

Moropoulou A, Labropoulos KC, Delegou ET, Karoglou M, Bakolas A (2013) Non-destructive techniques as a tool for the protection of built cultural heritage. Constr Build Mater 48:1222–1239. https://doi.org/10.1016/j.conbuildmat.2013.03.044

Murphy M, Mcgovern E, Pavia S (2009) Historic building information modelling (HBIM). Struct Surv 27:311–327. https://doi.org/10.1108/02630800910985108

NTC2018 (2018) D.M. 17/01/2018. Norme Tecniche per le Costruzioni (NTC2018). Gazzetta Ufficiale, n. 42 del 20/02/2018, Supplemento ordinario n.8 [in Italian]

Osello A, Davardoust S, Semeraro F, Lucibello G, Barone L, Dellosta M, Ronzino A, Del Giudice M, Fonsati A, De Luca D, Ugliotti FM, Rapetti Niccollo' (2016) La centralità della Rappresentazione con il BIM. In: Disegnare con... Roberto Mingucci, pp 5.1–5.16

Paoletti D, Ambrosini D, Sfarra S, Bisegna F (2013) Preventive thermographic diagnosis of historical buildings for consolidation. J Cult Herit 14:116–121. https://doi.org/10.1016/j.culher.2012.05.005

Pepe M, Garofalo AR, Costantino D, Tana FF, Palumbo D, Alfio VS, Spacone E (2024) From point cloud to BIM: a new method based on efficient point cloud simplification by geometric feature analysis and building parametric objects in rhinoceros/grasshopper software. Remote Sens (Basel) 16. https://doi.org/10.3390/rs16091630

Quattrini R, Pierdicca R, Morbidoni C, Malinverni ES (2017) Conservation-oriented hbim. The bimexplorer web tool. In: International archives of the photogrammetry, remote sensing and spatial information sciences—ISPRS archives. International Society for Photogrammetry and Remote Sensing, pp 275–281

Saygi G, Remondino F (2013) Management of architectural heritage information in BIM and GIS: state-of-the-art and future perspectives. Int J Heritage Digital Era 2:695–713. https://doi.org/10.1260/2047-4970.2.4.695

Schops T, Sattler T, Hane C, Pollefeys M (2015) 3D Modeling on the Go: interactive 3D reconstruction of large-scale scenes on mobile devices. In: Proceedings—2015 International Conference on 3D vision, 3DV 2015. Institute of Electrical and Electronics Engineers Inc., pp 291–299

Scott J, Laing R, Hogg G (2013) Built heritage digitization: opportunities afforded by emerging cloud based applications. In: Proceedings of the international conference on cloud computing technology and science, cloudcom. IEEE Computer Society, pp 88–93

Simeone D, Cursi S, Toldo I, Carrara G (2014) B(H)IM-Built Heritage Information Modelling Extending BIM approach to historical and archaeological heritage representation. In: eCAADe.

Department of Architecture and Built Environment, Faculty of Engineering and Environment, Newcastle upon Tyne, pp 613–622

Smits J (2011) Application of 3D terrestrial laser scanning to map building surfaces. J Archit Conserv 17:81–94. https://doi.org/10.1080/13556207.2011.10785083

Stefani C, Brunetaud X, Janvier-Badosa S, Beck K, De Luca L, Al-Mukhtar M (2014) Developing a toolkit for mapping and displaying stone alteration on a web-based documentation platform. J Cult Herit 15:1–9. https://doi.org/10.1016/j.culher.2013.01.011

Sun C, Jiang F, Jiang S (2013) Research on RFID applications in construction industry. J Netw 8:1221–1228. https://doi.org/10.4304/jnw.8.5.1221-1228

Vähä P, Heikkilä T, Kilpeläinen P, Järviluoma M, Gambao E (2013) Extending automation of building construction—survey on potential sensor technologies and robotic applications. Autom Constr 36:168–178

Valero E, Adán A, Cerrada C (2015) Evolution of RFID applications in construction: a literature review. Sensors (Switzerland) 15:15988–16008. https://doi.org/10.3390/s150715988

Valero E, Bosché F, Forster A (2018) Automatic segmentation of 3D point clouds of rubble masonry walls, and its application to building surveying, repair and maintenance. Autom Constr 96:29–39. https://doi.org/10.1016/j.autcon.2018.08.018

Volk R, Stengel J, Schultmann F (2014) Building Information Modeling (BIM) for existing buildings—literature review and future needs. Autom Constr 38:109–127

Xi XH, Wang C, Wan YP, Khuon KN (2015) Ta keo temple reconstruction based on terrestrial laser scanning technology. In: International archives of the photogrammetry, remote sensing and spatial information sciences—ISPRS archives. International Society for Photogrammetry and Remote Sensing, pp 449–452

Chapter 4
Multi-scale and Multi-risk Analysis of Built Heritage

Abstract This chapter provides an in-depth exploration of seismic and climate-related risk assessment for built heritage through a multi-scale and multi-risk perspective. It introduces the four-level approach for seismic vulnerability, covering methods ranging from preliminary visual assessments and Masonry Quality Index calculations to advanced analytical and vulnerability index techniques, applied both at territorial individual buildings scales. The chapter analyses local and global collapse mechanisms, modelling strategies, and risk estimation, with special attention to the derivation of fragility curves and damage distributions. Additionally, it broadens the risk framework to include climate change impacts, examining the effects of temperature fluctuations, moisture, precipitation, wind, and solar radiation on the structural integrity and material durability of heritage buildings. In this context, the Urban Heat Island (UHI) effect is also examined, as its combination with global warming and heat waves can lead to elevated urban ambient temperatures, potentially compromising thermal comfort and increasing energy demand during the summer season. The calculation of UHI intensity is presented at both the urban and building-scale levels. This combined approach supports integrated conservation strategies to address the challenges posed by seismic events and long-term climatic impacts.

4.1 Seismic Vulnerability Levels: Four-Level Approach

The assessment of a heritage building is based on the knowledge of its historical and constructive evolution, which forms the basis for restoration efforts, the evaluation of structural safety, and the design of strengthening interventions. This knowledge is essential not only for appreciating the building's historical and architectural significance but also for conducting vulnerability and risk assessments at different scales and levels (Cantagallo et al. 2020), allowing public authorities to make informed decisions regarding its conservation, restoration, and management. These decisions are made at different levels, either in terms of territorial planning or the restoration of individual buildings, depending on the type of vulnerability and risk assessment.

© The Author(s), under exclusive license to Springer Nature Switzerland AG 2025
C. Cantagallo et al., *Digitization of Built Heritage*, Digital Innovations in Architecture, Engineering and Construction, https://doi.org/10.1007/978-3-031-96598-2_4

The two critical factors that most influence the overall behavior of a masonry structure are the quality of the constituent materials and the wall-to-wall connections as well as the connections between decks and walls. These characteristics have a significant impact on the structural response of a building, particularly under seismic loading.

Assessing these two parameters is essential for understanding the seismic behavior of historic buildings and serves as the foundation for the four-level methodology discussed in the following sections. Each level contributes to a more comprehensive understanding of the structure's behavior under seismic actions, and is fundamental to ensuring a safer built environment, which in turn supports the sustainable development of tourism in the region (Borri et al. 2020; Cantagallo et al. 2024).

Figure 4.1 visually summarizes the multi-level methodology for assessing seismic vulnerability and planning interventions for historic masonry buildings. The approach is structured into four distinct levels (LV0 to LV3), each addressing a progressively detailed understanding of the building's structural behavior under seismic loading. The process begins with a **preliminary assessment (LV0)** based on the Masonry Quality Index (MQI), which evaluates the typological and constructive characteristics of masonry, such as its materials, texture, and structural arrangement. Buildings with inadequate masonry quality, determined when the MQI due to the horizontal out-of-plane actions O (MQI_O) is lower than 4, are immediately flagged for focused interventions. In such cases, further levels of assessment are unnecessary, as the structure's inability to withstand seismic forces has already been established. For buildings that meet the minimum requirements at LV0, the **first-level assessment (LV1)** involves rapid evaluations of the building's condition using simplified mechanical models and visual screening techniques. This level is particularly valuable in large-scale territorial analyses, where the aim is to prioritize resources and plan interventions efficiently across a wide range of structures. The LV1 analysis provides essential insights into the general vulnerabilities of the building, forming a foundation for more detailed assessments. The transition from broader territorial analyses to detailed individual building assessments occurs through the **Building-Level Prevention Policy**, which identifies heritage building requiring strengthening. This policy prioritizes buildings that require more in-depth evaluation, leading to the accurate acquisition of building-specific knowledge. This step involves detailed documentation of the structural and material characteristics, as well as an understanding of the building's historical evolution, architectural significance, and existing vulnerabilities. This phase ensures that subsequent interventions are based on a comprehensive and accurate understanding of the building's unique characteristics, allowing for the development of tailored conservation and strengthening strategies.

The next stage focuses on **local collapse mechanisms**, addressed in the **second-level vulnerability assessment (LV2)**. This phase involves a detailed evaluation of specific vulnerabilities, such as out-of-plane failures, which are prevalent in masonry buildings with weak connections between walls, floors, or roofs. By identifying these localized weaknesses, the LV2 analysis enables the development of targeted strengthening interventions to mitigate potential failures at the component level. Finally,

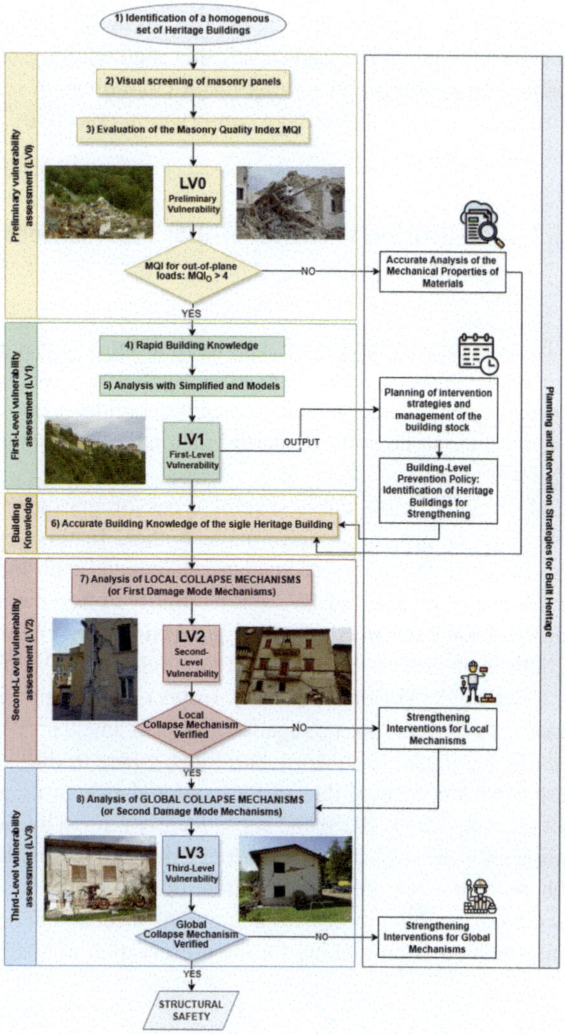

Fig. 4.1 Multi-level methodology for seismic vulnerability assessment and intervention planning for historic masonry buildings, illustrating the progression from preliminary evaluations (LV0) to detailed global analyses (LV3) and their integration into broader conservation strategies

the **third-level assessment** (**LV3**) provides a comprehensive analysis of the entire building by simulating its global structural behavior under seismic forces. Advanced techniques, including finite element modeling, are used to assess overall structural integrity. This level offers the most detailed understanding of the building's seismic performance, which is crucial for designing holistic conservation and restoration strategies. By addressing global collapse mechanisms, the LV3 assessment ensures the building's **structural safety**.

Figure 4.1 highlights the practical applications of this **multi-level approach**, linking the assessment process to broader planning and management strategies. The results support the development of building-level prevention policies and inform decisions on resource allocation for strengthening interventions. By integrating the insights gained at each level, this methodology offers a systematic and replicable framework for preserving historic buildings in seismic-prone areas, balancing structural safety with heritage conservation objectives.

4.2 Preliminary Vulnerability Assessment and Masonry Quality Index

The structural behavior of historic buildings subjected to seismic events varies depending on the quality of the masonry. Surveys conducted after recent earthquakes revealed that, in areas near the epicenter, buildings constructed with rubble masonry or irregular stone suffered partial or total collapse due to masonry disaggregation (Sisti et al. 2019).

During an earthquake, several types of masonry are unable to form macro-elements or behave as deformable bodies. In this case, a different failure mode occurs earlier than the mechanisms associated with local or global collapse, investigated in the LV1 and LV2 Vulnerability Assessments, respectively. This failure mode, known as "**masonry disaggregation**" or "**crumbling collapse**" (Borri et al. 2020) is specifically addressed in the Preliminary Vulnerability Assessment'(LV0).

The MQI (**Masonry Quality Index**) method provides a robust framework for assessing masonry disaggregation and mechanical properties. This approach is particularly significant for historic buildings, where traditional structural testing methods are often impractical due to the potential risk of damage. Initially introduced by Borri et al. (2015, 2018) and subsequently refined by Borri et al. (2020), the method allows for the evaluation of masonry disaggregation, highlighting cases where this phenomenon requires careful attention. By using this method, professionals gain access to essential information about the mechanical behavior of masonry, ensuring informed decision-making for the preservation of historic buildings.

4.2.1 Masonry Quality Index

Masonry Quality Index (**MQI**), proposed and refined in several literature works (Borri and De Maria 2009; Marino et al. 2014; Borri et al. 2015, 2018, 2020; Castori et al. 2018), is a fundamental parameter for a simplified, visual and semi-quantitative assessment of the structural and mechanical characteristics of masonry. Masonry, particularly in historical structures, is often a highly heterogeneous composite material, resulting from the combination of elements such as bricks or stones—whether regular, irregular, or roughly hewn—bound with mortar joints of varying thickness and often constructed with multiple layers or wythes. To develop effective interventions for such structures, an **initial visual inspection** is crucial to evaluate the mechanical behavior of the masonry, focusing on parameters such as the **level of interconnection between wall components**, the **distribution of internal voids**, and the **mechanical and physical properties of the materials**. The MQI method identifies the key construction parameters that influence the masonry's response to applied loads, yielding a numeric value called MQI. The numerical value of this index is derived based on compliance with specific construction standards known as the **"rules of the art"**, which include all construction techniques that ensure the wall's integrity, compactness, and monolithic behavior. Since these rules affect the masonry's response differently depending on the direction of the applied force, the MQI is further divided into:

1. MQI_V for vertical actions V,
2. MQI_O for horizontal out-of-plane actions O
3. MQI_I for horizontal in-plane actions I.

For each load application direction, the MQI is determined by assigning a qualitative rating (**Fulfilled-F**, **Partially Fulfilled-PF**, or **Not Fulfilled-NF**) based on the evaluation of seven constructive parameters:

- Characteristics and Quality of Masonry Units **SM** (Table 4.1)
- Dimensions of Masonry Units **SD** (Table 4.2)
- Shape of Masonry Units **SS** (Table 4.3)
- Connection Between Wall Leaves **WC** (Tables 4.4 and 4.5)
- Horizontality of Mortar Joints **HJ** (Table 4.6)
- Staggering of Vertical Joints **VJ** (Tables 4.7 and 4.8)
- Mortar Quality and Unit Contact **MM** (Table 4.9).

After having assigned the qualitative evaluation to each parameter according to criteria shown from Tables 4.1, 4.2, 4.3, 4.4, 4.5, 4.6, 4.7, 4.8 and 4.9, these ratings

Table 4.1 Qualitative evaluation of the characteristics and quality of masonry units SM (Borri et al. 2020)

Degraded/damaged elements (>50% of total number of elements)	**NF**
Hollow-core bricks (solid < 30%)	
Mud bricks	
Unfired bricks	
Degraded/damaged elements (≥10%; ≤ 50%)	**PF**
Hollow bricks (55% ≥ solid ≥ 30%)	
Sandstone or tuff elements	
Undamaged elements or degraded/damaged elements <10%	**F**
Solid fired bricks	
Hollow bricks (55% < solid)	
Concrete units	
Hardstones	

Table 4.2 Qualitative evaluation of the dimensions of masonry units SD (Borri et al. 2020)

Predominance of blocks with larger dimension <20 cm	**NF**
Header bond (no stretchers)	
Predominance of blocks with larger dimension of 20–40 cm	**PF**
Copresence of elements of different dimensions	
Predominance of blocks with larger dimension >40 cm	**F**

Table 4.3 Qualitative evaluation of the shape of the masonry units SS (Borri et al. 2020)

Rubble, or rounded or pebble stonework (predominant) on both masonry leaves	**NF**
Copresence of rubble, or rounded or pebble stonework; barely/perfectly cut stones; and bricks on both masonry leaves	**PF**
For double-leaf walls: squared blocks or bricks for one leaf and pebbles or irregular masonry for the adjacent wall leaf	
Masonry made of irregular (rubble, rounded, and pebble) stones, with pinning stones	
Barely cut stones or perfectly cut stones on both masonry leaves (predominant)	**F**
Brickwork masonry	

Table 4.4 Qualitative evaluation of the connection between wall leaves WC, for non-accessible vertical section of the wall (Borri et al. 2020)

Small stones	**NF**
No headers	
For double-leaf walls	**PF**
Limited number of stone headers	
Wall thickness larger than stone larger dimension	
Wall thickness similar to stone larger dimension	**F**

(**F**, **PF**, **NF**) are converted into numerical values according to the direction of load application, as specified in Table 4.10.

The quantitative values shown in Table 4.10 lead to a **global score (MQI) ranging between 0 and 10** for each direction of applied force. The formulas used to calculate MQI vary depending on the type of masonry.

For **solid brick or stone block masonry**:

$$\begin{aligned} MQI_V &= m \cdot g \cdot r_V \cdot SM_V \cdot (SD_V + SS_V + WC_V + HJ_V + VJ_V + MM_V) \\ MQI_O &= m \cdot g \cdot r_O \cdot SM_O \cdot (SD_O + SS_O + WC_O + HJ_O + VJ_O + MM_O) \\ MQI_I &= m \cdot g \cdot r_I \cdot SM_I \cdot (SD_I + SS_I + WC_I + HJ_I + VJ_I + MM_I) \end{aligned} \quad (4.1)$$

For other types of masonry, where thick bed joints or other structural features are not considered, Eq. (4.1) simplifies to:

Table 4.5 Evaluation of the connection between wall leaves WC, for accessible vertical section of the wall (Borri et al. 2020)

Single-leaf wall: $M_l^* < 1.4$	**NF**
Double-leaf wall: $M_l < 1.4$ (on a wall leaf) and $M_l < 1.6$ (on the other leaf)	
Header bonding (brickwork masonry)	
Walls made of small stones	
Unconnected wall leaves	
Single-leaf wall: $1.4 < M_l < 1.6$	**PF**
Double-leaf wall:	
(a) both wall leaves with $1.4 < M_l < 1.6$	
(b) $M_l > 1.6$ for only one wall leaf, $M_l < 1.4$ for the other	
(c) $M_l > 1.6$ for only one wall leaf, $1.4 < M_l < 1.6$ for the other	
Single-leaf wall: $M_l > 1.6$	**F**
Double-leaf wall: $M_l > 1.6$ for both wall leaves	

*where M_l is the "minimum distance" passing between mortar joints of a vertical section of a wall with height equal to 1 m

Table 4.6 Qualitative evaluation of the horizontality of the mortar joints HJ (Borri et al. 2020)

Bed joints not continuous	**NF**
Bed joints partially continuous	**PF**
For double-leaf wall: only one leaf with continuous bed joints	
Bed joints continuous	**F**
Stone masonry wall with bricks courses (distance between courses <60 cm)	

Table 4.7 Evaluation of the staggering of vertical joints VJ with a qualitative analysis (Borri et al. 2020)

Vertically aligned head joints	**NF**
Vertically aligned head joints for at least 2 large stones	
Header bond (no stretchers)	
No mechanical interlocking between the stones	
Partially staggered head joints (head joints in successive courses are not offset by one-half the unit length)	**PF**
Properly staggered head joints (head joints in successive courses are offset by one-half the unit length)	**F**

$$
\begin{aligned}
MQI_V &= m \cdot SM_V \cdot (SD_V + SS_V + WC_V + HJ_V + VJ_V + MM_V) \\
MQI_O &= m \cdot SM_O \cdot (SD_O + SS_O + WC_O + HJ_O + VJ_O + MM_O) \\
MQI_I &= m \cdot SM_I \cdot (SD_I + SS_I + WC_I + HJ_I + VJ_I + MM_I)
\end{aligned}
\quad (4.2)
$$

Table 4.8 Evaluation of the staggering of vertical joints VJ with a quantitative analysis (Borri et al. 2020)

Single-leaf wall: $M_l < 1.4$	**NF**
Double-leaf wall: $M_l > 1.4$ for a wall leaf, $M_l < 1.6$ for the other	
Header boding pattern (brickwork masonry)	
Walls made of pebbles or small stones	
No mechanical interlocking between the stones	
Single-leaf wall: $1.4 < M_l < 1.6$	**PF**
Double-leaf wall:	
(a) both wall leaves $1.4 < M_l < 1.6$	
(b) $M_l > 1.6$ for a wall leaf, $M_l < 1.4$ for the other	
(c) $M_l > 1.6$ for a wall leaf, $1.4 < M_l < 1.6$ for the other	
Single-leaf wall: $M_l > 1.6$	**F**
Double-leaf wall: $M_l > 1.6$ for both wall leaves	

*where M_l is the "minimum distance" passing between mortar joints of a wall panel with length and height equal to 1 m (Fig. 4.2)

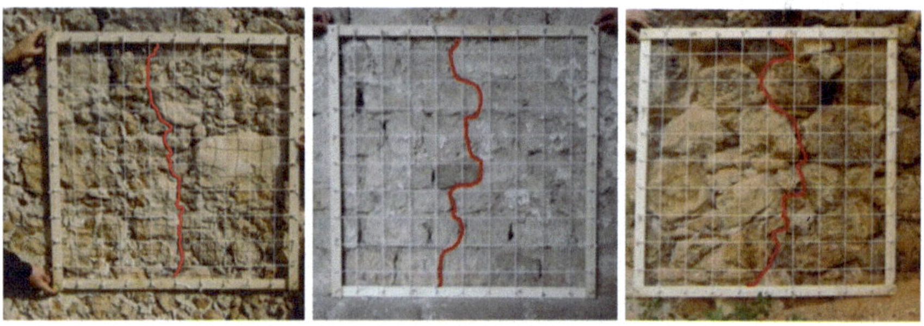

Fig. 4.2 Identification of the M_l length on wall panels measuring 1 square meter

Table 4.9 Qualitative evaluation of mortar properties MM (Borri et al. 2020)

Very weak mortar and dusty mortar with no cohesion	**NF**
No mortar (dry rubble or pebble stonework)	
Thick bed joints made of weak mortar (thickness comparable to stone/brick thickness)	
Porous stones/bricks with weak bond to mortar	
Medium quality mortar, with bed joints not largely notched	**PF**
Masonry made of irregular (rubble) stones and weak mortar, and pinning stones	
Good-quality and non-degraded mortar, regular bed joints, or thick bed joints made of very good-quality mortar	**F**
Masonry made of large perfectly cut stones with no mortar (dry) or very thin bed mortar joints	

Table 4.10 Coefficients applied to the 7 parameters used to calculate the MQI (Borri et al. 2015, 2020)

	Vertical loading (V)			Horizontal in-plane loading (I)			Horizontal out-of-plane loading (O)		
	NF	**PF**	**F**	**NF**	**PF**	**F**	**NF**	**PF**	**F**
HJ	0	1	2	0	0.5	1	0	1	2
WC	0	1	1	0	1	2	0	1.5	3
SS	0	1.5	3	0	1	2	0	1	2
VJ	0	0.5	1	0	1	2	0	0.5	1
SD	0	0.5	1	0	0.5	1	0	0.5	1
MM	0	0.5	2	0	1	2	0	0.5	1
SM	0.3	0.7	1	0.3	0.7	1	0.5	0.7	1

where m, g, and r are corrective coefficients that account for the quality of mortar and bed joints. Specifically, the coefficient m reflects the influence of weak mortar quality (for very-weak mortar, having a compressive strength f_m lower than 0.7 MPa, $m = 0.7$, otherwise $m = 1.0$), g accounts for the thickness of the bed joints in brickwork or block masonry (for thick bed joints, greater than 13 mm, $g = 0.7$, otherwise $g = 1$), r represents a set of correction factors (r_V, r_O and r_I) varying in function of the quality of the mortar MM and the specific loading direction (vertical, out-of-plane, or in-plane), as indicated below:

- MM not fulfilled (NF): $r_V = 0.2$; $r_O = 1$; $r_I = 0.1$
- MM partially fulfilled (PF): $r_V = 0.6$; $r_O = 1$; $r_I = 0.85$
- MM fulfilled (F): $r_V = 1$; $r_O = 1$; $r_I = 1$.

Equations (4.1) or (4.2) yields three distinct MQI values, each ranging between 0 and 10, depending on the direction of the applied load (vertical, in-plane horizontal, or out-of-plane horizontal) and the type of analyzed masonry (e.g., solid brick, stone block, or other masonry types). A higher MQI value indicates better performance of the masonry under the given loading condition, reflecting greater compliance with the "rules of the art" in the construction of the masonry assemblage.

The research team that developed the MQI method (Borri et al. 2015, 2018, 2020) also established correlations between MQI values and the typical mechanical properties of different masonry types, as defined in Table C8.5.I of the Italian Guidelines to the Construction Regulatory Code (Circolare 7/2019 2019). Using the multiplication factors provided in Table C8.5.II of Circolare 7/2019 (2019), they simulated a total of 74 virtual walls by exploring different combinations of masonry types and their possible variations. The calculated MQI values for these walls were then correlated to their corresponding mechanical properties. This work provided a set of equations allowing user to numerically calculate the **masonry properties starting from the evaluation of MQI**. Equations (4.3)–(4.7) report the estimation of the masonry shear strength τ_0, the Elastic Modulus E, the masonry compressive strength f, the Shear

Modulus G and the Shear Strength f_{v0}, respectively. Since Table C8.5.I. of Circolare 7/2019 (2019) specifies three values for each masonry type and mechanical parameter—minimum, average, and maximum—three corresponding equations are reported for each mechanical parameter.

$$\begin{aligned} \tau_{0,\min} &= 0.0004(MQI_I)^2 + 0.0055(MQI_I) + 0.0173 \ R^2 = 0.825 \\ \tau_{0,\text{mean}} &= 0.0006(MQI_I)^2 + 0.0075(MQI_I) + 0.0224 \ R^2 = 0.913 \\ \tau_{0,\max} &= 0.0008(MQI_I)^2 + 0.0094(MQI_I) + 0.0275 \ R^2 = 0.843 \end{aligned} \tag{4.3}$$

$$\begin{aligned} E_{\min} &= 608\, e^{0.154(MQI_V)} \ R^2 = 0.709 \\ E_{\text{mean}} &= 742.1\, e^{0.153(MQI_V)} \ R^2 = 0.720 \\ E_{\max} &= 876\, e^{0.151(MQI_V)} \ R^2 = 0.724 \end{aligned} \tag{4.4}$$

$$\begin{aligned} f_{\min} &= 1.055\, e^{0.193(MQI_V)} \ R^2 = 0.803 \\ f_{\text{mean}} &= 1.444\, e^{0.182(MQI_V)} \ R^2 = 0.824 \\ f_{\max} &= 1.892\, e^{0.175(MQI_V)} \ R^2 = 0.828 \end{aligned} \tag{4.5}$$

$$\begin{aligned} G_{\min} &= 201.4\, e^{0.142(MQI_I)} \ R^2 = 0.652 \\ G_{\text{mean}} &= 254.6\, e^{0.141(MQI_I)} \ R^2 = 0.680 \\ G_{\max} &= 298.7\, e^{0.141(MQI_I)} \ R^2 = 0.695 \end{aligned} \tag{4.6}$$

$$\begin{aligned} f_{v0,\min} &= 0.03\, MQI_I^{0.909} \ R^2 = 0.893 \\ f_{v0,\text{mean}} &= 0.0475\, MQI_I^{0.854} \ R^2 = 0.928 \\ f_{v0,\max} &= 0.0654\, MQI_I^{0.822} \ R^2 = 0.918 \end{aligned} \tag{4.7}$$

Digitization Notes

The use of digital technologies offers new perspectives in automating the Masonry Quality Index (MQI). One of the most recent advancements in this field is the development of a method to digitalize the MQI within **Heritage Building Information Modelling (HBIM)**. The **Visual Programming Language (VPL)**, as demonstrated by Calvano et al. (2022), provides a user-friendly, node-based interface for integrating complex datasets into HBIM. Using VPL tools such as **Grasshopper**, the MQI can be dynamically linked to geometric and material data within an HBIM model, enabling real-time adjustments and simulations. For instance, modifications to a masonry wall's geometry or material properties automatically update related MQI calculations, providing immediate feedback on the wall's seismic performance.

This parametric modelling process also enhances information flow efficiency. VPL automates data transfer between software platforms, **reducing**

manual input errors and ensuring consistency across disciplines. The adoption of these methods highlights the transformative role of digitization in heritage conservation, offering a **scalable and replicable framework** for the systematic analysis of historic masonry structures.

4.2.2 Preliminary Vulnerability Assessment and Masonry Disaggregation

The phenomenon of masonry disaggregation (Fig. 4.3), commonly referred to as "crumbling," is a critical collapse mechanism in historic unreinforced masonry buildings subjected to seismic forces. While mechanisms involving masonry macro-elements have been extensively studied, disaggregation remains underexplored despite its significant implications for structural safety. This mode of failure can start with lower seismic forces than those required to activate macro-element or global mechanisms and typically occurs in localized portions of buildings.

MQI, particularly the **MQI_O** component, serves as a valuable tool for assessing the susceptibility of walls to disaggregation. This visual method evaluates masonry against the "rules of the art," considering parameters such as material properties, wall geometry, and mortar characteristics. Borri et al. (2020) demonstrated that **walls with MQI_O values below a threshold of 4 are more likely to experience disaggregation during seismic events**. Their findings are supported by analyses of masonry types documented in the Italian Seismic Code (NTC2018 2018) and calibrated against real case studies from past earthquakes.

The most vulnerable masonry typologies include irregular stone arrangements, often characterized by weak mortar, unconnected wall leaves, and irregular bonding patterns. Conversely, masonry with horizontal bed joints, adequate mortar quality, and proper connections exhibit greater resilience. Disaggregation is less common in squared hardstone or brick masonry, which tend to perform satisfactorily under seismic loads unless weakened by poor mortar quality or lack of transversal connections.

4.3 Seismic Vulnerability and Risk for Large-Scale Analyses

Large-scale seismic vulnerability and risk assessment represents a fundamental tool for territorial risk management and the strategic planning of mitigation interventions. This approach enables understanding how an entire urban area or region might respond to a seismic event, providing crucial information for the protection

Fig. 4.3 Examples of masonry disaggregation occurred during the 2016 Central Italy Earthquake

of built heritage and the safeguarding of human lives. Unlike analyses at the single building level, large-scale assessments focus on identifying the most vulnerable areas and systemic issues that could compromise the overall resilience of a historic center. This is made possible through the use of **expeditious surveys**, **simplified indicators** and historical data, which help build an overview of seismic vulnerability at territorial, urban or neighborhood level.

The main goal of these analyses is to provide an efficient tool that allows public decision-makers to develop **prevention plans** and **prioritize interventions**. By identifying high-risk areas, public authorities can **optimize available resources** and **plan reinforcement measures** and targeted preventive actions, thereby minimizing the destructive effects of future seismic events as much as possible. In this perspective, the large-scale approach serves as an essential starting point for subsequent, more detailed analyses and more targeted local interventions, ensuring the effective use of available economic and technical resources.

Fig. 4.4 Seismic Risk as combination of three factors: hazard, vulnerability and exposure

The Italian and European territories are characterized by significant seismic activity. The existing building stock is often highly vulnerable, as it includes a large number of historic structures that lack modern seismic protection measures. Consequently, the seismic risk in these areas is significantly high.

Seismic risk is defined as a function of three factors: (a) seismic **vulnerability** of buildings, which depends on the intrinsic characteristics of the structures; (b) seismic **hazard** of the construction site, which is influenced by its location; and (c) **exposure**, which quantifies the assets at risk in terms of users, functions, value, and contents within the structures (Fig. 4.4).

Research on large-scale seismic vulnerability and risk assessment began in the early 1970s and has since evolved through various methodological approaches (D'Ayala and Novelli 2010):

- **Empirical Methods**: These methods are based on Damage Probability Matrices (DPMs).
- **Analytical/Mechanical Methods**: Also referred to as Capacity Curve-based (CCs) methods, they allow for evaluations of structural behavior using various models and analysis techniques.
- **Vulnerability Index Methods**: These methods are based on qualitative observations and are particularly useful for assessing large numbers of buildings where detailed analyses are impractical.

4.3.1 Empirical Methods

One of the earliest approaches to rapid seismic vulnerability assessment was the **Damage Probability Matrix** (**DPM**) method, introduced in the 1970s by Whitman et al. (1973), which used damage probability matrices derived from observations of damage to over 1600 buildings affected by the 1971 San Fernando earthquake in California ($M_w = 6.5$) to evaluate the vulnerability of the area.

This approach was subsequently used by Grünthal (1998) to define the **European Macroseismic Scale EMS-98**, which directly links construction types to six specific vulnerability classes. These classes, labelled A to F, represent decreasing levels of vulnerability, ranging from the weakest structures, such as adobe or unreinforced masonry (URM), to the most resistant, such as those with advanced earthquake-resistant design (ERD).

- **Classes A to C** correspond to typical building types with increasing structural strength, including adobe, brick, and reinforced concrete (RC) structures.
- **Classes D and E** reflect progressively lower vulnerability due to enhanced seismic design, incorporating materials like timber, confined masonry, and steel, known for their earthquake resilience.
- **Class F** represents the highest level of earthquake resistance, reserved for structures incorporating advanced seismic design principles.

According to Grünthal (1998), to evaluate a building's vulnerability, the first step is to identify its **construction type**. This determines its basic **vulnerability class**, which can then be adjusted based on specific construction details or seismic retrofitting. The Vulnerability Table (Fig. 4.5) lists the most prevalent European building types, indicating their most likely vulnerability classification along with the potential range of variability.

The EMS-98 also establishes a **damage scale**, widely used by researchers worldwide, to classify observed damage for each structural typology into five damage grades (DGs). These grades range from slight damage (DG1) to complete collapse (DG5). Table 4.11 provides this classification for unreinforced masonry (URM) buildings, the structural typology that typically characterizes heritage buildings.

The DPM methodology categorizes urban areas into building groups based on predefined qualitative descriptors, such as material and construction age (e.g., old reinforced concrete buildings, unreinforced masonry buildings, etc.). For each DG, the percentage of buildings observed to have sustained that level of damage is recorded and a histogram displaying the frequencies observed for each DG is

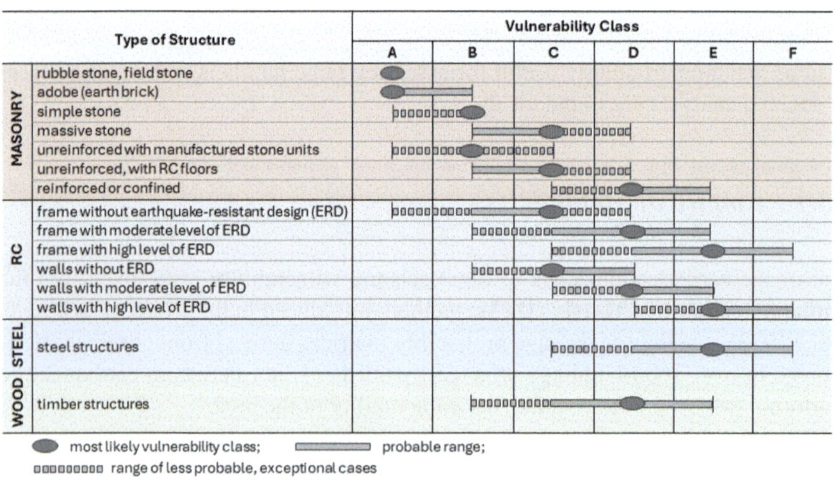

Fig. 4.5 Vulnerability classes defined by Grünthal (1998) for each structural typology

Table 4.11 Classification of damage to masonry buildings (Grünthal 1998)

Damage grade DG	Effects	Image
DG 1: Negligible to slight damage (no structural damage, slight non-structural damage)	• Hair-line cracks in very few walls • Fall of small pieces of plaster only • Fall of loose stones from upper parts of buildings in very few cases	
DG 2: Moderate damage (slight structural damage, moderate non-structural damage)	• Cracks in many walls • Fall of fairly large pieces of plaster • Partial collapse of chimneys	
DG 3: Substantial to heavy damage (moderate structural damage, heavy non-structural damage)	• Large and extensive cracks in most walls • Roof tiles detach. Chimneys fracture at the roof line; failure of individual non-structural elements (partitions, gable walls)	
DG 4: Very heavy damage (heavy structural damage, very heavy non-structural damage)	• Serious failure of walls; partial structural failure of roofs and floors	
DG 5: Destruction (very heavy structural damage)	• Total or near total collapse	

constructed. Numerous studies have adopted this type of representation to empirically analyse seismic damage in historical centres of small villages (Tavani et al. 2017; Formisano 2017a, b; Brando et al. 2017), in entire cities or parts of them (Eleftheriadou and Karabinis 2011, 2013; Rosti et al. 2018), or within an entire diocese (De Matteis et al. 2016).

As an example, the two histograms in Fig. 4.6 illustrate example damage probability matrices derived from observations of damage in 100 buildings. Figure 4.6a shows that 25% of buildings sustained DG 0, 32% DG 1, 27% DG 2, 12% DG 3, 3% DG 4, and 1% DG 5. Meanwhile, Fig. 4.6b indicates that 4% of buildings experienced DG 0, 20% DG 1, 44% DG 2, 26% DG 3, 5% DG 4, and 1% DG 5.

Several scientific studies (Braga and Liberatore 1982; Brando et al. 2017; De Matteis et al. 2016; Rosti et al. 2018) indicate that the **binomial distribution** is generally representative of the observed damage level distribution. The binomial distribution p_k is defined based on the average observed damage level μ_D and the number of observed samples k, as shown in the equation below.

$$p_k = \frac{5!}{k!(5-k)!}\left(\frac{\mu_D}{5}\right)^k \cdot \left(1 - \frac{\mu_D}{5}\right)^{5-k} \tag{4.8}$$

The average observed damage level μ_D can be calculated using the following formula:

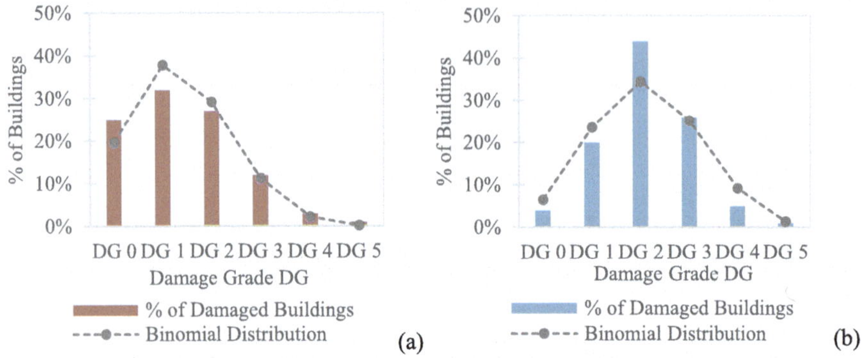

Fig. 4.6 Representation of two example of damage probability matrices with the corresponding binomial distributions

$$\mu_D = \frac{\sum_{k=0}^{5} D_{ki}}{N} = \frac{\sum_{k=0}^{5} k \cdot n_k}{N} \tag{4.9}$$

where N is the number of observed buildings and D_{ki} is the score k assigned to each building, ranging from 0 to 5, and equal to the product of the damage score k and the number of buildings in damage state n_k.

The application of the binomial distribution to the example shown in Fig. 4.6 generates the curve represented with the grey dashed line. The comparison between the empirical DPM and the binomial distribution confirms that this type of model is well-suited for evaluating damage distribution after a seismic event, provided that the average damage level is known.

4.3.2 *Analytical/Mechanical Methods*

The analytical/mechanical methods identify key **geometric and mechanical characteristics** of buildings during field inspections and correlate them with **analytical models**. These models are used to carry out **simplified nonlinear analyses**, predicting potential damage levels at varying seismic intensities.

Significant applications of the analytical/mechanical methods include the FAMIVE procedure (D'Ayala and Speranza 2003) and other studies (Lang and Bachmann 2004; Formisano 2017a, b).

The Guidelines for the Seismic Risk Evaluation and Reduction of Cultural Heritage—Alignment to the Technical Standards for Construction (MiBACT 2011) also propose **simplified mechanical models** (**LV1**) for assessing the seismic performance of the built heritage at the territorial scale, aimed at a preliminary risk evaluation. Specifically, for palaces, villas, and other structures with spine walls and intermediate floors, the simplified mechanical model enables a quantitative evaluation of the return period corresponding to reaching the Life Safety Limit State (LSLS) and the associated peak ground acceleration (PGA). This evaluation assumes that there is an **in-plane wall collapse** as part of the structure's **global behaviour**.

The model is based on a **few geometric and mechanical parameters** that are easy to collect during the initial stages of investigation, and it does not account for possible interactions with adjacent buildings. According to the Guidelines (MiBACT 2011), the evaluation of the acceleration corresponding to the LSLS should follow a simplified nonlinear model of the structure through the following steps:

1. Calculation of shear strengths at each storey level and along the building's principal axes (x and y).
2. Identification of the most vulnerable storey and direction based on the minimum obtained values.
3. Assessment of the acceleration a_{LSLS} for the weakest storey and direction.

The seismic capacity assessment simulates the global structural behaviour and the damage or collapse of the in-plane walls due to shear or bending.

Formisano (2017a) and Casapulla et al. (2018) compared the capacity curves obtained using the simplified model proposed by MiBACT (2011) with those derived from detailed three-dimensional nonlinear models and advanced LV3 evaluation. The results indicate a good agreement in identifying the weaker direction (aligned with the building's smaller dimension in plan), the prevalent failure mechanism of masonry piers. However, also in this case, significant differences emerged in the calculation of base shear capacity and the corresponding ground acceleration.

The comparisons between the results obtained by the LV1 and LV3 methods reveal that the two methods are not alternatives but belong to different evaluation levels. The LV1 method provides only partial insights into the structural behaviour of buildings and provides more conservative results compared to the more detailed LV3 approach.

Digitization Notes

The procedure proposed by MiBACT (2011) for the LV1 vulnerability assessment was implemented in the **online platform** of the Italian Ministry of Culture named Information System of Protected Assets—SIVARS, Podestà et al. (2010), DG ABAP (2017). Figure 4.7 shows the page of this platform that summarizes the characteristic geometric features of the masonry walls, entered by drawing them.

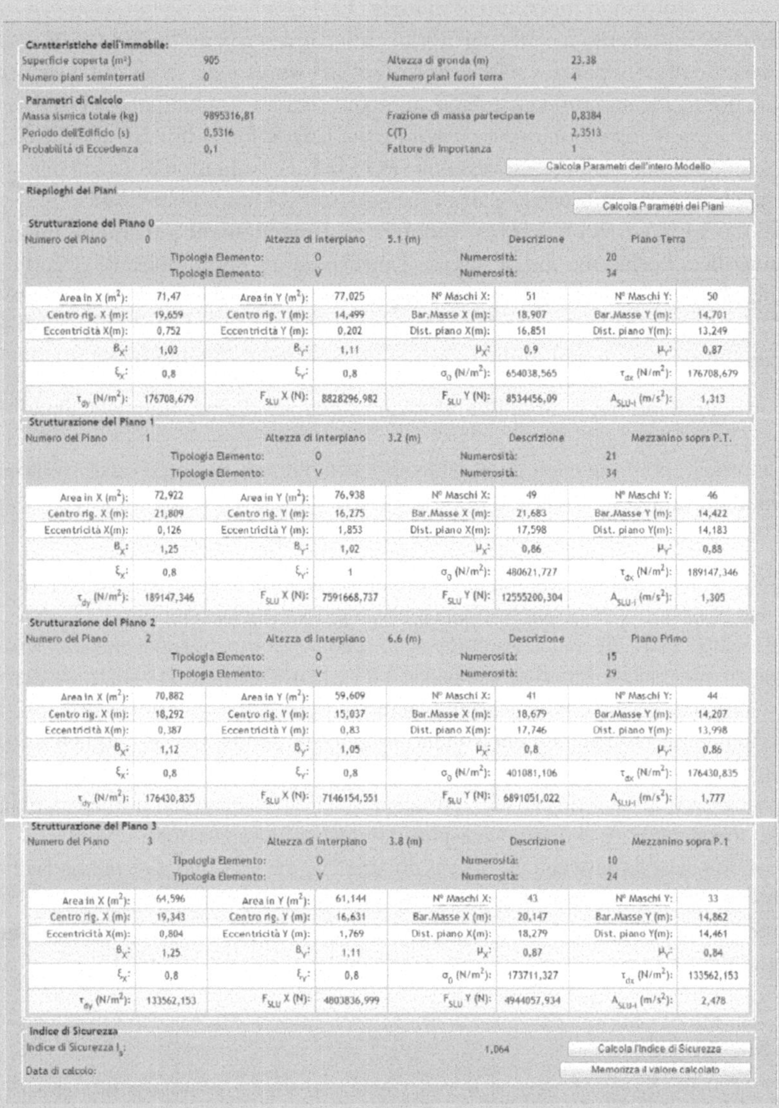

Fig. 4.7 Screenshot of the calculation of security indices (in Italian) in the SIVARS information system (Podestà et al. 2010; DG ABAP 2017)

4.3.3 *Vulnerability Index Methods*

The **Vulnerability Index** method assesses the vulnerability of a structure (such as a building, church, palace, or other type of construction) by defining a numerical indicator, called Vulnerability Index (i_v) obtained through a suitable combination of scores and weights assigned to various vulnerability factors (such as the lack of box-like behavior, material characteristics, or irregular geometries) and seismic prevention measures (such as the presence of tie rods or buttresses). **Expert judgment** plays a key role in this process.

This section explores three distinct approaches to seismic vulnerability assessment, each characterized by different methodologies, focuses, and objectives: the **GNDT framework** as a foundational tool, the **macro-element method** for evaluating churches vulnerability and their artistic and architectural value, and the **predictive approach** aimed at large-scale analyses of historic centers.

4.3.3.1 Initial Seismic Vulnerability Framework

The VI method was originally proposed by Benedetti and Petrini (1984) and subsequently adopted by the National Italian Group for Earthquake Defense GNDT (GNDT 1993) and other researchers worldwide. The GNDT method is a tool often used for **urban-scale vulnerability studies**, as it does not require detailed modelling but rather relies on the geometric survey of the units under study, combined with thorough visual inspection (Vicente et al. 2011; Ferreira et al. 2013; Hamidatou and Sbartai 2017; Cara et al. 2018; Sbrogiò et al. 2019; Ramírez Eudave et al. 2024). This method assesses structural deficiencies through field surveys and calculates a vulnerability index (i_v) based on a weighted scoring system. Buildings are then classified into **vulnerability levels** (low, medium, high) according to their respective i_v values. The form developed by GNDT for masonry buildings (Fig. 4.8) is organized into **11 sections**, each corresponding to a specific parameter to be assessed. These parameters are as follows:

1. Type and organization of the load-bearing system
2. Quality of the load-bearing system
3. Conventional strength
4. Building position and foundation
5. Horizontal structures (floors)
6. Planimetric configuration
7. Elevation configuration
8. Maximum wall distance (D_{max})
9. Roof
10. Non-structural elements
11. Current state.

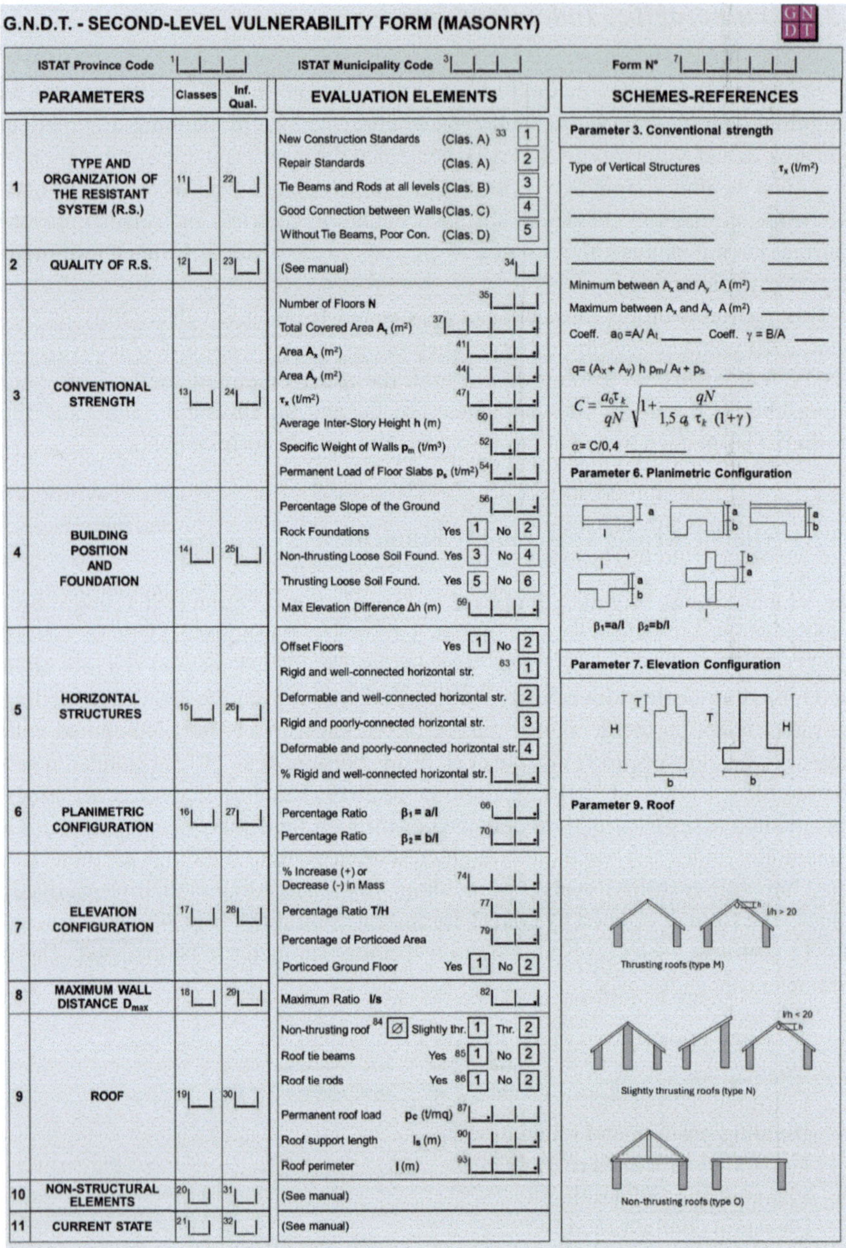

G.N.D.T. - SECOND-LEVEL VULNERABILITY FORM (MASONRY)

ISTAT Province Code [1] — ISTAT Municipality Code [3] — Form N° [7]

	PARAMETERS	Classes	Inf. Qual.	EVALUATION ELEMENTS
1	TYPE AND ORGANIZATION OF THE RESISTANT SYSTEM (R.S.)	11	22	New Construction Standards (Clas. A) 33 [1]; Repair Standards (Clas. A) [2]; Tie Beams and Rods at all levels (Clas. B) [3]; Good Connection between Walls (Clas. C) [4]; Without Tie Beams, Poor Con. (Clas. D) [5]
2	QUALITY OF R.S.	12	23	(See manual) 34
3	CONVENTIONAL STRENGTH	13	24	Number of Floors N 35; Total Covered Area A_t (m²) 37; Area A_x (m²) 41; Area A_y (m²) 44; τ_x (t/m²) 47; Average Inter-Story Height h (m) 50; Specific Weight of Walls p_m (t/m³) 52; Permanent Load of Floor Slabs p_s (t/m²) 54
4	BUILDING POSITION AND FOUNDATION	14	25	Percentage Slope of the Ground 56; Rock Foundations Yes [1] No [2]; Non-thrusting Loose Soil Found. Yes [3] No [4]; Thrusting Loose Soil Found. Yes [5] No [6]; Max Elevation Difference Δh (m) 59
5	HORIZONTAL STRUCTURES	15	26	Offset Floors Yes [1] No [2]; Rigid and well-connected horizontal str. 63 [1]; Deformable and well-connected horizontal str. [2]; Rigid and poorly-connected horizontal str. [3]; Deformable and poorly-connected horizontal str. [4]; % Rigid and well-connected horizontal str.
6	PLANIMETRIC CONFIGURATION	16	27	Percentage Ratio β_1 = a/l 66; Percentage Ratio β_2 = b/l 70
7	ELEVATION CONFIGURATION	17	28	% Increase (+) or Decrease (-) in Mass 74; Percentage Ratio T/H 77; Percentage of Porticoed Area 79; Porticoed Ground Floor Yes [1] No [2]
8	MAXIMUM WALL DISTANCE D_{max}	18	29	Maximum Ratio l/s 82
9	ROOF	19	30	Non-thrusting roof 84 [Ø] Slightly thr. [1] Thr. [2]; Roof tie beams Yes 85 [1] No [2]; Roof tie rods Yes 86 [1] No [2]; Permanent roof load p_c (t/mq) 87; Roof support length l_s (m) 90; Roof perimeter l (m) 93
10	NON-STRUCTURAL ELEMENTS	20	31	(See manual)
11	CURRENT STATE	21	32	(See manual)

SCHEMES-REFERENCES

Parameter 3. Conventional strength

Type of Vertical Structures τ_k (t/m²)

Minimum between A_x and A_y A (m²)

Maximum between A_x and A_y A (m²)

Coeff. $a_0 = A/A_t$ Coeff. $\gamma = B/A$

$q = (A_x + A_y)\, h\, p_m / A_t + p_s$

$$C = \frac{a_0 \tau_k}{qN}\sqrt{1 + \frac{qN}{1{,}5\, a_0\, \tau_k\, (1+\gamma)}}$$

α = C/0,4

Parameter 6. Planimetric Configuration

β_1=a/l β_2=b/l

Parameter 7. Elevation Configuration

Parameter 9. Roof

Fig. 4.8 The GNDT form for the evaluation of the vulnerability index i_v (version translated into English by the authors of this monograph)

Digitization Notes

The GNDT form (GNDT 1993), used for calculating i_v of individual buildings, can be **digitized** or **integrated into an online platform** to simplify the work of surveyors. These professionals are required to assess the 11 parameters above described for numerous buildings through visual inspections, which may also involve on-site visits. **Storing information directly using digital tools** such as tablets or laptops can significantly facilitate this process.

To this end, a **digital version of the GNDT form**, equivalent to the one shown in Fig. 4.8, should be developed. In this digital format, the evaluator could assign a class ranging from A to D to each of the 11 selected parameters. The platform, using predefined **algorithms**, could then automatically calculate the vulnerability level of the historic building based on the classes assigned to each parameter. The following sections provide a detailed explanation of the methods for defining these classes and the procedure to obtain i_v, which is essential for automating the vulnerability assessment process.

Parameter 1 (type and organization of the load-bearing system) evaluates the degree of **organization of vertical elements**, regardless of the material or the specific characteristics of the masonry. The key element here is the presence and **effectiveness of connections between orthogonal walls**, ensuring the box-like behaviour of the structure. To assess the quality of this parameter, the evaluator must select one of the following classes:

- **Class A**: Buildings constructed in compliance with seismic regulations for new constructions; buildings with masonry reinforced and/or repaired according to regulatory requirements.
- **Class B**: Buildings that feature connections at all levels and on all free sides, achieved using perimeter tie beams, rods or anchorages capable of transferring vertical shear forces.
- **Class C**: Buildings that, while not having tie beams or rods at all levels, consist of orthogonal walls well-anchored to each other.
- **Class D**: Buildings with orthogonal walls that are not effectively tied.

Parameter 2 (quality of the load-bearing system) takes into account the type of masonry, qualitatively defining its strength to evaluate its efficiency. The evaluator selects one of four possible strength classes, based on two factors: first, the **type of material** and the shape of the masonry components, and second, the **homogeneity of the material** and the consistency of its elements across the entire wall. Based on these factors, the evaluator indicates one of the following classes:

- **Class A**: Masonry in high-quality bricks, or squared stone or tuff, provided they are homogeneous throughout; well-bonded rubble masonry, homogeneous and with connections between the two wythes.

- **Class B**: Masonry in bricks, stone, or squared tuff, even if not homogeneous, as long as it has connections between the two wythes.
- **Class C**: Roughly squared stone masonry or low-quality brickwork with irregularities; well-bonded rubble masonry in tuff or stone without connections between the two wythes.
- **Class D**: Irregular stone masonry; low-quality brickwork with cobbles included; poorly bonded rubble masonry lacking connections between the two wythes.

Parameter 3 focuses on assessing the conventional strength of a building through a simplified evaluation process. This requires three key data points:

- **N**, the number of floors above the verification level, including the floor being evaluated.
- $\mathbf{A_t}$, the average covered area above the verification floor.
- $\mathbf{A_x}$ and $\mathbf{A_y}$, the total area of the structural elements resisting loads in two orthogonal directions.
- From these values, two ratios are calculated to describe the structural characteristics:
- **A**, the smaller of A_x and A_y, and **B**, the larger of the two.
- $\mathbf{a_0}$, the ratio of A/A_t, and γ, the ratio of B/A.

Using these ratios, the **strength coefficient C** is determined through the following formula, which considers both the reference shear strength (τ_k) and the average weight per unit area (q).

$$C = \frac{a_0 \cdot \tau_k}{q \cdot N} \sqrt{1 + \frac{q \cdot N}{1.5 \cdot a_0 \cdot \tau_k (1 + \gamma)}} \tag{4.10}$$

Specifically, q should reflect the weight of a typical floor slab and the masonry within an inter-story section and can be estimated as follows:

$$q = \frac{(A + B) \cdot h}{A_t} \cdot P_m + P_s \tag{4.11}$$

where P_m is the specific weight of the masonry, P_s is the average weight of the floor slabs per unit area and h is the average height of the inter-story sections.

If direct experimental data on $\boldsymbol{\tau_k}$ is unavailable, approximate values can be obtained from Table 4.12, which is based on the D.M. 02/07/1981. The ratio $\boldsymbol{\alpha}$, calculated as the value of parameter C divided by the reference value $C' = 0.4$, enables the user to determine one of the following vulnerability classes.

Table 4.12 Value of the reference shear strength τ_k (GNDT 1993)

(a) Unreinforced, undamaged masonry	τ_k	τ_k*
Solid bricks—lime mortar, 6–12,12	6–12	12
Modular blocks (compliant with DM 24.1.1986)—Mixed mortar (lime and cement-based)	8	8
Expanded clay or concrete blocks—Mixed mortar (lime and cement-based)	18	18
Stone masonry (with brick courses across the wall thickness, τ_k can be increased by 30%)		
– Poor-quality rubble stone	2	2
– Dressed and well-organized stone	7–9	7
– Sack masonry in good condition	4	4
– Tuff blocks	2–10	10
(b) New masonry		
Solid bricks with circular holes - cement mortar—Rm not less than 1450 t/m^2	20	20
UNI double hollow bricks (void/solid = 40.10) —cement mortar—Rm not less than 1450 t/m^2	18	18
(c) Reinforced masonry		
Solid brick masonry squared off and consolidated with two reinforced concrete slabs (minimum thickness 3 cm)	11	11
Injected rubble masonry—Sack masonry consolidated with two reinforced concrete slabs (minimum thickness 3 cm)	11	11
τ_k* values recommended by D.M. 02/07/1981		

Class A: $\alpha \leq 1$ (least vulnerable).
Class B: $0.6 \leq \alpha < 10.6$
Class C: $0.4 \leq \alpha < 0.60$
Class D: $\alpha < 0.4$ (most vulnerable).

This classification should be recorded in field (13) of the form shown in Fig. 4.8.

Parameter 4 assesses the influence of terrain and foundations on a building's seismic performance. Based on the interaction of terrain slope p (%), terrain type, and difference in foundation depth Δh, for foundation classes (field 14 in Fig. 4.8) can be defined as indicated in Table 4.13.

Table 4.13 Definition of the foundation classes according to GNDT (1993)

Building position and foundation			
Soil and foundation	Field 56	Field 59	Field 14
	Terrain Slope p (%)	Difference in foundation depth Δh	Class
1. Rock soil with foundation	$p \leq 10$	–	**A**
	$10 < p \leq 30$	–	**B**
	$30 < p \leq 50$	–	**C**
	$p > 50$	–	**D**
2. Rock soil without foundation	$p \leq 10$	–	**A**
	$10 < p \leq 30$	–	**B**
	$30 < p \leq 50$	–	**C**
	$p > 50$	–	**D**
3. Loose soil with foundation and without unbalanced thrusts caused by embankments	$p \leq 10$	$\Delta h = 0$	**A**
	$p \leq 10$	$0 < \Delta h \leq 1$	**B**
	$10 < p \leq 30$	$\Delta h \leq 1$	**B**
	$30 < p \leq 50$	$\Delta h \leq 1$	**C**
	$p > 50$	–	**D**
	–	$\Delta h > 1$	**D**
4. Loose soil without foundation and unbalanced thrusts caused by embankments	$p \leq 10$	$\Delta h = 0$	**A**
	$p \leq 10$	$0 < \Delta h \leq 1$	**B**
	$10 < p \leq 30$	$\Delta h \leq 1$	**B**
	$30 < p \leq 50$	$\Delta h \leq 1$	**C**
	$p > 50$	–	**D**
	–	$\Delta h > 1$	**D**
5. Loose soil with foundation and unbalanced thrusts caused by embankments	$p \leq 50$	$\Delta h \leq 1$	**C**
	$p > 50$	–	**D**
	–	$\Delta h > 1$	**D**
6. Loose soil without foundation and with unbalanced thrusts caused by embankments	$p \leq 30$	$\Delta h \leq 1$	**C**
	$p > 30$	–	**D**
	–	$\Delta h > 1$	**D**

Parameter 5 (horizontal structures) evaluates the quality of horizontal structural elements in buildings. The integrity of horizontal structures is particularly significant, as their collapse can lead to severe damage and loss of life. Horizontal structures are classified into four classes (field 15 in Fig. 4.8) based on their performance:

- **Class A**: Horizontal elements meet all three conditions:
  a. Negligible in-plane deformability (Code 1 or 3 in field 63).
  b. Effective connections between horizontal and vertical structures (Code 1 or 2 in field 63).
  c. Absence of staggered floors (Code 2 in field 62).
- **Class B**: Similar to Class A, but with staggered floors present.
- **Class C**: Horizontal elements exhibit significant in-plane deformability but are well-connected to walls.
- **Class D**: Horizontal elements, regardless of type, are poorly connected to vertical walls or frames.

Parameter 6 evaluates the **plan irregularity** of a building, as the seismic performance is significantly influenced by the shape of the building's plan. The evaluation considers both the proportion of the plan's dimensions and deviations from a regular rectangular shape. Regularity for **rectangular plans** is measured using the parameter $\beta_1 = (a/l) \times 100$, where a is the shorter side and l is the longer side of the rectangle. A higher β_1 value indicates a more compact and regular rectangular plan. For **irregular plans**, the parameter $\beta_2 = (b/l) \times 100$ accounts for the complexity of extensions or recesses. Buildings are classified based on the most unfavourable combination of β_1 and β_2, as indicated in Fig. 4.9.

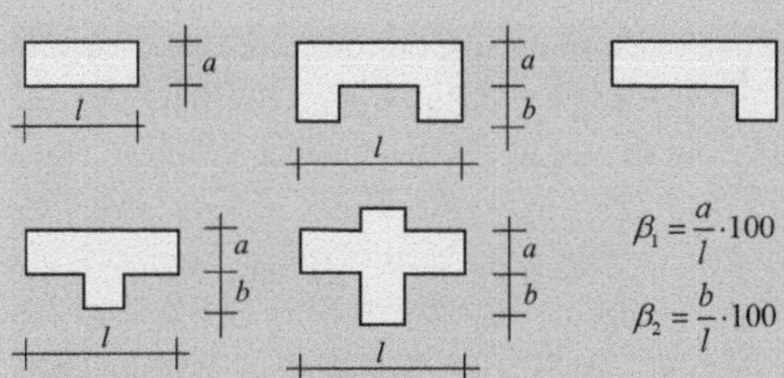

Class A:	$\beta_1 \geq 80$	$\beta_2 \leq 10$
Class B:	$60 \leq \beta_1 < 80$	$10 < \beta_2 \leq 20$
Class C:	$40 \leq \beta_1 < 60$	$20 < \beta_2 \leq 30$
Class D:	$\beta_1 < 40$	$\beta_2 > 10$

Fig. 4.9 Parameters used for the evaluation of the plan irregularity in the GNDT form (GNDT 1993)

Parameter 7 evaluates the **height irregularity of the building**. This parameter is particularly relevant for heritage masonry buildings, where irregularities are often introduced by porticoes, loggias, or towers. Key elements for the definition of the height irregularity are the following:

- **Porticoes and Loggias** (fields 79–80): Irregularities caused by porticoes or pilotis are evaluated as the percentage ratio between the portico area and the total floor area of the most unfavorable floor.
- **Towers or Turrets** (fields 77–78): The irregularity introduced by towers or turrets is assessed based on the ratio T/H, where T is the height of the tower and H is the total building height. Minor elements, such as chimneys, are not considered.
- **Mass Variations** (fields 74–76): Irregularities due to mass changes are evaluated using the ratio $\pm \Delta M/M$, where ΔM is the mass variation between two consecutive floors (sign + for increases, − for decreases) and M is the mass of the lower floor. Variations below 10% are considered negligible. Alternatively, the ratio $\pm\Delta A/A$ (floor area variations) may be used as a simplified proxy, where A and ΔA are the floor's covered area and its variation, respectively.

Taking into account for these parameters and material variations across vertical elements, buildings are categorized into the following four distinct classes (to be recorded in field 17):

- **Class A:** (a) Uniform mass and structural element distribution across the height; (b) Gradual and continuous decrease in mass and structural elements; (c) Recesses reducing the floor area by less than 10%.
- **Class B**: (a) Small porticoes or loggias affecting up to 10% of the total floor area; (b) Recesses reducing the floor area by more than 10% but not exceeding 20%; (c) Towers or turrets with heights less than 10% of the total building height.
- **Class C**: (a) Porticoes or loggias affecting more than 10% but up to 20% of the total floor area; (b) Recesses reducing the floor area by more than 20%; (c) Towers or turrets with heights between 10 and 40% of the total building height.
- **Class D**: (a) Porticoes or loggias affecting more than 20% of the total floor area; (b) Towers with heights exceeding 40% of the total building height. If different materials are used for vertical structural elements, leading to significant changes in stiffness or strength, the building's class is downgraded as follows: (1) Buildings classified as A or B based on geometry are downgraded to Class C and (2) Buildings classified as C are downgraded to Class D.

Parameter 8 evaluates the spacing between load-bearing walls and intersecting transverse walls, focusing on cases where the distance between them is excessive. The classification is determined by the ratio l/s between the center-to-center spacing of transverse walls and the thickness of the main load-bearing wall (field 82). Based on this ratio, buildings are categorized as follows (field 18):

- **Class A**: $l/s \leq 15$;
- **Class B**: includes those with a ratio greater than $15 < l/s \leq 18$;
- **Class C**: includes buildings with a ratio greater than $18 < l/s \leq 25$;
- **Class D**: $l/s \geq 25$.

Parameter 9 classifies roof types based on their impact on seismic behaviour. The evaluation criteria include: (a) the most critical roof type present—thrusting, slightly thrusting, or non-thrusting (field 84); (b) the presence or absence of ring beams (field 85); (c) the presence or absence of tie rods (field 86); (d) the roof's permanent load (field 87); and (e) the ratio between the supported length of the roof (l_s)—calculated as the perimeter minus the unsupported portions—and the total perimeter (l). The classification is defined as follows (field 19):

- **Class A**: non-thrusting roofs equipped with ring beams and/or tie rods.
- **Class B**: non-thrusting roofs lacking both ring beams and tie rods, slightly thrusting roofs that have either beams or tie rods.
- **Class C**: slightly thrusting roofs lacking both ring beams and tie rods, thrusting roofs equipped with either ring beams or tie rods.
- **Class D** thrusting roofs lacking both ring beams and tie rods.

Parameter 10 evaluates the presence of fixtures, appendages, and projections that may fall during seismic events, posing risks to people and property. The classification (field 20) is based on the condition and connection of these elements.

- **Class A and B**: buildings without fixtures, appendages, projections, or suspended ceilings, or buildings well-secured fixtures, small and lightweight chimneys, anchored suspended ceilings, and balconies integrated into the horizontal structures.
- **Class C**: buildings with small, poorly secured exterior fixtures or signs, small unanchored suspended ceilings, or well-secured large-suspended ceilings.
- **Class D** includes buildings with poorly anchored chimneys or roof appendages, inadequately constructed parapets, or heavy elements prone to collapse, as well as balconies or projections added after the construction of the main structure and weakly connected to it, or large, poorly secured suspended ceilings.

Parameter 11 assesses the **state of conservation of buildings**, classifying them based on the condition of their masonry and the presence of visible damage. The classes (field 21) are defined as follows:

- **Class A**: masonry in good condition and no visible cracks.
- **Class B**: localized hairline cracks, not derived from earthquakes.

- **Class C**: moderate cracks (2–3 mm in width) or hairline cracks caused by seismic events, as well as no visible cracks with a masonry whose state of conservation causes a significant loss of strength.
- **Class D**: walls out of plumb and/or severe cracks, even if not widespread, or severely deteriorated materials, no visible cracks but with a state of conservation that causes a substantial and severe loss of strength.

Each box in fields 22–32 of Fig. 4.8 must be filled with a letter indicating the **quality of the information** of each parameter:

- **E—High Quality**: Direct information (e.g., on-site measurements, readings from reliable graphic documents, or direct observation of elements), with a reliability level close to certainty.
- **M—Medium Quality**: Information mainly deduced (e.g., indirect readings from photographs, measurements from non-executive drawings, or low-reliability non-destructive tests, direct observations of analogous situations or oral information from trusted sources).
- **B—Low Quality**: Information mainly assumed (e.g., estimates based on reasonable assumptions, common practices, or frequent design choices).
- **A—Absent Information**: Information with a reliability level close to random selection. In such cases, the evaluator's assessment is purely indicative.

The **vulnerability index** $\boldsymbol{i_v}$ of each considered building is calculated using the following equation:

$$i_v = \sum_{i=1}^{11} \omega_i \cdot p_i \tag{4.12}$$

where p_i represents the score assigned to each of the 11 parameters based on their class, and ω_i is the weight associated with each parameter, as specified in Table 4.14. The vulnerability index i_v varies within the range $0 \leq i_v \leq 100$. The value of $i_v = 0$ indicates no seismic vulnerability, whereas the value of $i_v = 100$ indicates the maximum vulnerability (Meroni et al. 2000).

The GNDT method is a well-established tool for collecting and analysing structural data, offering an initial framework for assessing the seismic vulnerability index. While effective for many building types, its application to complex structures like **historical or religious buildings** often requires further refinement. To address these challenges, subsequent studies have expanded the GNDT approach, developing tailored methodologies. Notable examples include the methods introduced in MiBACT (2011) and other studies (De Matteis et al. 2016, 2019; Sangiorgio et al. 2021), specifically designed for evaluating the seismic vulnerability of **churches**, and the simplified procedure proposed by Brando et al. (2017) for **historic centers**. These approaches build on the GNDT framework, introducing modifications that enhance its applicability.

Table 4.14 Numerical values of t $\boldsymbol{i_v}$. The weights for parameters 5, 7, and 9 vary within a range of 0.131–0.261, depending on following factors: the percentage of rigid and well-connected horizontal structures, the presence of porticoed floors, and the weight of the roof (Benedetti and Petrini 1984; GNDT 1993; Meroni et al. 2000)

i	Parameters	Classes (p_i)				ω_i
		A	**B**	C	**D**	
1	Type and organization of the resistant system	0	0	20	45	0.261
2	Quality of the resistant system	0	5	25	45	0.065
3	Conventional strength	0	5	25	45	0.392
4	Building position and foundation	0	5	25	45	0.196
5	Horizontal structures	0	5	15	45	Var
6	Planimetric configuration	0	5	25	45	0.131
7	Elevation configuration	0	5	25	45	Var
8	Maximum wall distance D_{max}	0	5	25	45	0.065
9	Roof	0	15	25	45	Var
10	Non-structural elements	0	0	25	45	0.065
11	Current state	0	5	25	45	0.261

4.3.3.2 Macro-Element Method for Church Vulnerability

The Guidelines for Cultural Heritage (MiBACT 2011) introduced a simplified framework for assessing the seismic vulnerability of churches based on a **macro-element approach**. This methodology consists of identifying specific structural components (facade, nave, apse, bell tower, etc.) and evaluating their susceptibility to **28 damage mechanisms**. Table 4.15 provides an overview of these mechanisms grouping them into macro-elements.

The seismic vulnerability of the church is assessed using the **vulnerability index** $\boldsymbol{i_v}$ shown in Eq. (4.13). The index i_v ranges between 0 and 1 and is defined as a weighted average of the behavior of the different parts of the church:

$$i_v = \frac{1}{6}\frac{\sum_{k=1}^{28} \rho_k\left(v_{ki} - v_{kp}\right)}{\sum_{k=1}^{28} \rho_k} + \frac{1}{2} \tag{4.13}$$

For each mechanism k: v_{ki} and v_{kp} are the scores assigned to the indicators related to the vulnerability and seismic reinforcement, respectively (Table 4.16); ρ_k is the weight corresponding to the mechanism (it is 0 for mechanisms that cannot be activated in the church, due to the lack of the macro-element, and between 0.5 and 1 in other cases).

First, it is necessary to verify whether certain macro-elements are absent, i.e. if the mechanisms cannot occur in the church during the earthquake and assign them

Table 4.15 The damage mechanisms considered in MiBACT (2011) for the evaluation of the vulnerability index of churches

Facade	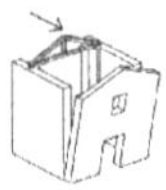	**1. Overturning of the facade**: Detachment of the facade
		2. Damage at the top of the facade: Overturning of tympanums, cracking of gables, and failure of roof connections
		3. In-plane mechanisms of the facade: Diagonal and vertical cracking due to shear and rotation forces
Nave and Aisles	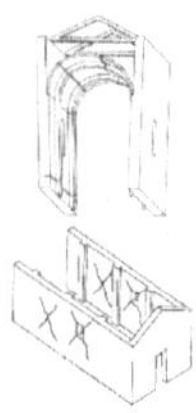	**4. Prothyrum (Narthex)**: Damage to arches or beams due to rotation or impacts
		5. Transverse response of the nave: Out-of-plane movements of walls, shear cracks in vaults, and column crushing
		6. Shear mechanisms in lateral walls: Diagonal cracking due to longitudinal forces
		7. Longitudinal response of colonnades: Cracking at the base of columns or vaults of lateral aisles
		8. Vaults of the central nave: Disconnection or collapse due to concentrated roof loads
		9. Vaults of lateral aisles: Similar vulnerabilities as in the central nave
Transept	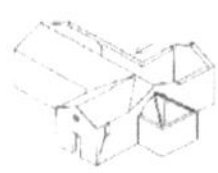	**10. Overturning of transept end walls**: Detachment or failure of gables due to poor connections
		11. Shear mechanisms in transept walls: Diagonal cracking and failures at material discontinuities
		12. Vaults of the transept: Cracking and disconnection under roof-induced thrusts
Apse and Presbytery	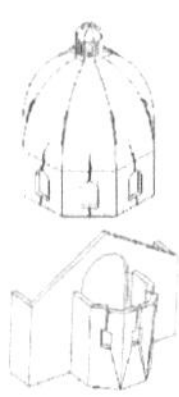	**13. Triumphal arches**: Sliding or crushing at the base of piers, and cracking in arches
		14. Dome and drum (tambour): Cracks at drum openings or disconnection from arches
		15. Lantern: Damage due to slenderness or excessive openings in the lantern structure
		16. Overturning of the apse: Weak connections and thrust from vaults
		17. Shear mechanisms in the presbytery or apse: Failure in masonry discontinuities
		18. Vaults of the presbytery or apse: Damage from concentrated loads or irregular profiles
Roof		**19. Failures in roofing over the nave**: Disconnection of beams, roof movement, and insufficient ties
		20. Failures in roofing over the transept: Similar vulnerabilities of nave
		21. Failures in roofing over the apse or presbytery: Localized failures

(continued)

Table 4.15 (continued)

Chapels and Bell Tower	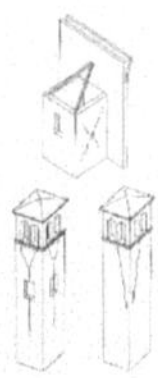	**22. Overturning of chapel walls**: Detachment from lateral walls
		23. Shear mechanisms in chapel walls: Cracking due to shear forces
		24. Vaults of chapels: Similar vulnerabilities as other vaults
		25. Irregularities in plan or elevation: Cracking and instability due to geometric discontinuities
		26. Protruding elements (pinnacles, spires, statues): Rotations or cracking due to slenderness or asymmetric loads
		27. Bell tower: Shear, cracking, and overturning due to its verticality and lack of proper ties
		28. Bell chamber: Cracks or disconnections at arches and support

Table 4.16 Vulnerability score assessment for each damage mechanism (MiBACT 2011)

Number of vulnerability indicators or anti-seismic measures	Effectiveness judgment	v_k
At least 1	3	3
At least 2	2	
1	2	2
At least 2	1	
1	1	1
None	0	0

a value of $\rho_k = 0$. For all other mechanisms, $\rho_k = 1$ should be assigned, except for mechanisms 4 and 15, where $\rho_k = 0.5$. For specific mechanisms (10, 11, 12, 18, 20, 22, 23, 24, 25, 26), a value between $0.5 < \rho_k < 1$ must be selected, depending on the importance of the element within the context of the structure.

For each mechanism, MiBACT (2011) suggests the potential anti-seismic measures that the structure could possess and vulnerability indicators. Each identified measure or vulnerability indicator must be assigned a level of effectiveness or severity, with a score ranging from 1 to 3. From these scores are then obtained the values of v_{ki} and v_{kp}, which are subsequently applied in Eq. (4.13) to calculate i_v, as detailed in Table 4.16.

The evaluation of i_v can be carried out after expeditious surveys, which involves completing specific forms that allow the evaluator to report all the church characteristics, as well as the surveyed damage. One of these forms was formulated by the Italian Department of the Civil Protection and is named "*Second-Level Survey Form for the Assessment of Damage and Vulnerability*" (DPC and MiBAC 2001; Lagomarsino et al. 2005).

During the surveys, evaluators must document all observed damage to the church. This information can be summarized using a damage index i_d (Molise et al. 2005), which ranges from 0 to 1 and is defined by the following equation:

$$i_d = \frac{1}{5} \frac{\sum_{k=1}^{28} \rho_k d_k}{\sum_{k=1}^{28} \rho_k} \tag{4.14}$$

where d_k is the level of damage of the kth mechanism (from 0 to 5).

The damage index reaches a value of 1 when the damage level corresponds to the collapse (total or partial) of all the macro-elements in the church. It represents a synthetic parameter that allows the comparison of the damage level of churches of different sizes and configurations, enabling the definition of a hierarchy of structural damage severity. For each mechanism, six levels of damage $d_k(d_0 - d_5)$ must be defined according to Grünthal (1998) and (De Matteis et al. 2016), which specializes the procedure according to the specific characteristics of churches.

Since the damage score i_d is a continuous variable ($0 < i_d < 1$) rather than discrete, Lagomarsino and Podestà (2004a) and Lagomarsino and Podestà (2004b) converted it into a discrete form associating specific ranges of i_d with the six damage levels d_k derived from the EMS-98 Scale (Grünthal 1998), as shown in Table 4.17.

Based on a statistical analysis of observed damage across various types of churches following numerous earthquakes, it was possible to correlate seismic intensity, expressed in terms of peak ground acceleration, with the vulnerability index i_v. This correlation enabled the calculation of the seismic intensity corresponding to the observed vulnerability index for each church at both the **Damage Limit State (DLS)** and the **Life Safety Limit State (LSLS)**.

$$\begin{aligned} a_{DLS} S &= 0.025 \cdot 1.8^{2.75-3.44 i_v} \\ a_{LSLS} S &= 0.025 \cdot 1.8^{5.1-3.44 i_v} \end{aligned} \tag{4.15}$$

Table 4.17 Correlations between damage levels and damage score (Lagomarsino and Podestà 2004a)

Damage level k	Damage score i_d	Description
0	$i_d \leq 0.05$	**No damage**: light damage only in one or two mechanism
1	$0.05 < i_d \leq 0.25$	**Negligible to slight damage**: light damage in some mechanisms
2	$0.25 < i_d \leq 0.4$	**Moderate damage**: light damage in many mechanisms, with one or two mechanisms activated at medium level
3	$0.4 < i_d \leq 0.6$	**Substantial to heavy damage**: many mechanisms have been activated at medium level, with severe damage in some mechanisms
4	$0.6 < i_d \leq 0.8$	**Very heavy damage**: severe damage in many mechanisms, with the collapse of some macro-elements of the church
5	$i_d > 0.8$	**Destruction**: at least 2/3 of the mechanisms exhibit severe damage

where a_{DLS} and a_{LSLS} are the peak ground accelerations corresponding to the damage and safety limit states, respectively, and S is the soil factor.

4.3.3.3 Macro-Element Method for Artistic and Architectural Value

While the MiBACT (2011) guidelines provide a robust method for quantifying structural vulnerability, recent developments in the scientific literature have emphasized the importance of integrating architectural and artistic values into the assessment process. Sangiorgio et al. (2021) propose an enhanced methodology that combines structural vulnerability evaluation with the **artistic and architectural significance** of each macro-element. This is obtained through an *Index of Artistic and Architectural Assets* ($\boldsymbol{i_a}$), which is based on the classic vulnerability index ($\boldsymbol{i_v}$) assigning different weights to macro-elements based on their cultural and historical importance.

$$i_a = \frac{1}{6}\frac{\sum_{k=1}^{28} \alpha_k \left(v_{ki} - v_{kp}\right)}{\sum_{k=1}^{28} \alpha_k} + \frac{1}{2} \tag{4.16}$$

where v_{ki} and v_{kp} are respectively the vulnerability score and the anti-seismic score of the macro-element, variable from 1 to 3 (MiBACT 2011) and α_k is the artistic and architectural importance of the macro-element in relation to the damage mechanisms.

The α_k value is obtained using the **Analytic Hierarchy Process (AHP)**. The first step in AHP consists of the hierarchic analysis of the church according to nine criteria i (with $i = 1, \ldots, 9$), grouped into three macro-criteria: the church, its macro-elements, and the associated artistic assets. For each criterion a set of alternatives j (with $j = 1, \ldots, n_i$) is defined to characterize these components, as shown in Fig. 4.10.

The second AHP step individually analyses each aspect of the problem to weight the involved parameters. The criteria and alternative weights are defined as follows:

- v_i is the weight associated with each ith criterion (whose sum is equal to 1).
- w_{ij} is the weight associated with each jth alternative related to the ith criterion.

For each criterion, the **judgment matrix A** of size n_i x n_i is constructed (where n_i is the number of alternatives of the considered criterion). Each upper diagonal element $a_{ij} > 0$ is obtained by comparing the ith element of the matrix with the jth element through using the Saaty (1987) scale, which convert the verbal judgement into numerical value (Table 4.18). The weights v_i and w_{ij} and are then obtained by solving the following eigenvector equation for the matrix A:

$$Aw = \lambda_{\max} w \tag{4.17}$$

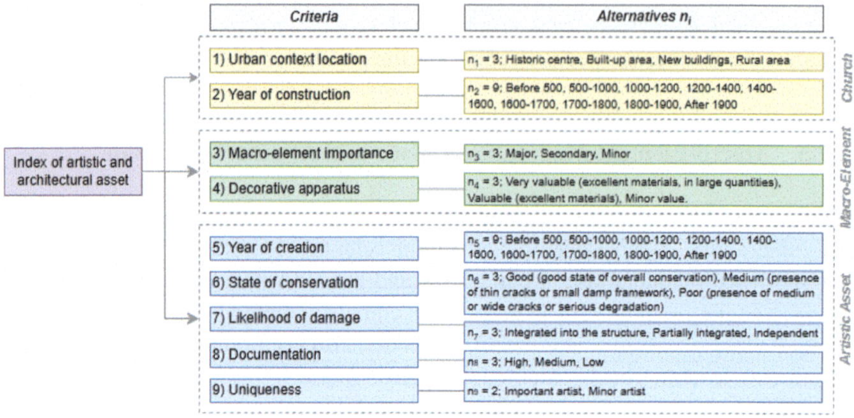

Fig. 4.10 Criteria and alternatives for calculating the α_k value representative of the artistic and architectural importance of the macro-element (Sangiorgio et al. 2021)

Table 4.18 Conversion of the verbal judgement in numerical values (Saaty 1987)

a_{ij}	Verbal scale
$a_{ij} = 1$	Equal importance
$a_{ij} = 3$	Moderate importance of one over another
$a_{ij} = 5$	Strong importance
$a_{ij} = 7$	Very strong importance
$a_{ij} = 9$	Extreme importance
1.5–4–6–8	Intermediate value
1/9, 1/8,…,1/2	The reciprocal expresses an opposite judgment

where w is the eigenvector and λ_{max} is the principal eigenvalue (Sangiorgio et al. 2021).

As an example, Table 4.19 reports the judgment matrix A and the weights, obtained for the alternatives related to Urban context location.

Finally, as indicated by the classical AHP procedure, the global weights associated with each alternative are obtained by multiplying the criteria weight by the alternative weight:

Table 4.19 Judgment matrix A and weights for the Urban context location (Sangiorgio et al. 2021)

a_{ij}	(a)	(b)	(c)	(d)	$w_{1,j}$
Historic centre (a)	1.0	5.0	8.0	3.0	1.00
Built-up area (b)	0.2	1.0	2.0	0.7	0.22
New buildings (c)	0.1	0.5	1.0	0.5	0.13
Agricultural area (d)	0.3	1.5	2.0	1.0	0.30

$$w_{ij}\prime = v_i \times w_{ij} \quad (4.18)$$

After having defined the weights w'_{ij}, the α_k value are calculated using the following equation:

$$\alpha_k = \sum\nolimits_{k=1}^{9} w_{ij}\prime \quad (4.19)$$

where k is the considered mechanism and w'_{ij} is the global weight associated with the artistic assets of the macro element. The parameter α_k varies between 0.25 and 1 and serves as a numerical tool to assess the artistic value associated with each macro-element, while also offering insights into the overall artistic and architectural heritage of the church. The average of the α_k values derived from all 28 mechanisms analyzed in MiBACT (2011), referred to as α_{med}, along with the vulnerability index i_a, enables professionals and researcher to quantitatively assess both the church's overall artistic value and its vulnerability.

Digitization Notes

The **MaChro form** represents a significant advancement in the rapid seismic vulnerability assessment of masonry churches. Developed by De Matteis et al. (2017), it is a standardized, **automated tool** aimed at reducing subjectivity in the evaluation of vulnerability and enhancing scalability for territorial analyses. The MaChro form is therefore a key element in the digitization of vulnerability assessments in historic centers and can be easily integrated into other digital platforms. From a digitization perspective, the inherent and potential features of this form are as follows:

D1. **Digitization and Automation of the Vulnerability Assessment**

The MaChro form provides a structured, automated approach by capturing essential data such as macro-element dimensions, material quality, and structural irregularities. This data, collected through the form, can be directly integrated into **Geographic Information Systems (GIS)** and other specialized platforms, enhancing its usability. The MaChro form's ability to link vulnerability data to precise spatial coordinates allows for detailed analyses that incorporate additional geographic features, such as fault lines or historical earthquake data. This makes it a cornerstone for identifying and addressing vulnerabilities in historic structures.

The interoperability of GIS platforms has the potential to enable the integration of MaChro form data with complementary datasets, including **seismic hazard maps**, **soil characteristics**, and **urban density distributions**. Such integration would allow for complex spatial analyses, which are crucial for prioritizing interventions. **Centralized databases** could further enhance accessibility, making the data readily available for researchers, engineers, and policymakers. This improved access would support emergency planning, structural interventions, and academic research.

Automation is another key feature of the MaChro form as it automatically calculates vulnerability indices (i_v) using an Excel-based system. Future advancements could enable this system to migrate to **cloud-based platforms** or mobile applications, enhancing accessibility and operational efficiency for a broader user base.

D2. **Digitization of Macro-Element Analysis**

The MaChro form adopts a macro-element approach, segmenting churches into structural units like façades, transepts, and apses. This division enables detailed assessments of potential collapse mechanisms for each element.

In addition, the form incorporates photographic documentation for verifying macro-element conditions. Enhancing this feature with advanced techniques such as 3D laser scanning or UAV-generated orthophotos would improve spatial accuracy.

D3. **Standardization and Data Sharing**

The MaChro form currently standardizes the collection of seismic vulnerability data by using a structured format ensuring consistency in the data gathered across different churches and facilitating data comparisons.

However, the potential for further development lies in adopting globally recognized standards, such as the **FAIR principles** (Findability, Accessibility, Interoperability, and Reusability). By aligning with these principles, MaChro form data could be more easily shared among stakeholders, including researchers, engineers, and policymakers. Future enhancements could also involve creating centralized databases where processed indices, raw input data, and fragility curves are stored, enabling reproducibility, extended analyses, and more efficient collaboration across disciplines (for further details, see Chap. 6).

D4. **Predictive Modeling and Scenario Analysis**

The MaChro form currently generates outputs that allow for localized assessments of seismic risk and help prioritize interventions for individual structures.

In the future, the MaChro form could be integrated with advanced **predictive modelling tools**, such as seismic risk analysis software, to simulate potential damage scenarios more accurately. Additionally, incorporating **interactive dashboards** could provide visual representations of risk distributions, enabling stakeholders to assess large-scale impacts and make informed decisions about mitigation strategies (for further details, see Chap. 6).

4.3.3.4 Predictive Approach for Historic Centers

In recent years the vulnerability approach suggested by the Italian Guidelines for churches (MiBACT 2011) has been adapted to the urban centers (Brando et al. 2017). According to this methodology, **for each structural unit**, the vulnerability index i_v (whose values range between 0 and 1) is defined as follows:

$$i_v = \frac{1}{6} \cdot \frac{\sum_{k=1}^{m} \rho_k \cdot \left(v_{kf} - v_{kp}\right)}{\sum_{k=1}^{m} \rho_k} + \frac{1}{2} \quad (4.20)$$

In Eq. (4.19), m represents the total number of **potential failure sources** that may affect each building. The coefficient ρ_k, ranging from 0 to 1.5, is used to assess the influence of a given failure source (k) on the overall structural stability (a value of 0 indicates no impact on the building's stability, while a value of 1.5 denotes maximum influence). The term v_{kf} describes the structural characteristics that maximize a specific vulnerability, while v_{kp} indicates the presence of anti-seismic measures or devices capable of mitigating the vulnerability associated with the failure source.

Brando et al. (2017) define **fourteen m_k vulnerability parameters** (Table 4.20) and the associated coefficients ρ_k based on the damage observed after the 2009 and 2016 earthquakes in the historic centres of the Abruzzo Region.

Table 4.20 Potential failure sources and associated coefficients ρ_k (Brando et al. 2017)

Failure sources	Vulnerability type	ρ_k
m_1	Position (in the cluster)	1.5
m_2	Number of storeys	1.5
m_3	1st mode mechanism	1.5
m_4	2nd mode mechanisms	1.0
m_5	Arches	1.0
m_6	Vaults	1.0
m_7	Slabs	1.0
m_8	Thrusting forces	0.8
m_9	Presence of added structures	0.5
m_{10}	Stairs	1.0
m_{11}	Irregularities	0.8
m_{12}	Non-structural elements	0.5
m_{13}	Site effects	1.5
m_{14}	Non-seismic external hazard	0.3

The score v_{kf} and v_{kp} ranges between 0 and 3 and are calculated as indicated in Eq. (4.21):

$$\begin{aligned} v_{kf} &= w \cdot z \cdot f \\ v_{kp} &= w \cdot z \cdot \eta \end{aligned} \tag{4.21}$$

where z indicates the presence or absence (1/0) of the m_k indicators and w represents an "importance factor" that can be equal to 1 or 2. The coefficient w increases proportionally to the number of factors contributing to v_{kf} and the effectiveness of protection measures, associated with v_{kp}. The parameter f is a "fragility factor" that ranges between 0 and 1.5 and quantifies the impact of factors contributing to m_k (0, indicates no influence on the activation of failure, to 1.5, representing full vulnerability with respect to failure initiation). Finally, η (ranging from 0 to 1.5) evaluates the effectiveness of the protection systems. Specific details regarding the evaluation of these parameters are can be found in Brando et al. (2017).

After calculating i_v for each building of the considered historic center, the aggregate vulnerability index $i_v{}^*$ can be obtained as either a simple or a weighted average of the individual i_v values. The mean damage μ_D of the group of the considered buildings is then estimated as a function of the vulnerability factor V, the expected earthquake intensity I and the ductility factor Q, using the formula proposed by Sandi and Floricel (1995):

$$\mu_D' = 2.5 \cdot \left[1 + \tanh\left(\frac{I + 6.25 \cdot V - 13.1}{Q}\right)\right] \tag{4.22}$$

where Q is typically equal to 2.3, as suggested by Lagomarsino and Giovinazzi (2006).

According to Brando et al. (2017), the vulnerability factor V can be calculated as follows:

$$V = a + b \cdot i_v^* + c \cdot i_v^{*2} + d \cdot i_v^{*3} = 0.53 + 1.16 \cdot i_v^* - 4.00 \cdot i_v^{*2} + 4.21 \cdot i_v^{*3} \tag{4.23}$$

where the coefficient a, b, c and d are calculated by Brando et al. (2017) from Eq. (4.22) imposing four conditions leading to the solution of four equations in four unknowns:

(i) $I = 0; i^*{}_v = 0$. Minimum intensity and vulnerability;

$$\mu_D'\left(I = 0; i_v^* = 0\right) = 2.5 \cdot \left[1 + \tanh\left(\frac{6.25 \cdot a - 13.1}{2.3}\right)\right] = 0.001 \tag{4.24}$$

(ii) $I = 11; i^*{}_v = 1$. Maximum intensity and vulnerability);

$$\mu'_D(I = 11; i_v^* = 1) = 2.5 \cdot \left[1 + \tanh\left(\frac{11 + 6.25 \cdot (a + b + c + d) - 13.1}{2.3}\right)\right] = 4.999 \tag{4.25}$$

(iii) $I = 8.5; i^*_v = 0.591$. Intensity and vulnerability corresponding to the case study of Poggio Picenze analyzed in Brando et al. (2017);

$$\mu'_D(I = 8.5; i_v^* = 0.591) = 2.5 \cdot \left[1 + \tanh\left(\frac{8.5 + 6.25 \cdot \left(a + b \cdot 0.591 + c \cdot 0.591^2 + d \cdot 0.591^3\right) - 13.1}{2.3}\right)\right] = 2.17 \tag{4.26}$$

(iv) $I = 7; i^*_v = 0.68$. Intensity and vulnerability corresponding to the case study of Goriano Sicoli analyzed in Brando et al. (2017);

$$\mu'_D(I = 7; i_v^* = 0.68) = 2.5 \cdot \left[1 + \tanh\left(\frac{7 + 6.25 \cdot \left(a + b \cdot 0.68 + c \cdot 0.68^2 + d \cdot 0.68^3\right) - 13.1}{2.3}\right)\right] = 1.35 \tag{4.27}$$

Following the indications provided in Brando et al. (2017), Eq. (4.23) indicates that a historic center with no vulnerability would not experience damage if subjected to an earthquake with a minimal intensity. Conversely, Eq. (4.24) predicts the occurrence of maximum damage for an earthquake with an intensity of $I = 11$, and maximum vulnerability ($i_v^* = 1$). Equations (4.25) and (4.26) describe the damage scenarios observed in the historic centers of Poggio Picenze ($I = 8.5$, $\mu_D = 2.17$; $i_v^* = 0.591$) and Goriano Sicoli ($I = 7$, $\mu_D = 1.35$; $i_v^* = 0.68$), respectively.

In general, the implementation of Eq. (4.22) using the formulation of V provided in Brando et al. (2017) allows for the estimation of the **mean damage** (μ_D) of an historic centre based on its vulnerability. This formulation was obtained and validated for the buildings of central Italy, and for other types of constructions further studies or other approaches are recommended. Note that, in this context, the vulnerability is estimated using an **expeditous approach**, without an accurate knowledge or advanced simulations carried out with Finite Element Models.

4.3.4 Seismic Risk Assessment for Large-Scale Analyses

The evaluation of seismic risk at a large scale involves the analysis of observed damages in multiple buildings and consists of three main steps: determination of damage distributions, derivation of fragility curves, and estimation of seismic risk.

4.3.4.1 Determination of Damage Distributions

The first step in evaluating the seismic risk of a historic center or more general of a group of heritage buildings is to collect data on the damage sustained by buildings after an earthquake. The **mean damage level** (μ_D) can be calculated using various formulations, depending on the method employed for vulnerability assessment. For example, if empirical methods are used, the damage surveyed on the group of buildings is categorized into discrete states, such as **no damage**, **slight damage**, **moderate damage**, **severe damage**, and **collapse**. These categories are numerically represented by damage scores (k) ranging from 0 (No Damage) to 4 (Collapse). The scores are then used to calculate μ_D by using Eq. (4.9), which is presented below, accounting for a different number of k_i scores.

$$\mu_D = \frac{\sum_{k=0}^{4} k \cdot n_k}{N} \tag{4.28}$$

where k is the damage score n_k is the number of buildings in damage state k and N is the total number of buildings.

The **binomial distribution** is often used to model damage distributions due to its simplicity and effectiveness. Its primary advantage is being defined solely by μ_D. For a specific damage score k, its probability of occurrence p_k is expressed by Eq. (4.8). The **standard deviation** of this damage distribution is given by:

$$\sigma_D = \sqrt{\mu_D \cdot \left(1 - \frac{\mu_D}{5}\right)} \tag{4.29}$$

The standard deviation of the damage distribution provides a measure of the dispersion of data around μ_D. In large-scale analyses, this parameter is essential for quantifying the uncertainty associated with the observed damage distribution across multiple buildings. A high standard deviation indicates greater variability in the seismic response of buildings, suggesting the influence of local factors such as different construction typologies or specific vulnerability conditions. Conversely, a low standard deviation implies a more uniform damage distribution, facilitating simplified modeling and the calibration of fragility curves that are representative of the analyzed building stock.

4.3.4.2 Derivation of Fragility Curves

Fragility curves represent the **probability of exceeding a specific damage state** (e.g., Collapse) as a function of seismic intensity, such as Peak Ground Acceleration (PGA) or Modified Mercalli Intensity (MMI). The development of fragility curves involves the following steps:

(a) **Linking observed damage to intensity**: Each observed damage level is associated with the seismic intensity recorded in the affected area. For example, in an area with $I = 7$, 5% of buildings exceeded "Collapse"; in an area with $I = 8$, 10% exceeded "Collapse."
(b) **Cumulative probabilities**: To calculate the **probability of exceeding a specific damage state** the probabilities derived from the binomial distribution function p_k are used, according to the formulation reported in Eq. 4.25. If the damage states are defined as (0) *No Damage*, (1) *Slight Damage*, (2) *Moderate Damage*, (3) *Severe Damage* and (4) *Collapse*, the probability of exceeding *Collapse* ($k = 4$) is equal to p_4, the probability of exceeding *Severe Damage* ($k = 3$) is equal to $p_3 + p_4$, the probability of exceeding *Moderate Damage* ($k = 2$) is equal to $p_2 + p_3 + p_4$ and so forth.
(c) **Curve fitting**: The cumulative probabilities are fitted using a statistical model, typically a **lognormal distribution**, to produce smooth fragility curves.

4.3.4.3 Estimation of Seismic Risk

Seismic risk quantifies the expected losses based on fragility curves and regional seismic hazard data. Specifically, this is carried out by first **combining fragility curves with seismic hazard models** and then **estimating losses** by multiplying the probability of exceeding each damage state (from fragility curves) for the associated loss values (e.g., repair costs or total reconstruction costs). These associated loss values can be determined from various sources depending on the damage state and the context of the analysis:

1. **Cost of Repairs**: For minor damage states (e.g., slight or moderate damage), repair costs are often estimated as a percentage of the building's replacement value. Typical values can be obtained from:

   - **Construction cost databases**: National or regional databases, such as RSMeans (USA), BCIS (UK), or ISTAT (Italy), which provide detailed construction and repair cost estimates.
   - **Historical earthquake studies**: Reports on past earthquakes often document repair costs for specific damage levels, e.g., the EERI (Earthquake Engineering Research Institute) case studies.
   - **Building codes and guidelines**: Seismic assessment standards (FEMA P-1 2012a; FEMA P-2, 2012b) may include approximate cost models for repair.

2. **Replacement Cost of Damaged Elements**: For higher damage states, such as severe damage, loss values can be derived from the replacement cost of individual structural and non-structural components. Sources include:
   - **Standardized component costs**: Manuals and regional price lists can provide detailed cost breakdowns for components such as walls, roofs, and windows.
   - **Local market rates**: Regional construction markets often publish current prices for materials and labour.
3. **Total Reconstruction Costs**: For the collapse state, the loss value is typically equal to the full reconstruction cost of the building. This can be estimated using:
   - **Building valuation records**: Official property valuations or insurance appraisals.
   - **National or regional statistics**: Government data on average construction costs per square meter or per building type.
   - **Post-disaster reports**: Organizations such as the World Bank or UNISDR (United Nations Office for Disaster Risk Reduction) publish damage assessments that include reconstruction costs.

For example, repair costs may be estimated as a percentage of the building's value for lower damage states, while the cost of collapse may be considered equal to the total reconstruction value. For example, if the annual probability of an I = 9 earthquake is 2%, and the expected cost of collapse damage is €1,000,000, the annual expected loss is equal to 0.02·1,000,000 = €20,000.

4.3.4.4 Example

This section provides a practical example illustrating how damage distributions observed in a **sample of 100 buildings** for different seismic intensities are used to calculate **fragility curves** and **evaluate seismic risk**. The analysis is based on both observed and fitted probabilities for different damage states, enabling a deeper understanding of the process.

Table 4.21 presents the observed damage distribution for seismic intensities ranging from $I = 4$ to $I = 10$. The damage states are categorized as follows: $k = 0$ No Damage; $k = 1$ Slight Damage, $k = 2$ Moderate Damage, $k = 3$ Severe Damage and $k = 4$ Collapse. The distribution of damage shown in Fig. 4.11 indicates a progressive increase in damage state as seismic intensity rises, and a higher number of damaged buildings as seismic intensity decreases.

Table 4.22 shows the mean damage (μ_D) and the corresponding standard deviation (σ_D) calculated for each intensity level I. As an example, Eq. (4.26) reports the explicit calculation of μ_D for I = 6.

$$\mu_D(I=6) = \frac{\sum_{k=0}^{4} k \cdot n_k}{N} = \frac{(0 \cdot 70) + (1 \cdot 20) + (2 \cdot 8) + (3 \cdot 2) + (4 \cdot 0)}{100} = 0.42 \tag{4.30}$$

Table 4.21 Damage distribution of a sample of 100 buildings

Intensity *I*	No damage ($k = 0$)	Slight damage ($k = 1$)	Moderate damage ($k = 2$)	Severe damage ($k = 3$)	Collapse ($k = 4$)
4	4	95	5	0	0
5	5	85	10	5	0
6	70	20	8	2	0
7	50	30	15	5	0
8	30	30	25	10	5
9	10	20	30	25	15
10	5	10	20	30	35

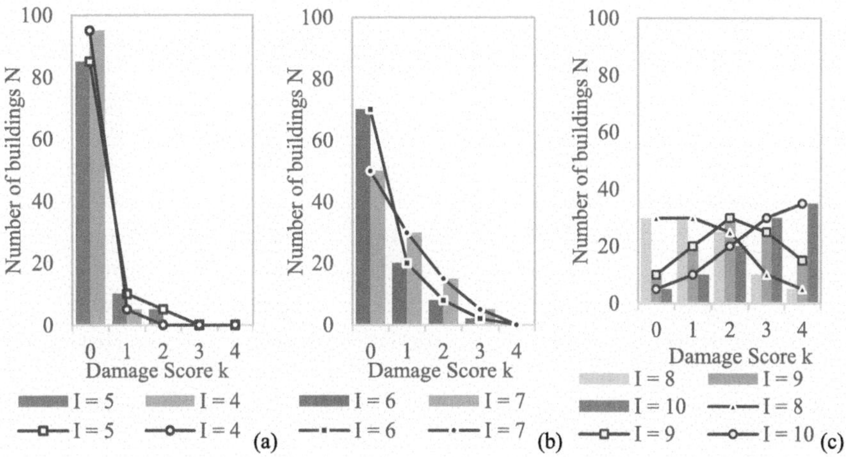

Fig. 4.11 Observed damage distribution for a sample of 100 buildings: **a** at I = 5 and I = 6; **b** at I = 6 and I = 7; **c** at I = 8, I = 9 and I = 10

Table 4.22 also includes the occurrence probability of each damage state p_k for each intensity level. These probabilities are derived using the binomial distribution formula shown in Eq. (4.9):

The cumulative **probability of exceeding each damage state** P ($k \geq$ X) is then calculated from the observed distributions (Table 4.23) summing the probability of the current state to those of the subsequent damage states as indicated below:

- Probability of Exceeding *Collapse* = $P\ (k \geq 4) = p_4$
- Probability of Exceeding *Severe Damage* = $P\ (k \geq 3) = p_3 + p_4$
- Probability of Exceeding *Moderate Damage* = $P\ (k \geq 2) = p_2 + p_3 + p_4$
- Probability of Exceeding *Slight Damage* = $P\ (k \geq 1) = p_1 + p_2 + p_3 + p_4$

Table 4.22 Mean and Standard Deviation of the observed damages for each Intensity Level and their Probability of Occurrence calculated according to the binomial distribution function

Intensity I	μ_D mean damage	σ_D standard deviation	p_0 no damage	p_1 slight damage	p_2 moderate damage	p_3 severe damage	p_4 collapse
4	0.05	0.22	0.951	0.048	0.001	0.000	0.000
5	0.20	0.44	0.815	0.170	0.014	0.001	0.000
6	0.42	0.62	0.645	0.296	0.054	0.005	0.000
7	0.75	0.80	0.444	0.392	0.138	0.024	0.002
8	1.30	0.98	0.222	0.390	0.274	0.096	0.017
9	2.15	1.11	0.060	0.227	0.342	0.258	0.097
10	2.80	1.11	0.016	0.105	0.267	0.340	0.216

Table 4.23 Probability of exceedance of each observed damage state *k*

Intensity *I*	Observed $P(k \geq 0)$	Observed $P(k \geq 1)$	Observed $P(k \geq 2)$	Observed $P(k \geq 3)$	Observed $P(k \geq 4)$
4	1.0000	0.0490	0.0010	0.0000	0.0000
5	1.0000	0.1846	0.0148	0.0006	0.0000
6	1.0000	0.3551	0.0594	0.0052	0.0002
7	0.9999	0.5562	0.1647	0.0265	0.0022
8	0.9988	0.7769	0.3871	0.1132	0.0169
9	0.9853	0.9251	0.6982	0.3558	0.0974
10	0.9449	0.9284	0.8235	0.5564	0.2164

- Probability of Exceeding *No Damage* $= P(k \geq 0) = p_0 + p_1 + p_2 + p_3 + p_4.$

Plotting for each damage state the seismic intensities I on the X-axis and the corresponding probabilities of exceedance $P(k \geq X)$ on the Y-axis, the fragility curves of the observed probabilities are obtained.

To derive continuous fragility curves, the observed probabilities are fitted to specific functions for enhanced representation and generalization. In this example the following function is used:

$$P_{fit}(k \geq X) = \frac{1}{1 + \exp[-\beta \cdot (I - I_0)]} \tag{4.31}$$

The fitted probabilities are reported in Table 4.24. The parameters β and I_0 are estimated by minimizing the error between the observed and fitted probabilities. This error is quantified as the sum of squared residuals computed for each considered intensity level:

$$Error = \sum_{i=I_{\min}}^{i=I_{\max}} \left[P(k \geq X) - P_{fit}(k \geq X)\right]^2 \quad (4.32)$$

The minimization is performed using numerical optimization techniques.

Figure 4.12 shows the fragility curves derived from observed (Table 4.23) and fitted (Table 4.24) exceedance probabilities. These curves represent the **probability of exceeding specific damage states** as the seismic intensity increases. For instance, in this example, they indicate that, for an earthquake with $I = 6$, the analyzed set of buildings has a probability of exceeding the Slight Damage state equal to 34%. Note that in a complete distribution, $P\ (k \geq 0)$ must be 1 for all intensities, as it includes all buildings (100% of cases).

The fragility curves can also be expressed in terms of the **expected PGA on rigid soil** (Fig. 4.13). This transformation is achieved by converting the macro-seismic intensity I into PGA by using the correlation proposed by Margottini et al. (1992).

$$PGA = c_1 \cdot c_2^{(I-5)} \quad (4.33)$$

Table 4.24 Fitted probability of exceedance of each observed damage state k

Intensity I	Fitted P_{fit} $(k \geq 4)$	Fitted P_{fit} $(k \geq 3)$	Fitted P_{fit} $(k \geq 2)$	Fitted P_{fit} $(k \geq 1)$	Fitted P_{fit} $(k \geq 0)$
4	0.0004	0.0020	0.0069	0.0739	1.0000
5	0.0011	0.0058	0.0211	0.1699	1.0000
6	0.0033	0.0171	0.0627	0.3442	1.0000
7	0.0099	0.0492	0.1720	0.5739	1.0000
8	0.0295	0.1335	0.3918	0.7755	1.0000
9	0.0847	0.3147	0.6665	0.8986	1.0000
10	0.2195	0.5777	0.8611	0.9579	1.0000

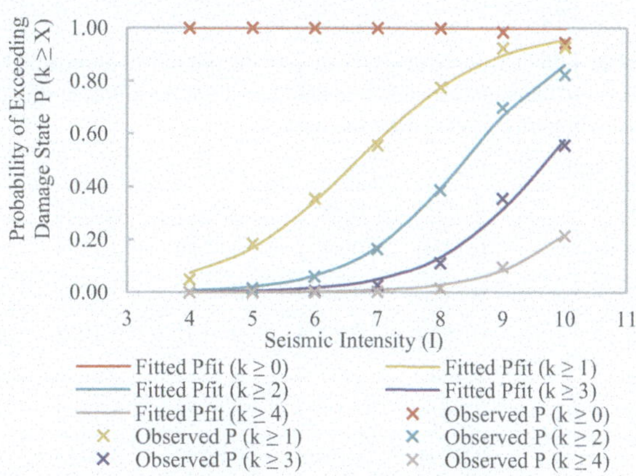

Fig. 4.12 Fragility curves in terms of Seismic Intensity (I) for a sample of 100 buildings

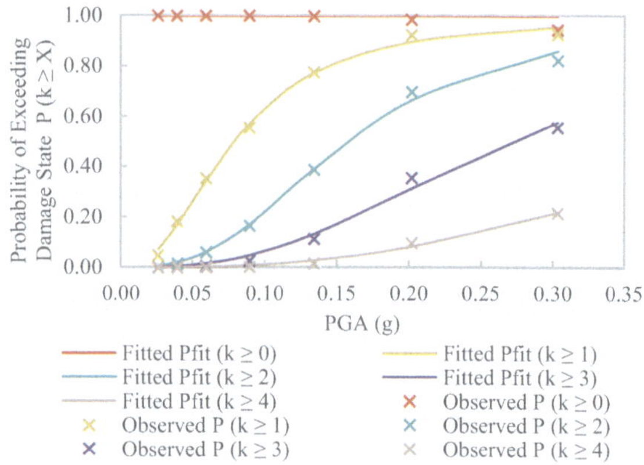

Fig. 4.13 Fragility curves in terms of Peak Ground Acceleration (PGA) on rigid soil

where $c_1 = 0.04$ and $c_2 = 1.5$ (Margottini et al. 1992).

An alternative method for deriving fragility curves is the **predictive approach** proposed by Brando et al. (2017). This methodology calculates the mean damage value μ_D as a function of seismic intensity I and a **vulnerability factor** V, providing a useful framework particularly suited for historic centers where detailed empirical data may be unavailable.

Assuming a dataset of 100 vulnerability indices i_v, one for each analyzed building, the **mean vulnerability index** (i_v^*) is calculated as the arithmetic mean of these 100 values, resulting in $i_v^* = 0.495$. The i_v^* value is then used to derive the **vulnerability factor** (V) using Eq. (4.23). Subsequently, for each seismic intensity I, the mean damage μ_D is computed using Eq. (4.22), as shown in Table 4.25. The other steps needed for the construction of the fragility curve are the same described previously. Specifically, the probabilities of observing different damage states are derived based on μ_D for each intensity level, using the binomial distribution (Eq. 4.9).

The exceedance probabilities $P\ (k \geq X)$ are then calculated by summing the probabilities of the current and previous damage states starting from p_4, as illustrated

Table 4.25 Example of application of the predictive approach: Mean and Standard Deviation of the observed damages for each Intensity Level and their Probability of Occurrence calculated according to the binomial distribution function

Intensity I	μ_D	σ_D	p_0	p_1	p_2	p_3	p_4
4	0.0807	0.2817	0.9219	0.0756	0.0025	0.0000	0.0000
5	0.1760	0.4121	0.8359	0.1525	0.0111	0.0004	0.0000
6	0.3756	0.5894	0.6768	0.2748	0.0446	0.0036	0.0001
7	0.7654	0.8051	0.4357	0.3938	0.1424	0.0257	0.0023
8	1.4343	1.0114	0.1844	0.3710	0.2985	0.1201	0.0241
9	2.3618	1.1163	0.0409	0.1831	0.3278	0.2934	0.1313
10	3.3291	1.0548	0.0042	0.0415	0.1654	0.3296	0.3284

in Table 4.26. The calculated exceedance probabilities are finally fitted to obtain continuous fragility curves (Table 4.27). The curves shown in Fig. 4.14 report the probability of exceeding specific damage states as a function of seismic intensity according to the predictive approach, based on the predictive approach. Note that, according to this approach, μ_D depends only on i_v^*. In cases where data is unavailable or it is not feasible to assess individual buildings in a historic center, i_v^* can be estimated approximately through interviews with local experts. In this context, the prevailing building typology within the historic center (e.g., two-story masonry buildings with reinforced concrete gable roofs) is identified, and the 14 reference parameters needed to calculate i_v^* are determined based on the input provided from local technicians.

The fragility curves provide the probability of exceeding specific damage states $P\ (k \geq \mathrm{X})$ as a function of seismic intensity (I). These probabilities can further be used to calculate the **expected economic losses for each intensity level**, leading to the derivation of **seismic risk curves**.

In this example, the expected loss (L) for each seismic intensity was calculated by combining the probabilities of exceedance of Table 4.26 with the economic value

Table 4.26 Example of application of the predictive approach: probability of exceedance of each observed damage state k

Intensity (MMI)	Observed P ($k \geq 0$)	Observed P ($k \geq 1$)	Observed P ($k \geq 2$)	Observed P ($k \geq 3$)	Observed P ($k \geq 4$)
4	1.0000	0.0781	0.0025	0.0000	0.0000
5	1.0000	0.1641	0.0115	0.0004	0.0000
6	1.0000	0.3232	0.0484	0.0038	0.0001
7	0.9999	0.5642	0.1704	0.0281	0.0023
8	0.9981	0.8136	0.4427	0.1442	0.0241
9	0.9765	0.9356	0.7525	0.4248	0.1313
10	0.8691	0.8650	0.8235	0.6580	0.3284

Table 4.27 Example of application of the predictive approach: fitted probability of exceedance of each observed damage state k

Intensity (MMI)	Fitted P_{fit} ($k \geq 4$)	Fitted P_{fit} ($k \geq 3$)	Fitted P_{fit} ($k \geq 2$)	Fitted P_{fit} ($k \geq 1$)	Fitted P_{fit} ($k \geq 0$)
4	0.0015	0.0031	0.0084	0.0739	1.0000
5	0.0038	0.0089	0.0253	0.1699	1.0000
6	0.0096	0.0255	0.0741	0.3442	1.0000
7	0.0242	0.0708	0.1977	0.5739	1.0000
8	0.0596	0.1815	0.4315	0.7755	1.0000
9	0.1392	0.3924	0.7003	0.8986	1.0000
10	0.2921	0.6529	0.8780	0.9579	1.0000

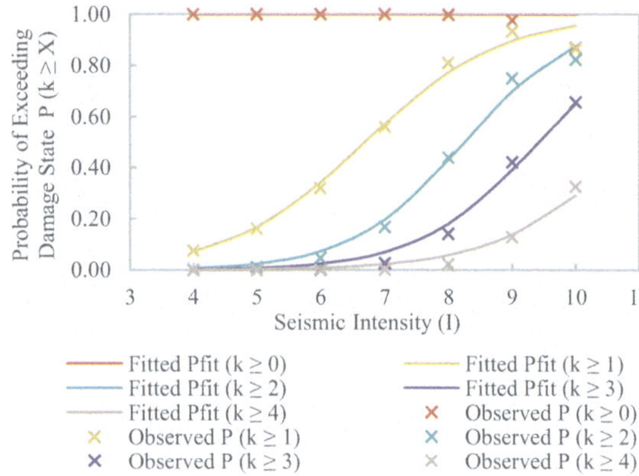

Fig. 4.14 Fragility curves calculated according to the predictive approach

associated with each damage state. The formula used for this calculation is:

$$L(I) = \sum_{k=0}^{4} P(k \geq X) \cdot C_k \tag{4.34}$$

where C_k are the economic cost associated with damage state k, expressed as a fraction of the total building value.

The damage state costs were assigned as follows:

- $C_0 = 0$: No Damage,
- $C_1 = 0.1$: Slight Damage,
- $C_2 = 0.3$: Moderate Damage,
- $C_3 = 0.6$: Severe Damage,
- $C_4 = 1.0$: Collapse.

Figure 4.15 shows the expected loss values for a single building, based on an assumed average economic value of €100,000, across different seismic intensity levels. The total expected loss for the entire set of 100 buildings, as a function of seismic intensity level, is summarized in Table 4.28.

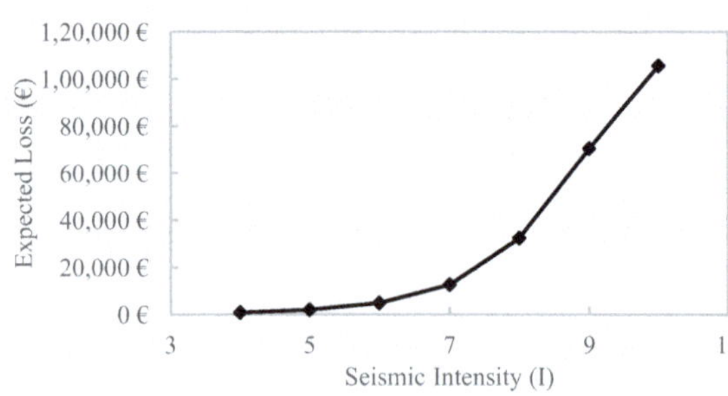

Fig. 4.15 Expected loss values for a representative building and the total for the entire set of 100 buildings

Table 4.28 Expected loss values of the representative building and the overall set of buildings

Intensity I	Expected loss per building (€)	Total expected loss for 100 buildings (€)
4	856	85,600
5	2,010	201,000
6	4,922	492,200
7	12,670	1,267,000
8	32,479	3,247,900
9	70,549	7,054,900
10	105,675	10,567,500

In the context of the developed example, the expected losses $L(I)$ represent the costs associated with a single seismic event of a specific intensity (I) for a single building or the entire set of buildings.

If the goal is to calculate the **annual expected losses** (*AEL*), it is necessary to consider the annual probability of a seismic event occurring at a given intensity using a **seismic hazard model**. For this purpose, the annual probability of exceeding a given PGA is derived from the **seismic hazard curves** provided by the national governments. In the Italian case, the National Institute of Geophysics and Volcanology (INGV), in collaboration with the National Department of the Civil Protection, provides the seismic hazard curves [http://ess1.mi.ingv.it/] for each point of a regular grid covering the entire Italian territory. As an example, for the latitude 42.335 and 13.423° longitude point, corresponding to L'Aquila (AQ), the Italian seismic hazard curve indicates that there is annual probability of 3.3% that an earthquake with $PGA = 0.08\ g$ occurs, i.e. P (PGA = 0.08 g) = 0.033 (Fig. 4.16).

As previously indicated, the conversion between I and PGA values, for the existing Italian building can be carried out using the formulations proposed in Margottini et al. (1992). Specifically, by applying the inverse formula of Eq. (4.29), the following relationship is obtained, where c_1 and c_2 are equal to 0.04 and 1.5, respectively:

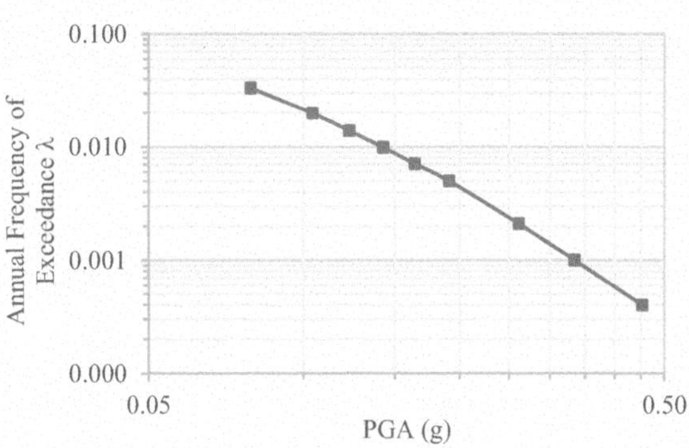

Fig. 4.16 Acceleration values for annual exceedance frequencies (lat. 42.335, long. 13.423—ID 26,528)

$$I = 5 + \frac{\log_{10}(PGA/c_1)}{\log_{10} c_2} \tag{4.35}$$

For each PGA value shown in Fig. 4.16 and listed in Table 4.28, the corresponding seismic intensity (I) is calculated using Eq. (4.31). Each PGA value is therefore associated with an Annual Exceedance Frequency, as derived from the seismic hazard curve (Fig. 4.16). Based on the expected loss per building reported in Table 4.27, the losses corresponding to the calculated intensities are obtained using a linear interpolation. For $I \geq 10$, the expected losses for $I = 10$ are considered. Finally, the Annual Expected Loss per Building is computed by multiplying the Annual Exceedance Frequency with the corresponding expected loss per building (Table 4.29). Figure 4.17 illustrates the risk curve of a representative building, showing the relationship between the Annual Frequency of Exceedance and the Annual Expected Loss per Building.

Table 4.29 Annual expected loss values per building based on the seismic hazard curve and the expected loss per building for each intensity level

Annual exceedance frequency	PGA (g)	Seismic intensity I	Expected loss per building (€)	Annual expected loss per building (€)
0.0004	0.45260	10.9836	105,675	42
0.0010	0.33430	10.2364	105,675	106
0.0021	0.26100	9.6259	92,534	194
0.0050	0.19080	8.8532	64,961	325
0.0071	0.16400	8.4799	50,750	360
0.0099	0.14240	8.1316	37,490	371
0.0139	0.12260	7.7624	27,772	386
0.0199	0.10410	7.3590	19,780	394
0.0332	0.07890	6.6754	10,155	337

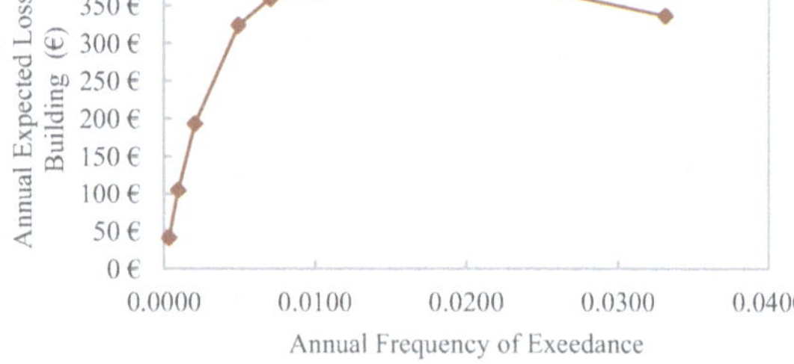

Fig. 4.17 Annual expected loss per building as a function of the annual frequency of exceedance

4.4 Seismic Vulnerability and Risk for Individual Heritage Buildings

4.4.1 Seismic Behavior of Masonry Heritage Buildings

Understanding the seismic behavior of masonry heritage buildings is crucial for preserving both their structural integrity and cultural significance. Unlike modern structures, heritage buildings often **lack the rigid connections** and **uniformity** required to resist seismic forces effectively. Their performance depends on the interaction of individual components and their degree of connection within the structural system. This section introduces first- and second-mode collapse mechanisms alongside their corresponding first- and second-level vulnerability assessments, focusing on how and when these mechanisms are likely to activate.

First-mode mechanisms, triggered by **out-of-plane seismic actions**, such as forces perpendicular to walls, can cause overturning, flexural failure, or connection failures (Fig. 4.18a). These mechanisms are considered the **first damage mode** of existing masonry buildings. Second-mode mechanisms, by contrast, activate only when first-mode mechanisms are effectively restrained.

In heritage masonry buildings, first-mode mechanisms are particularly prevalent due to weak or **missing connections between walls and floors**. These deficiencies often cause walls to act independently, leading to localized collapse rather than a coordinated structural response. Mitigating these vulnerabilities may require the installation of **tie-rods, anchors, horizontal diaphragms, or ring beams**. The absence of such reinforcements often results in dominant out-of-plane responses, preventing the activation of second-mode mechanisms and compromising the building's overall stability.

Second-mode mechanisms develop when **in-plane actions** dominate the response of the masonry walls. These failures are typically associated with global seismic behavior, where walls interact to resist lateral forces. Seismic damage observations and laboratory tests have demonstrated that masonry piers subjected to in-plane loading exhibit two main behavioral patterns, associated with different failure modes: **flexural behaviour** and **shear behaviour**.

Flexural failure is characterized by the rocking motion of a masonry wall within its plane, leading to a loss of load-bearing capacity due to progressive cracking in the tensile region and/or crushing in the compressive region (Fig. 4.19a).

Shear failure is associated with two failure modes: (a) **sliding shear failure**, which consists of sliding along a horizontal mortar joint, typically at the pier's ends

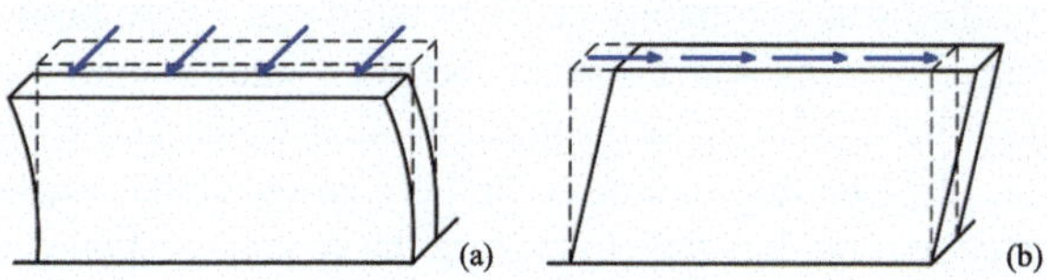

Fig. 4.18 First Mode (**a**) and Second Mode (**b**) mechanisms

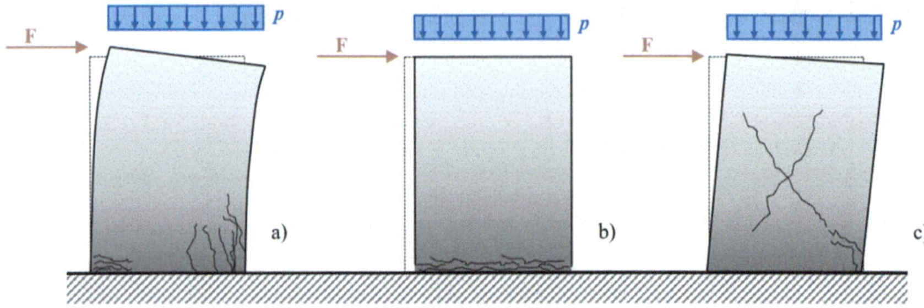

Fig. 4.19 In-plane failure mechanisms of a masonry wall. **a** flexural failure; **b** shear-sliding failure; **c** shear-diagonal failure

(Fig. 4.19b); (b) **diagonal cracking**, that occurs through a diagonal crack, starting at the pier's centre and propagating toward the corners (Fig. 4.19c). Cracks may follow mortar joints (stair-stepped pattern in regular masonry) or pass through the blocks.

The global stability of a heritage masonry building relies on achieving **box-like behaviour**, where walls and diaphragms act cohesively to resist seismic loads. Key factors influencing this behaviour include:

- **Wall Connections**: Rigid connections between orthogonal walls are essential to transfer loads and prevent independent movement.
- **Horizontal Diaphragms**: Floors and roofs must be rigid in their plane to distribute lateral forces effectively and minimize out-of-plane displacements.
- **Vertical Load Stabilization**: The presence of stabilizing vertical loads, such as those from upper floors, enhances the behaviour of masonry walls.

For heritage buildings, the implementation of these strategies is often challenging due to preservation constraints. However, targeted interventions, such as installing tie-rods or reinforcing existing connections, can significantly enhance global stability without compromising the building's cultural value.

4.4.2 Vulnerability Assessment for First-Mode Mechanisms (LV2)

First-mode mechanisms correspond to out-of-plane kinematics affecting individual walls or parts of a structure. These phenomena, such as out-of-plane rocking and out-of-plane bending, lead to structural collapse due to a loss of equilibrium. These mechanisms commonly occur in buildings where **poor or ineffective connections** between walls and floors, or between intersecting walls, prevent cohesive structural behaviour. As a result, individual walls tend to act independently, activating out-of-plane failure modes. In these cases, global analysis using three-dimensional structural models fails to capture the seismic response unless specific measures are first implemented to mitigate these mechanisms.

(a) (b)

Fig. 4.20 Simple wall overturning mechanisms observed during the 2016 Central Italy Earthquake

4.4.2.1 Simple Wall Overturning

The simple overturning of external walls due to seismic forces is one of the most frequent and hazardous damage scenarios. This failure mode typically occurs when the affected wall lacks adequate connection to orthogonal walls. The presence of structural elements such as **tie rods** or **ring beams** generally prevents this mechanism by providing restraint. Key indicators of activated overturning include vertical cracks at wall intersections, out-of-plumb conditions of the overturning wall, and the detachment of beams from floors (Fig. 4.20). Overturning can occur in various configurations, influenced by the arrangement of openings such as doors and windows. It can also affect one or more stories, depending on the level of connection between floors and walls at different heights.

4.4.2.2 Vertical Out-of-Plane Bending

A common scenario in masonry buildings involves **walls constrained at their ends but free in the central portion**, such as a wall restrained at the roof with a ring beam but lacking intermediate floor connections. Vertical out-of-plane bending occurs when walls are well-anchored at their ends, weakly tied at the sides, and unrestrained in the middle. Excessive slenderness, poorly bonded masonry wythes, or intermediate horizontal thrusts (e.g., from arches or vaults) further exacerbate this behaviour. This failure mode is characterized by the **formation of a horizontal hinge dividing the wall into two blocks** that rotate relative to one another. Symptoms include bulging and out-of-plumb conditions, horizontal and vertical cracking, and detachment of floor beams (Fig. 4.21).

(a) (b)

Fig. 4.21 Vertical out-of-plane bending mechanisms observed during the 2016 Central Italy Earthquake

4.4.2.3 Horizontal Out-of-Plane Bending—Arch Mechanism

Horizontal out-of-plane bending affects the **upper portions of walls** when they are poorly tied at the top but well-anchored to adjacent walls. This mechanism is triggered by horizontal thrusts perpendicular to the wall, often due to roof or floor loads. Material expulsion from the upper section of the wall and detachment of wedge-shaped blocks frequently accompany this failure mode. Horizontal hinges, both oblique and vertical, may form. The behaviour can be conceptualized as the formation of a three-hinge resisting arch within the wall thickness. This arch sustains horizontal forces until the masonry crushes. Typical symptoms include vertical and diagonal cracks and roof beam detachment (Fig. 4.22).

(a) (b)

Fig. 4.22 Horizontal out-of-plane bending mechanisms observed during the 2016 Central Italy Earthquake

Fig. 4.23 Complex overturning mechanisms observed during the 2016 Central Italy Earthquake

4.4.2.4 Complex Overturning

Complex overturning involves the **rotation of a wall perpendicular to seismic forces**, accompanied by the dragging of adjoining wall sections. This occurs in cases where the overturning wall lacks top restraints but is well-connected to orthogonal walls. Indicators include diagonal cracking in bracing walls, out-of-plumb conditions, and beam detachment (Fig. 4.23).

4.4.2.5 Hammering of Floors and Roofs

Mixed damage mechanisms often arise from the hammering of roof structures or thrust from diagonal braces, exacerbating local failure modes (Fig. 4.24).

Fig. 4.24 Hammering of roof mechanisms observed during the 2016 Central Italy Earthquake

4.4.2.6 Analysis of First-Mode Mechanisms

Before employing advanced software for global structural analysis, engineers must ensure that such damage mechanisms will not activate. In the post-earthquake scenario, these mechanisms are often visible, allowing for relatively reliable identification of remedial solutions. However, assessing seismic vulnerability in undamaged buildings requires extensive experience, especially regarding historical construction and the interpretation of diagnostic results from structural and material surveys.

The analysis typically adopts a **static or dynamic kinematic approach** to evaluate the equilibrium of a labile structure (kinematic chain) composed of rigid masonry segments. This approach considers **local mechanisms** under proportional force distributions (λW) or height-variable distributions scaled by a multiplier (λ). The critical task is identifying significant local mechanisms based on the building's configuration, prior seismic damage, or potential crack patterns. Factors such as wall connectivity, masonry texture, tie rods, and interactions with adjacent structures must also be considered. For further details on the analysis type, see for example Curti et al. 2007, Andreotti et al. 2015, Fabbrocino et al. 2019, Casapulla 2021.

4.4.3 Vulnerability Assessment for Second-Mode Mechanisms (LV3)

4.4.3.1 Building Behaviour

The seismic behavior of monumental buildings and historic structures presents significant challenges due to their architectural and structural complexity, as well as the intrinsic constraints posed by conservation requirements. These buildings are often characterized by irregular geometries, discontinuous construction techniques, and a lack of standardized design principles, reflecting their evolution over centuries.

Post-seismic evidence and numerous experimental campaigns demonstrate that when masonry disaggregation is prevented and the connections between walls and between walls and floors are sufficient to inhibit local mechanisms, masonry walls may experience two modes of collapse: **flexural failure (in-plane bending)** and **shear failure** (Fig. 4.25). These mechanisms define the critical modes of collapse and are influenced by the geometry, material properties, and boundary conditions of the walls.

In-plane flexural failure of a wall occurs when horizontal seismic forces, combined with vertical loads, induce significant eccentricities (Fig. 4.19). This process develops in two main stages:

(a) **Partial Compression of the Section**: Low axial loads or high eccentricities substantially reduce the compression zone, leading to uneven stress distribution and the initiation of tensile cracks at the tensioned edge.

(a) (b) (c)

Fig. 4.25 In-plane failure mechanisms of unreinforced masonry walls

(b) **Ultimate Collapse**: As compression concentrates in a reduced zone, crushing occurs in the compressed area, often accompanied by rigid-body rotation of the wall around its compressed edge.

This mechanism is particularly common in slender walls, characterized by a high height-to-thickness ratio (h/t). These walls are especially vulnerable to bending moments caused by horizontal forces, which promote early flexural cracking and ultimately lead to crushing failure.

Shear failure in masonry walls subjected to in-plane forces can occur in two primary modes (Fig. 4.19):

- **Sliding Shear Failure**: Governed by the inability of mortar joints to resist horizontal shear stresses. Sliding typically occurs along a horizontal plane, such as a mortar joint, and is common in walls with weak bonding.
- **Diagonal Cracking**: This mode is dominated by tensile stresses acting diagonally across the panel. Cracks propagate from the centre of the panel toward its corners, following a stair-stepped pattern in regular masonry or passing through blocks in weaker masonry (Fig. 4.25).

The mode of collapse depends on several critical factors:

- **Axial Load Level**: Low vertical loads Favor flexural failures due to reduced compression zones, while higher loads enhance shear resistance but may increase the risk of crushing.
- **Wall Geometry**: Slender walls ($h/t > 1.5$) are more susceptible to flexural failure, while squat walls ($h/t \approx 1$) typically fail in shear.
- **Boundary Conditions**: Effective anchoring and interlocking of walls enhance their ability to resist in-plane forces and reduce vulnerability to localized failures.

For specific details about the bending and shear capacity of the existing masonry walls, see the Italian and European Regulatory codes (EN 1998-3 2005; NTC2018 2018; Circolare 7/2019 2019; EN 1998-1 2004).

4.4.3.2 Modelling Strategies

Historic masonry buildings often exhibit significant architectural complexity, characterized by multiple structural units resulting from extensions built over different historical periods. These buildings frequently include complex structural systems such as vaulted systems, misaligned openings, unsupported walls, lack of floor diaphragms, etc. Such complexities necessitate structural models that account for these variables to provide a realistic representation of the building's response to seismic actions.

Nonlinear **modeling of heritage masonry buildings** can be performed using various approaches that account for stiffness degradation under increasing displacements, resistance to horizontal forces, and the progressive deterioration of the structure up to collapse. Structural element models, which describe the nonlinear response of masonry at the panel scale using generalized stress characteristics, are widely applied in practice. These models are advantageous because they enable nonlinear analysis with manageable computational effort, even for highly complex structures. Additionally, they are typically based on mechanical parameters that can be directly derived from commonly used diagnostic surveys (when feasible and necessary) or from literature and standards. Within this modeling framework, **equivalent frame models** are the most widely adopted in practice.

Specifically, equivalent frame models involve the identification of vertical elements (**masonry piers**) and horizontal elements (**spandrels**) within a wall, with spandrels located between two vertically aligned openings. The piers serve as the primary load-bearing components, supporting both static and seismic actions, while the spandrels act as secondary resisting elements that significantly influence the coupling and redistribution of forces within the piers (Fig. 4.26).

Nonlinearity in the structural response is concentrated in the piers and spandrels, while the intersections between these elements (**node panels**) are considered rigid, modelled through the introduction of kinematic constraints. This schematization is supported by post-seismic damage observations, which consistently show that damage is predominantly concentrated in the piers and spandrels.

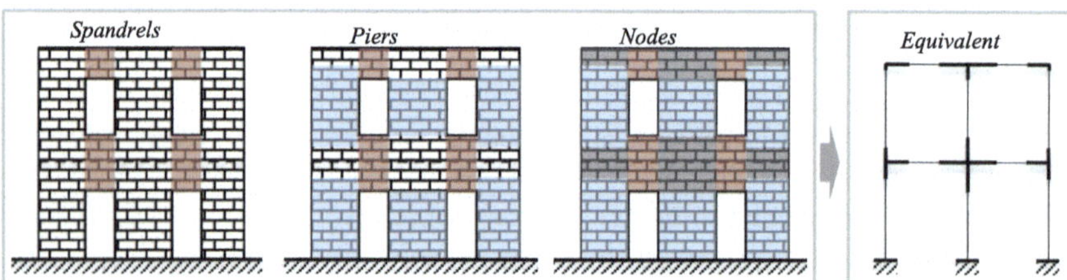

Fig. 4.26 Identification of structural elements (spandrels, piers, and nodes) and their transformation into an equivalent frame model

The degree of coupling between the vertical and horizontal elements—often determined by the presence or absence of tie beams (ring beams) or other connections—plays a critical role in defining the structural response. Depending on the **level of connection** between the masonry piers and spandrels, various analytical models can be developed, which will be detailed throughout this chapter. These models provide a basis for understanding how the building's structural components interact and contribute to its global seismic performance.

The structural model of a masonry building can vary depending on the degree of coupling provided by the spandrels. Specifically, when the spandrels lack stiffness, the wall behaves as a series of isolated cantilevers composed solely of masonry piers. In this case, the wall model can be characterized as follows:

- When **spandrels lack both extensional and flexural stiffness**, the wall behaves as a series of isolated cantilevers made up solely of masonry piers. The structural model is characterized by masonry piers extending over the full height of the wall, while node panels are absent. The spandrels only support vertical loads and are considered non-existent in resisting horizontal actions. This behaviour is typical of buildings without perimeter connection elements, such as ring beams or ties, which are essential to ensuring horizontal coupling.
- When **spandrels have extensional stiffness**, typically provided by a single tensile-resistant element such as a tie rod, ring beam, or steel lintel, the wall functions as a system of vertical cantilevers connected by horizontal tensile members. The structural model is characterized by masonry piers extending to the full height of the inter-story, while node panels are essentially reduced to negligible lines. The spandrels function as single-tie horizontal elements, providing primarily axial resistance without contributing significantly to the structural response. The masonry piers are continuous vertical cantilevers (from foundation to roof) and are coupled translationally through floors.
- When spandrels have both **significant extensional and flexural stiffness**, typically due to the presence of continuous ring beams along the entire wall length, the wall behaves like a rigid-jointed frame under horizontal forces. In this case, the spandrels prevent rotations at the ends of masonry piers and the structural model consists of piers that correspond to the height of the openings and are bound by two spandrels. Node panels are fully enclosed, except at the extremities of the wall, while the spandrels are rigidly connected to the node panels and ensure the deformation compatibility.

The equivalent frame schematization is well-suited for structures with regular openings and architectural features typical of ordinary masonry construction. However, for buildings belonging to the historic built heritage, this schematization is more challenging to apply. If not properly calibrated, it can lead to results that significantly deviate from the actual structural behaviour. Therefore, its application requires a highly skilled and cautious approach.

The most common challenges associated with this modelling approach include:

- **Idealization of walls into an equivalent frame**: Issues arise with irregular openings, filled-in openings, very small openings, and overestimation of stiffness due to the inclusion of rigid node panels.
- **Modelling of masonry panels**, particularly the spandrels.
- **Representation of specific construction systems**, such as overhanging walls, arches, and misaligned inter-story levels in adjacent spaces or buildings.
- **Modelling of floor diaphragms**, including assumptions about their stiffness and the calculation of forces transmitted by vaulted systems.

These challenges can be addressed through the following methods:

- **Sensitivity Analysis**: Sensitivity analysis helps evaluate the impact of model uncertainties arising from the approximations and assumptions of the equivalent frame approach. For example, in cases of highly irregular opening layouts, alternative geometries for piers and spandrels can be considered. Based on the results, further investigations—such as diagnostic surveys or more detailed numerical analyses—can be conducted to determine the need for alternative models that provide a more reliable safety assessment.
- **Detailed Analyses of Specific Structural Portions**: For localized structural features or unique construction systems, more refined models (e.g., Finite Element Models-FEM) can be used to analyse specific portions of the structure. The results from these detailed studies can then be employed to appropriately calibrate the response of structural elements within the equivalent frame model.

In the following sections, examples and possible alternatives are presented to address these challenges effectively and enhance the reliability of the equivalent frame approach in the context of heritage masonry structures.

Misaligned Openings. In the presence of misaligned openings, the identification of masonry piers becomes challenging, as the piers on upper floors often align with the openings of lower floors (Fig. 4.27a). In an equivalent frame model, this results in vertical elements from upper floors resting on deformable horizontal elements, creating unrealistic concentrated stresses (Fig. 4.27b). To address this issue, comparative models using two-dimensional finite elements can help determine the most accurate strategy for representing the wall's structural behaviour. One possible solution involves modelling the portions of masonry piers that overlap with openings as deformable beams with dimensional characteristics similar to the pier (Fig. 4.27c). This approach simplifies and regularizes the structural model while accounting for the weight and mass of the masonry portions above the openings.

Infilled Openings. In historic buildings, the configuration of openings may have undergone many variations over the centuries. Depending on the degree of bond between the infill and adjacent masonry, infilled openings can be modelled as either "openings" (if the infill is unbonded) or as solid masonry panels (if continuity has been restored).

Small Openings. Small openings must be carefully assessed to determine whether they significantly affect force transmission and the continuity of masonry piers. If they do not, their inclusion in the model may result in excessively short piers (and

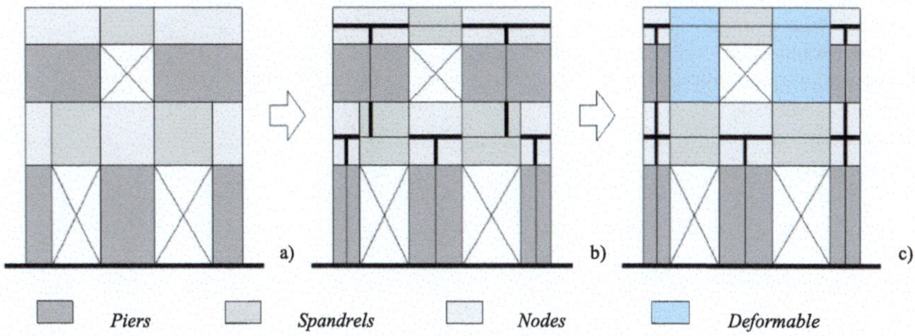

Fig. 4.27 Hypothesis of schematization of the equivalent Frame Model in case of misaligned openings

consequently squat elements), which could unrealistically affect the activation of weak-story mechanisms or localized failures. This issue arises because structural collapse criteria are typically drift-dependent, a function of panel height.

Overestimation of Stiffness Due to Rigid Node Panels. In cases of highly misaligned or very small openings, conventional methods for equivalent frame idealization may lead to disproportionately large rigid node panels. Ignoring the stiffness of these regions can significantly distort the overall structural stiffness. One solution is to model these portions as masonry panels with elastic behavior.

Overhanging Walls. Overhanging walls are often excluded from models by replacing their contribution with equivalent distributed or concentrated loads. However, in some cases, these walls may significantly affect the overall stiffness distribution and should be included in the global model. If the overhanging wall is supported by a floor, it can be modelled as part of the system, resting on supporting elements (e.g., beams representing the floor, or two-dimensional elements for advanced floor modelling).

Misaligned Interstory Levels. Misaligned interstory levels complicate equivalent frame modelling because masonry piers between misaligned floors are interrupted by horizontal elements. Comparative analysis of a limited section of the structure using more detailed models can help validate the results. Additionally, the applicability of nonlinear static analysis methods must be verified for such configurations.

Arched Systems (Porticos, Loggias, Arches). Arched systems are common in historic structures but are often not directly represented in equivalent frame models. When these systems play a significant role in structural safety, detailed modelling (e.g., finite element analysis) is recommended to calibrate elements that can emulate their effects within the equivalent frame model. For instance, detailed modelling can identify plastic hinge locations in the arch system and inform the calibration of hinged inclined beams to replicate the equivalent "strut effect".

Diaphragm Stiffness. Equivalent frame models often assume rigid floor diaphragms, but this assumption must be carefully evaluated based on the actual characteristics of the structure. While rigid diaphragms are valid for reinforced concrete

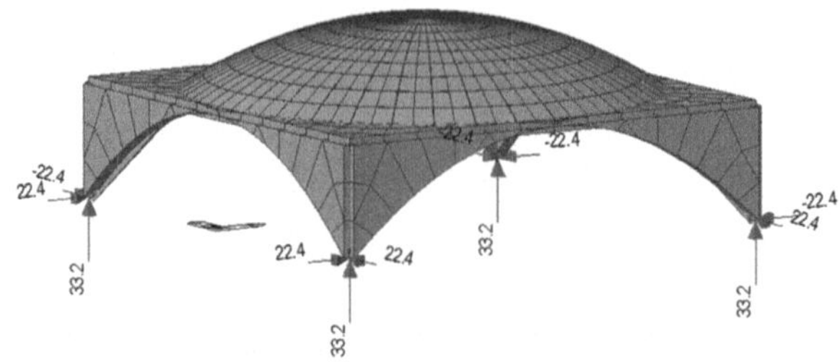

Fig. 4.28 Support reactions obtained from a two-dimensional finite element model of a dome on pendentives under the G_{kI} loading condition (permanent load)

floors or composite floors with sufficient stiffness, traditional floors in historic buildings (e.g., timber beams, vaulted systems, or metal beams with masonry infills) often require modelling with finite in-plane stiffness. Calculation methods for various floor types are provided in standards such as ASCE/SEI 41/06 and NZSEE 2006. When diaphragm stiffness cannot be directly assessed, employing rigid and infinitely flexible diaphragm models as alternative scenarios can help estimate epistemic modelling uncertainties.

Vaulted Systems. In one-dimensional finite elements, vaulted systems introduce additional complexity as their load distribution depends on the vault's shape and type. Vault loads can be approximated by equivalent loads derived from their distribution characteristics. Vaulted systems can be categorized as (a) **Point-loaded systems** (e.g., groin vaults, sail vaults, domes on pendentives), where loads are concentrated at specific points or (b) **Distributed-load systems** (e.g., barrel vaults, pavilion vaults, domes on drums), where loads are uniformly distributed along walls. In both cases, the thrust forces are applied at the vault's springing points, often requiring adjustments to apply equivalent forces at the floor level in the global model. Detailed finite element modelling of vaults can provide the necessary thrust forces for integration into the global structural model (Fig. 4.28).

Addressing these challenges through detailed modelling or comparative analyses ensures that the equivalent frame approach remains a reliable tool for assessing the safety of historic masonry structures.

The **FEM (Finite Element Method) model** provides a much more precise and detailed alternative to the equivalent frame approach for analyzing historic masonry buildings. Unlike the simplified schematization of walls into piers, spandrels, and rigid nodes, FEM modeling offers a refined representation of the structure, capturing the complex interactions between masonry components and their nonlinear behavior under seismic or static loads. However, this enhanced precision requires significantly higher computational resources and detailed input data.

FEM modelling can be divided into two primary approaches: **macro-modelling** and **micro-modelling** (Asteris et al. 2015; Lourenço 2015). Macro-modelling considers masonry as a homogenized material with averaged properties for the entire assemblage, offering computational efficiency and suitability for large-scale structural analysis (Pantò et al. 2015). However, it cannot accurately capture localized damage phenomena, such as crack propagation or crushing. In contrast, micro-modelling represents masonry in its actual configuration, where individual blocks and mortar joints are modelled separately, allowing for detailed investigation of specific failure mechanisms. While this approach provides a high level of accuracy, it is computationally intensive and demands extensive data on material properties and geometry (Petracca et al. 2017, 2023).

Compared to the equivalent frame method, FEM models are particularly advantageous in analysing **irregular structures**, such as those with misaligned openings, overhanging walls, or arched systems. They are also indispensable for assessing complex elements like vaulted systems or domes, where FEM modelling can capture the complex stress redistributions and thrust mechanisms that simplified models cannot consider.

Despite these advantages, FEM models are **computationally demanding** and require validation through diagnostic surveys or experimental data to ensure reliability (Kumar and Pallav 2020). For this reason, their use is often reserved for critical components of heritage structures or for situations where a high level of detail is necessary to understand structural behaviour and guide interventions.

In conclusion, while FEM modelling demands more resources than the equivalent frame approach, it offers unmatched accuracy for global analyses. Its application is particularly valuable for complex heritage structures, where the behaviour of unique elements like vaulted systems or arched constructions must be understood in detail. Integrating FEM modelling with other approaches ensures a comprehensive evaluation of structural safety and the development of effective preservation strategies.

4.4.3.3 Analysis Methods

The global seismic response analysis for existing masonry structures can be performed using the following methods:

Linear Static Analysis: This method involves applying equivalent static forces to the structure, representing seismic loads, and evaluating the induced stresses and displacements under the assumption of linear elastic behaviour. It is typically suitable for simple and regular structures.

Linear Dynamic Analysis: Utilizing modal analysis, this method considers the dynamic properties of the structure (e.g., natural frequencies and mode shapes) to calculate the seismic response. It accounts for higher modes of vibration and is more accurate than static analysis, especially for irregular or taller structures.

Nonlinear Static Analysis (Pushover Analysis): A monotonic loading process is applied to simulate the progression of damage under increasing seismic forces. This method provides insights into the capacity curve of the structure, relating base shear to roof displacement, and is particularly useful for identifying global and local failure mechanisms.

Nonlinear Dynamic Analysis (Time-History Analysis): This is the most detailed and accurate method, involving the application of ground motion records to simulate the time-dependent response of the structure. It captures the full nonlinear behaviour, including stiffness degradation and hysteresis, but requires significant computational resources and detailed input data.

Given that the expected nonlinear response of historic buildings to seismic actions can be significant, even under moderate seismic intensities or when evaluating ultimate limit states, **nonlinear static analysis** plays a critical role. Linear analysis, in contrast, is often insufficient because it cannot accurately capture redistribution effects or the limitations in defining appropriate reduction factors for structural behaviour. Recognizing these challenges, Italian regulatory codes (NTC2018 2018; Circolare 7/2019 2019) explicitly allow nonlinear analysis for masonry structures, regardless of the participating mass in the first mode of vibration. For instance, the standards lower the threshold for participating mass to 60% instead of the usual 75% and permit its use for existing buildings in all cases. European code for existing masonry buildings (EN 1998-3 2005) does not provide specific limitations for the use of nonlinear static analysis.

However, employing nonlinear analysis requires caution, especially for highly irregular structures or buildings with flexible diaphragms. Careful attention must be given to:

- The distribution of forces used in the analysis,
- The displacement representative of the global response on the pushover curve,
- The conversion of the original system to an equivalent oscillator.

For irregular buildings, where the participating mass in the first mode is relatively low, relying solely on the first modal shape for nonlinear static analysis may yield erroneous results.

For further details about the application of the method of analysis, see Fajfar (1999), Fajfar (2000), NTC2018 (2018), Circolare 7/2019 (2019), Cantagallo et al. (2023a) and Cantagallo et al. (2023b).

4.4.4 Risk Assessment for Individual Heritage Buildings

The seismic risk assessment of individual historic buildings, particularly monumental heritage, is a crucial process for ensuring the preservation of cultural, historical, and architectural values. Unlike ordinary structures, the assessment must go beyond traditional parameters of safety and economy to encompass the intrinsic heritage value, which includes **authenticity**, **cultural significance**, and **socio-economic impact**.

Seismic risk assessment integrates three fundamental components: hazard, vulnerability, and exposure. **Hazard** represents the probability of seismic events occurring in a given region, determined through geological and seismological studies. **Vulnerability** quantifies the susceptibility of a structure to sustain damage when subjected to seismic forces. **Exposure**, in the context of heritage buildings, extends beyond economic considerations to include cultural, artistic, and historical values.

4.4.4.1 Risk Assessment

Risk assessment for heritage buildings consists of identifying potential seismic threats, determining the structure's response to these hazards under different scenarios, and evaluating the potential losses. The assessment framework typically involves analysing the building's performance across various **limit states**, namely Operational, Damage, Life Safety, and Collapse (Cosenza et al. 2018; Ministry Decree n.58 28/02/2017 2017). These states represent a hierarchy of degradation levels due to seismic actions. The **Operational Limit State** (OLS) corresponds to minor damage with full functionality retained, the **Damage Limit State** (DLS) involves reparable damage that affects usage, the **Life Safety Limit State** (LSLS) indicates significant damage with life-threatening conditions and the **Collapse Limit State** (CLS) entails complete structural failure, risking lives and cultural loss.

Key indices used in this methodology are the **Life-Safety Index (SI-LS)** and the **Expected Annual Loss (EAL)**. The SI-LS is calculated as the ratio between the peak ground acceleration capacity of the structure at the Life Safety Limit State ($PGA_{C\text{-}LS}$) and the seismic demand at the same limit state ($PGA_{D\text{-}LS}$).

$$\mathrm{SI\text{-}LS} = \mathrm{PGA}_{\mathrm{C\text{-}LSLS}} / \mathrm{PGA}_{\mathrm{D\text{-}LSLS}} \tag{4.36}$$

SI-LS can be therefore obtained only after a careful seismic vulnerability analysis of the structure, and a higher SI-LS corresponds to a lower likelihood of significant damage. Ministry Decree n.58 28/02/2017 (2017) identifies seven different classes of SI-LS ($Class_{SI\text{-}LS}$), ranging from $A^{+}_{SI\text{-}LS}$ (very safe) to $F_{SI\text{-}LS}$ (unsafe), as indicated in Table 4.30.

The **Expected Annual Loss (EAL)** is a measure that quantifies the potential economic impact of seismic events on a heritage building. It combines the probabilities of reaching various limit states with the corresponding economic losses associated with each state. These limit states include Operational, Damage, Life Safety, and Collapse, representing progressive levels of structural degradation. The calculation of EAL involves integrating the **annual exceedance probabilities** of these limit states with their respective **economic loss estimates** over the building's lifespan (Fig. 4.29). The result is represented graphically as an EAL curve, where

Table 4.30 Life-safety index classes (Ministry Decree n.58 28/02/2017 2017)

Safety index	$Class_{SI\text{-}LS}$
100% ≤ SI-LS	$A^{+}_{SI\text{-}LS}$
80% ≤ SI-LS ≤ 100%	$A_{SI\text{-}LS}$
60% ≤ SI-LS ≤ 80%	$B_{SI\text{-}LS}$
45% ≤ SI-LS ≤ 60%	$C_{SI\text{-}LS}$
30% ≤ SI-LS ≤ 45%	$D_{SI\text{-}LS}$
15% ≤ SI-LS ≤ 30%	$E_{SI\text{-}LS}$
SI-LS ≤ 15%	$F_{SI\text{-}LS}$

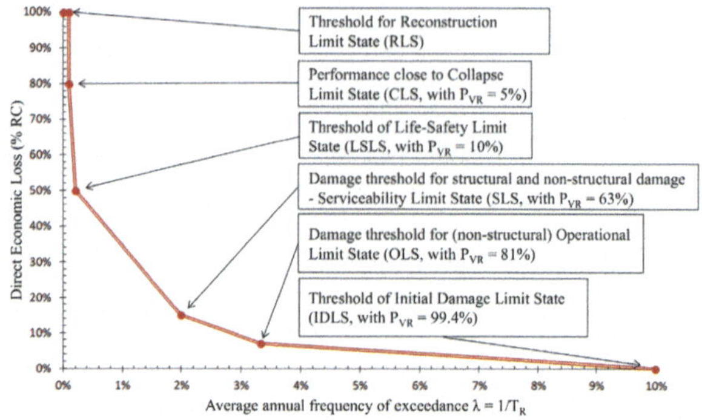

Fig. 4.29 EAL-curve for a residential building (use class II and nominal life of 50 years) according to the Ministry Decree n.58 28/02/2017 (2017)

the area under the curve corresponds to the total expected economic loss. The curve illustrates the thresholds for each limit state: Initial Damage Limit State (IDLS), Operational Limit State (OLS), Serviceability Limit State (SLS), Life-Safety Limit State (LSLS), Collapse Limit State (CLS), and Reconstruction Limit State (RLS). The x-axis indicates the average annual frequency of exceedance ($\lambda = 1/T_R$), while the y-axis represents the direct economic loss as a percentage of the Reconstruction Cost (%RC). Each point on the curve corresponds to a specific probability of reaching or exceeding a limit state (P_{VR}). In particular, the definition of the EAL-curve is carried out obtaining first the Peak Ground Acceleration Capacity (PGA_C) and the Peak Ground Acceleration Demand (PGA_D) related to DLS and LSLS. Subsequently, the return period capacity $T_{R,C}$ corresponding to the two considered limit states are obtained following the indications of the current regulatory codes (NTC2018 2018):

$$T_{R,C} = T_{R,D}\left(PGA_C / PGA_D\right)^{\eta} \tag{4.37}$$

where

- $T_{R,C}$ is the return period provided by the regulatory codes based on the number of years during which the structure is expected to maintain specific performance levels and the building's use class, defined with reference to the consequences of an operational interruption or a potential collapse of the building.
- $\eta = 1/0.49$ for $a_g > 0.25g$; $\eta = 1/0.43$ for $0.15g < a_g \leq 0.25g$; $\eta = 1/0.356$ for $0.05g < a_g \leq 0.15g$ and $\eta = 1/0.34$ for $a_g \leq 0.05g$ (with a_g equal to the seismic acceleration on rock soil corresponding to the considered site).

The inverse of each return period capacity $T_{R,C}$ provide therefore the values of the average annual frequency of exceedance λ of the EAL-curve. After having calculated λ for DLS (λ_{DLS}) and LSLS (λ_{LSLS}), the Italian guidelines (Ministry Decree n.58 28/02/2017 2017), suggest obtaining λ for OLS (λ_{OLS}) and CLS (λ_{CLS}) using the following simplified formulations:

$$\begin{aligned} \lambda_{OLS} &= 1.67\lambda_{DLS} \\ \lambda_{CLS} &= 0.49\lambda_{LSLS} \end{aligned} \tag{4.38}$$

The Italian guidelines (Ministry Decree n.58 28/02/2017 2017) establish the **repair costs** for each limit state based on evaluations of seismic damage and analyses of macro-seismic data observed after the recent earthquakes in Italy (Dolce and Manfredi 2015; Cosenza et al. 2018). These costs are expressed as percentages of the reconstruction costs (RC) and are assigned as indicated in Table 4.31.

The EAL index corresponds to the area beneath the EAL curve and can be calculated using the following formulation:

$$\mathrm{EAL} = \left\{ \sum_{i=2}^{5} [\lambda(\mathrm{LS}_{i-1}) - \lambda(\mathrm{LS}_i)] \cdot [\mathrm{RC}(\mathrm{LS}_i) + \mathrm{RC}(\mathrm{LS}_{i-1})]/[\mathrm{RC}(\mathrm{LS}_i) + \mathrm{RC}(\mathrm{LS}_{i-1})]/2 \right\} + \lambda(\mathrm{CLS}) \cdot \mathrm{RC}(\mathrm{RLS}) \quad (4.39)$$

where the index i is the limit state ($i = 5$ corresponds to CLS and $i = 1$ to IDLS), RLS denotes the Reconstruction Limit State. Italian guidelines associate the defined ranges of EAL index to specific classes ($\mathrm{Class}_{\mathrm{EAL}}$), as indicated in Table 4.32.

As a result, the seismic risk classification for the analysed building is determined by the worst-case scenario between $\mathrm{Class}_{\mathrm{SI\text{-}LS}}$ and $\mathrm{Class}_{\mathrm{EAL}}$.

For **heritage buildings**, the economic evaluation must account for factors beyond conventional reconstruction costs. For this reason, Diaferio et al. (2021) introduced the **Reconstruction Cost for Historic buildings** ($\mathrm{RC_H}$) defined as the sum of the conventional reconstruction cost (RC), calculated based on regional price lists, and

Table 4.31 Repair costs for each limit state (Ministry Decree n.58 28/02/2017 2017)

Limit state	% RC
Reconstruction (RLS)	100
Collapse (CLS)	80
Life Safety (LSLS)	50
Damage (DLS)	15
Operational (OLS)	7
Initial Damage (IDLS)	0

Table 4.32 Classes of expected annual losses (Ministry Decree n.58 28/02/2017 2017)

Expected Annual Loss (EAL)	$\mathrm{Class}_{\mathrm{EAL}}$
$\mathrm{EAL} \leq 0.50\%$	$\mathrm{A^{+}}_{\mathrm{EAL}}$
$0.5\% \leq \mathrm{EAL} \leq 1.0\%$	$\mathrm{A}_{\mathrm{EAL}}$
$1.0\% \leq \mathrm{EAL} \leq 1.5\%$	$\mathrm{B}_{\mathrm{EAL}}$
$1.5\% \leq \mathrm{EAL} \leq 2.5\%$	$\mathrm{C}_{\mathrm{EAL}}$
$2.5\% \leq \mathrm{EAL} \leq 3.5\%$	$\mathrm{D}_{\mathrm{EAL}}$
$3.5\% \leq \mathrm{EAL} \leq 4.5\%$	$\mathrm{E}_{\mathrm{EAL}}$
$4.5\% \leq \mathrm{EAL} \leq 7.5\%$	$\mathrm{F}_{\mathrm{EAL}}$
$7.5\% \leq \mathrm{EAL}$	$\mathrm{G}_{\mathrm{EAL}}$

additional parameters linked to the building's cultural, artistic, architectural, and socio-economic value. This relationship can be expressed as:

$$\mathrm{RC_H} = \mathrm{RC} + \sum_{i=1}^{n} C(k_i) \quad (4.40)$$

where k_i represents each parameter (e.g., cultural or historical value), $C(k_i)$ is the cost associated with it, and n is the total number of considered parameters. For instance, in the case of a historic church, k1 could represent the artistic value of frescoes or sculptures within the building, and its associated cost $C(k_1)$ would reflect the expense of restoring or preserving these elements. Similarly, k_2 might account for the architectural uniqueness of the structure, such as intricate vaults or rare construction techniques, with $C(k_2)$ capturing the specialized skills and materials required for reconstruction. Another parameter, k_3, could address the socio-economic impact, such as the loss of tourism revenue during the reconstruction period, with $C(k_3)$ calculated based on annual revenue losses and estimated recovery time.

As for ordinary buildings, repair costs of built heritage associated with specific limit states can be calculated as percentages of the reconstruction cost (RC), in line with existing guidelines (Table 4.30). The Expected Annual Loss of historic buildings ($\mathrm{EAL_H}$) is derived from plotting repair costs as percentages of $\mathrm{RC_H}$ against annual exceedance probabilities for each limit state λ and integrating repair costs over λ. Finally, the seismic risk class is identified as the worst class between $\mathrm{Class_{SI-LS}}$ $\mathrm{Class_{EAL\text{-}H}}$.

To clarify the concepts, a hypothetical numerical example is reported below:

4.4.4.2 Example

Assume a historic theatre has a conventional reconstruction cost (RC) of €2,000,000. Additional parameters reflecting the building's intrinsic values are introduced:

- k_1: Artistic value of murals and sculptures based on specialized restoration costs provided by regional price lists or consultation with art conservation experts. Assumed cost $C\ (k_1) =$ €200,000,
- k_2: Architectural uniqueness requiring specialized restoration based on the true expenses required to restore architecturally unique features authentically and in compliance with heritage standards. Assumed cost $C\ (k_2) =$ €300,000,
- k_3: Socio-economic impact due to loss of annual tourism revenue during reconstruction. It is calculated as annual Revenue Loss ($\mathrm{R_L}$) multiplied by the reconstruction duration ($\mathrm{T_r}$). If $\mathrm{R_L} =$ €100,000 and $\mathrm{T_r} = 1$ year, $C\ (k_3) = \mathrm{R_L} \cdot \mathrm{T_r} =$ €100,000.

Using Eq. (4.40), the reconstruction cost of the theatre is equal to

$$\mathrm{RC_H} = 2{,}000{,}000 + (200{,}000 + 300{,}000 + 100{,}000) = 2{,}600{,}000€ \quad (4.41)$$

A risk curve represents the relationship between the likelihood of a hazardous event occurring (typically plotted on the X-axis) and the resulting losses associated with that event (commonly represented on the Y-axis). To represent the likelihood of a seismic event, the annual frequency of exceedance (λ) is calculated. Specifically, the annual frequency λ for each limit state is determined as the inverse of the corresponding return period ($T_{R,C}$). As previously indicated, $T_{R,C}$ is specifically calculated for DLS and LSLS calculated while for the other limit states there are simplified formulations.

Assuming that, at LSLS, PGA_C (typically obtained by an accurate nonlinear analysis of the structural model of the theater) is equal to 0.25 g, PGA_D (that depends on the seismic hazard of the site) is equal to 0.30 g and $T_{R,D}$ (calculated as the product between the nominal life of the structure and its use class) equal to 75 years, $T_{R,C}$ is calculated as follows:

$$T_{R,C} = T_{R,D}\left(PGA_C/PGA_D\right)^{\eta} = 75\left(0.25/0.30\right)^{2.04} = 51.7\,\mathrm{yr} \tag{4.42}$$

If the same calculation is repeated at the DLS assuming that PGA_C is equal to 0.10 g and PGA_D is equal to 0.15 g, $T_{R,C}$ equal to 24.0 yr is obtained. Consequently, the λ values for DLS and LSLS are equal to 0.0416 and 0.0196, respectively. For OLS and CLS, the λ values are calculated according to Eq. (4.34) and the corresponding values are reported in Table 4.33.

Subsequently, to calculate the risk curve, the expected annual losses corresponding to each limit state are determined based on the Italian guidelines for seismic risk assessment (Ministry Decree n.58 28/02/2017 2017), which assign a specific percentage of RC_H to each limit state (as shown in Table 4.33).

Figure 4.30 shows the EAL_H curve obtained for the example considered in this study. The total EAL_H is calculated as the area under the curve using Eq. (4.35) and is equal to 2.73% of RC_H, as indicated in Eq. (4.39).

$$\begin{aligned} EAL_H = {} & \left[(10-6.95)\cdot(0.07+0)/2\right] + \left[(6.95-4.16)\cdot(0.07+0.15)/2\right] + \\ & \left[(4.16-1.93)\cdot(0.50+0.15)/2\right] + \left[(1.93-0.95)\cdot(0.50+0.80)/2\right] + \\ & +(0.95\cdot 1) = 2.73\% \text{ of } RC_H \end{aligned} \tag{4.43}$$

Table 4.33 Example values for the calculation of the seismic risk

Limit state	$T_{R,C}$ (years)	RC_H (%)	RC_H (€)	λ (%)
IDLS	10.0	0	0	10
OLS	14.4	7	182,000	6.95
DLS	24.0	15	390,000	4.16
LSLS	51.7	50	1,300,000	1.93
CLS	105.5	80	2,080,000	0.95
RLS	–	100	2,600,000	0

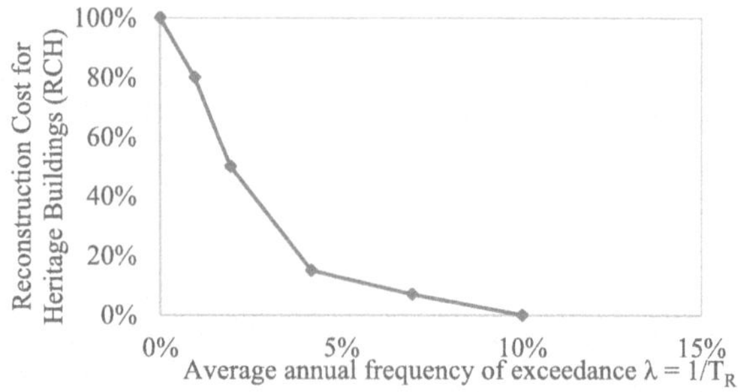

Fig. 4.30 EAL_H-curve for the considered example

Since $2.5\% \leq EAL_H \leq 3.5\%$, according to Italian guidelines for seismic risk assessment (Ministry Decree n.58 28/02/2017 2017) **$Class_{EAL}$** is classified as **D_{EAL}**.

The **Life-Safety Index (SI-LS)** is calculated as the ratio of PGA_C to PGA_D, resulting in a value of 0.833. According to the Italian guidelines (Ministry Decree n.58 28/02/2017 2017), this index corresponds to **$Class_{SI\text{-}LS}$ =** $A_{SI\text{-}LS}$.

Finally, the overall Risk Class is calculated as the worst-case scenario between SI-LS and EAL_H and is classified as D_{EAL}.

4.5 Climate Change Vulnerability and Risk for Large-Scale Analyses

Climate change vulnerability and related risk can be assessed at different scales, depending on the context of analysis and research objectives. In the current Sect. 4.5, a large-scale analysis approach is presented at the urban district level, followed by an examination of vulnerability and risk at the individual building scale in Sect. 4.6. This latter analysis explores the relationship between building pathologies and degradation, highlighting how climatic variables influence materials, structures, and conservation strategies for the built heritage.

4.5.1 Climate Change and Urban Risk: An Introductory Overview

The Mediterranean region is recognized as a "*hot spot*" for climate change, characterized by significant vulnerability due to rising temperatures and decreasing precipitation levels (Manigrasso 2022). Scientific studies indicate that climate change and in particular global warming effects are already manifesting, rather than being a distant future concern. Among these, the increasing frequency of extreme weather events, such as Mediterranean tropical-like cyclones (*medicanes*), highlights the urgency of adaptation strategies (Adamo 2016).

Advances in climate models have enabled more precise simulations at higher resolutions and increasingly accurate assessments of current and future climate variations. The Euro-Mediterranean Center on Climate Change (CMCC, https://www.cmcc.it/) has developed a regional model with an 8 km spatial resolution, optimized for Italy (Montesarchio et al. 2014; Russo et al. 2019), named COSMO model in CLimate Mode (COSMO-CLM). Using this model, some researchers (Jacob et al. 2014; Zollo et al. 2016) analyzed climate projections for the near future (2021–2050), and the results showed a general temperature increase of up to 2 °C across Italy, with significant seasonal variability. The summer months present the most pronounced changes, characterized by higher average temperatures and reduced precipitation levels. In addition, climate indicators reveal an increase in days exceeding 29 °C, a reduction in days with temperatures below freezing and an increasing frequency of drought periods (Mercogliano et al. 2017).

Increased occurrences of intense precipitation events, extreme heat waves, and reduced water availability make urban environments particularly vulnerable due to the direct health and safety risks they pose and their impact on air quality deterioration. Moreover, existing urban planning deficiencies—such as uncontrolled urban expansion, soil impermeabilization, and inadequate drainage systems—increase these climate-related hazards.

Most Italian cities are still inadequately prepared to deal with rising temperatures and shifting precipitation patterns. Extreme weather events are becoming increasingly frequent, accelerating hydrogeological instability. Studies conducted by institutions such as the Superior Institute for Environmental Protection and Research - ISPRA (Putaud et al. 2018; Munafò et al. 2024), the Italian National Agency for New Technologies—ENEA (Dalmastri and Uccelli 2024; Müller et al. 2024), Legambiente (Pietrapertosa et al. 2019), the Euro-Mediterranean Center on Climate Change (Nicolì et al. 2023) as well as the studies carried out from numerous worldwide universities, have provided extensive analyses of climate change impacts on urban environments. These studies highlight the urgent need for specific policies to adapt cities to climate change.

The United Nations Educational, Scientific and Cultural Organization (UNESCO) has highlighted that climate change poses a significant threat to cultural heritage worldwide sites, including those in Italy (Lafrenz Samuels and Platts 2022; Nguyen and Baker 2023). The increasing frequency of extreme weather events, such as intense rainfall and heatwaves, can cause physical damage to historical structures and accelerate material deterioration. Moreover, rising sea levels and increased humidity can lead to structural instability in coastal heritage sites. One of the most emblematic cases is **Venice**, a UNESCO World Heritage site that faces severe threats from climate change. Rising sea levels and increased storm surges have intensified the phenomenon of *acqua alta* (high water), causing frequent and severe flooding (Fig. 4.31). The city's historic architecture, built on wooden piles, is particularly vulnerable to prolonged water exposure, leading to structural weakening and material degradation. Furthermore, saltwater infiltration accelerates the erosion of masonry and frescoes, posing an existential risk to Venice's invaluable artistic and architectural heritage. The MOSE project, a system of movable barriers designed to protect

Fig. 4.31 Piazza San Marco, Venice. Acqua Alta Phenomenon: October 29, 2018 (**a**) and December 1, 2008 (**b**)

the Venetian Lagoon from extreme tidal events, represents a crucial adaptation effort, yet its long-term effectiveness remains debated. Venice is a typical example of how climate change directly endangers both tangible and intangible cultural heritage, necessitating comprehensive and urgent mitigation strategies.

Despite progress in climate adaptation efforts, both Europe and Italy continue to have significant gaps in effectively integrating climate risk assessments into spatial governance. The lack of a comprehensive and coordinated approach limits the effectiveness of long-term strategies. Strengthening the use of high-resolution climate projections and multi-risk assessment methodologies can provide policymakers with essential tools to enhance urban resilience, implement adaptive strategies, and safeguard cultural assets from the increasing threats posed by climate change.

4.5.2 Urban Heat Island Intensity Evaluation

One of the most evident and harmful effects of climate change is global warming, which not only increases average annual temperatures but also causes increasingly frequent heat waves during the summer season. In this context, the Urban Heat Island (UHI) phenomenon is particularly relevant, as it describes the tendency of cities to experience significantly higher temperatures than nearby rural areas, primarily due to human activities and urban morphology.

When heat waves combine with the Urban Heat Island (UHI) effect, temperatures in cities can reach extremely high levels, posing serious challenges for urban populations. Indeed, the UHI effect, when coupled with heat waves, exacerbates health risks, particularly for vulnerable groups, by increasing cases of heat stress, dehydration, and respiratory illnesses. Additionally, it leads to a significant rise in cooling energy demands, driving up air conditioning costs and intensifying urban energy consumption, further contributing to climate change. Consequently, the study of the

Urban Heat Island Intensity (UHII) phenomenon and its mitigation strategies has become a topic of extreme interest in the global scientific community.

The study by Sangiorgio et al. (2020) identifies the urban parameters that influence the intensity and manifestation of the phenomenon. By applying this framework to historic city centers, it becomes possible to pinpoint the key factors that contribute most significantly to the formation of the Urban Heat Island (UHI) effect. These factors include:

- **Albedo and Surface Materials**: The materials used in historic buildings, such as stone and brick, often exhibit high solar radiation absorption, which, in combination with their thermal inertia properties, allows them to store heat throughout the day and release it slowly at night. This prolongs periods of elevated temperatures, especially during the heatwaves.
- **Vegetation and Green Spaces**: Historic city centers often have limited green areas due to their dense urban fabric. The absence of parks, tree-lined streets, and vertical greenery exacerbates the lack of natural cooling mechanisms, making these areas more susceptible to extreme heat.
- **Building Density and Morphology**: The narrow streets and high buildings typical of historic centers create an enclosed urban canyon effect, reducing airflow and trapping heat within the built environment. This lack of ventilation further amplifies temperature increases.
- **Anthropogenic Heat Emissions**: The concentration of human activities, including tourism, commercial businesses, and the extensive use of air conditioning, leads to additional heat emissions. The energy demand for cooling historic buildings, which were not designed for high thermal efficiency, is particularly high, further contributing to urban overheating.

These factors exacerbate heatwaves and pose a severe threat to **cultural heritage preservation**. Increased temperatures and prolonged exposure to high heat accelerate material degradation, leading to the deterioration of frescoes, facades, and structural elements in historic buildings. Furthermore, high temperatures combined with air pollution contribute to chemical reactions that worsen the decay of historical monuments and artifacts.

In addition to thermal effects, climate change increases hydrological risks to historic cities. Extreme precipitation events, increasingly frequent due to climate change, overload inadequate drainage systems, causing urban flooding.

Addressing the impact of climate change in historic urban centers requires targeted adaptation strategies, including: (a) implementing **climate-responsive restoration techniques**, such as breathable protective coatings and sustainable materials that enhance resilience against heat and humidity; (b) increasing **urban greenery** to improve shading and cooling; (c) enhancing **traditional cooling strategies**, such as shaded courtyards and reflective surfaces, inspired by historical urban design practices that minimized heat retention; (d) developing **smart climate monitoring systems**, integrating temperature, humidity, and air quality sensors to provide real-time data for risk assessment and mitigation. Further details on the adaptation strategies on climate changes are described in Chap. 5.

Understanding the interactions between climate change and historic urban environments is fundamental to preserving cultural heritage while ensuring urban resilience. Section 4.5.3 specifically explores hazard assessment tools for large-scale urban climate adaptation.

4.5.3 A Quantitative Index for Assessing Urban Heat Island Intensity

The study of Sangiorgio et al. (2020) proposes a quantitative method for evaluating the **maximum potential of the Urban Heat Island Intensity (UHII)** in cities. This index is designed using a **Multi-Criteria Decision-Making (MCDM)** approach and has been calibrated with empirical data from **41 urban districts across 35 European cities**. The aim of the index is to provide a predictive tool for assessing the intensity of urban heat islands and identifying critical areas where mitigation strategies should be prioritized.

4.5.3.1 Calculation of the Urban Heat Island Intensity Index

The UHII index (Sangiorgio et al. 2020) is calculated in function of eleven key parameters, identified based on their influence on urban heat island intensity (Fig. 4.32). These parameters are the following:

Meteorological Variables

(1) Windless days: The number of days with minimal wind, reducing natural ventilation and enhancing heat accumulation.
(2) Average maximum summer temperature: A key meteorological factor indicating the baseline thermal conditions affecting urban overheating.
(3) Average summer thermal excursion: The difference between daily maximum and minimum temperatures, which affects heat dissipation at night.
(4) Clear sky days: The frequency of cloudless days, which increases direct solar radiation exposure and contributes to urban overheating.

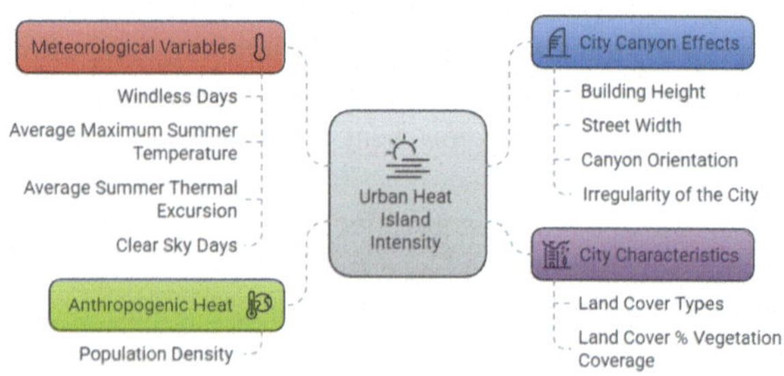

Fig. 4.32 Factors needed for calculating the UHI Index

Characteristics of the City

(5) Land Cover Types (Albedo of surfaces): The ability of urban surfaces to reflect solar radiation, influencing heat absorption and retention. High albedo materials reduce heat accumulation, whereas dark surfaces intensify heat retention.
(6) Land Cover (% vegetation coverage): The percentage of greenery within an urban district, which mitigates urban heat through shading and evapotranspiration processes.

Anthropogenic Heat

(7) Population density: The number of inhabitants per square kilometer, which affects anthropogenic heat emissions from buildings, vehicles, and human activities.

City canyon

(8) Building height: The vertical dimension of structures, which affects the urban canyon effect by trapping heat and reducing air circulation.
(9) Street width: The average width of roads, influencing airflow and the dissipation of heat within urban canyons.
(10) Canyon orientation: The alignment of streets relative to the sun's path, which influences solar radiation exposure and shading effects.
(11) Irregularity of the city: The arrangement of buildings and roads, which impacts wind flow patterns and localized heat retention.

The UHII is calculated using a weighted sum, in which each of the 11 parameters i is assigned a weight v_i and an intensity w_{ij}, determined through statistical analysis and optimization techniques.

$$UHII = \sum_{i=1}^{11} v_i \cdot w_{ij} \tag{4.44}$$

where v_i represents the weight associated with each parameter and w_{ij} corresponds to the corresponding intensity range

4.5.3.2 Calibration and Validation of the Urban Heat Island Intensity Index

UHII index is calibrated through an optimization procedure, leveraging a large-scale data acquisition conducted across 41 urban districts to ensure accuracy and reliability in its development.

To validate the UHII index, a **Jackknife resampling approach** was used, systematically removing one urban district at a time from the dataset, recalculating the index, and comparing it to the observed UHII values. This process helped assess the **stability** and **sensitivity** of the model by measuring the variation in results when specific data points were excluded. The consistency of the recalculated values confirmed that the index maintained a high predictive accuracy across different urban contexts, reinforcing its robustness in various climatic and morphological conditions. The results are measured on three indicators: (1) **Mean Absolute Error (MAE)**, (2) **Mean Relative Error (MRE)** and (3) **Pearson correlation coefficient**.

MAE is calculated as:

$$MAE = \frac{1}{n}\sum_{i=1}^{n}\left|UHII_{pred,i} - UHII_{obs,i}\right| \tag{4.45}$$

where $UHII_{pred,i}$ is the UHII value predicted by the model for urban district, $UHII_{obs,i}$ is the observed UHII value for the same district, n is the total number of urban districts analyzed. MAE calculated from Sangiorgio et al. (2020) on 41 urban districts is equal to 0.9 °C. This means that, on average, the UHII predictions are within 0.9 °C of the actual measured values.

MRE is given by:

$$MRE = \frac{1}{n}\sum_{i=1}^{n}\frac{\left|UHII_{pred,i} - UHII_{obs,i}\right|}{UHII_{obs,i}} \times 100 \tag{4.46}$$

MRE quantifies the relative deviation of predictions from observed values as a percentage. This metric provides a percentage-based evaluation of the model's accuracy, allowing for comparison across different UHII magnitudes. MRE calculated from Sangiorgio et al. (2020) is equal to 14.5%, a value that demonstrates a strong reliability of the model across different urban contexts.

The Pearson correlation coefficient (ρ) is computed as:

$$\rho = \frac{\sum_{i=1}^{n}\left(UHII_{pred,i} - \overline{UHII}_{pred}\right)\left(UHII_{obs,i} - \overline{UHII}_{obs}\right)}{\sqrt{\sum_{i=1}^{n}\left(UHII_{pred,i} - \overline{UHII}_{pred}\right)^2}\sqrt{\sum_{i=1}^{n}\left(UHII_{obs,i} - \overline{UHII}_{obs}\right)^2} \tag{4.47}$$

where $\overline{UHII}_{pred}$ and $\overline{UHII}_{obs}$ are the mean values of the predicted and observed UHII, $UHII_{pred}$ and $UHII_{obs}$ are the individual data points, and n is the total number of observations.

A Pearson correlation coefficient closer to 1 indicates a strong positive correlation, confirming that the model successfully predicts *UHII* values with high reliability.

In this case, Sangiorgio et al. (2020) found a high Pearson correlation coefficient ($\rho = 0.716$) between the predicted and observed *UHII* values, reinforcing the effectiveness of the model.

4.5.3.3 Application Example: Comparative Analysis of UHII in Different Urban Contexts

To illustrate the application of the UHII index, two hypothetical urban environments are analyzed: **a dense historic city center** and **a suburban area with higher greenery coverage**. These two scenarios represent extreme cases where urban morphology, land use, and microclimatic conditions significantly influence heat island intensity.

The historic city center is characterized by compact buildings, narrow streets, high population density, and minimal vegetation, leading to higher heat retention. Conversely, the suburban area is characterized by wider streets, lower building density, and increased vegetation, contributing to better heat dissipation and lower UHII values. To quantify the differences in Urban Heat Island Intensity (UHII) between these two environments, the **UHII index** is calculated.

To this purpose, in this example, a hypothetical yet plausible value is assigned to each of the eleven parameters described by Sangiorgio et al. (2020) for calculating the UHII index. Next, v_i weights, representing the importance of each parameter, are assumed, and w_{ij} weights, corresponding to their intensity, are determined. Specifically, the w_{ij} weights are assigned based on predefined intervals, as shown in Table 4.34.

Table 4.35 presents the assumed values for the eleven key UHII parameters in both scenarios, along with their respective weights (v_i) and intensity range weights (w_{ij}). The final UHII score is obtained by summing the products of these values according to Eq. (4.44).

Based on these calculations, the **UHII score for the historic city center is 8.24**, while the **UHII score for the suburban area is 3.31**, confirming the expected higher heat island effect in dense urban areas.

Figure 4.33 compares the contribution of each parameter to the UHII score for two urban settings: a historic city center (light blue) and a suburban area (grey). The values represent the weighted influence of each factor, illustrating how meteorological variables, characteristics of the city, anthropogenic heat and city canyon impact UHII. Notably, parameters such as albedo, street width and canyon orientation show significant differences between the two contexts.

4.5.4 Microclimate Analyses and Simulations

Understanding the effects of climate change and urbanization on the microclimate of **historic city centers** is essential for mitigating heat stress and preserving built heritage. Microclimatic alterations in **historic districts** are influenced by both **climate change**, which is increasing the frequency and intensity of **heatwaves**, and **urban morphology**, which amplifies the **UHI effect** due to high-density built environments, low vegetation cover, and the thermal inertia of traditional materials such as stone and brick. This combination of factors generates prolonged periods of **thermal**

Table 4.34 Intensity Ranges and Corresponding Weights for assessing UHII index (Sangiorgio et al. 2020)

	Parameter	Intensity range	Weight (w_{ij})
1	Windless days (%)	>15	1
		10–15	0.71
		5–10	0.43
		1–5	0.14
		<1	0.03
2	Average maximum summer temperature (°C)	>30	1
		28–30	0.85
		26–28	0.7
		24–16	0.55
		22–24	0.4
		<22	0.25
3	Average summer thermal excursion (°C)	>16	1
		14–16	0.85
		12–14	0.7
		10–12	0.55
		8–10	0.4
		<8	0.25
4	Clear sky days (%)	>70	1
		60–70	0.87
		50–60	0.73
		40–50	0.6
		30–40	0.47
		20–30	0.33
		10–20	0.2
		<10	0.07
5	Albedo of surfaces	0.10–0.12	1
		0.12–0.14	0.85
		0.14–0.16	0.73
		0.16–0.18	0.65
		0.18–0.20	0.58
		0.20–0.22	0.52
		0.22–0.24	0.48
		0.24–0.26	0.44
		0.26–0.28	0.41
		0.28–0.30	0.38

(continued)

Table 4.34 (continued)

	Parameter	Intensity range	Weight (w_{ij})
6	Vegetation coverage (%)	<5	1
		5–10	0.9
		10–15	0.8
		15–20	0.7
		20–25	0.6
		25–30	0.5
		30–35	0.4
		35–40	0.3
		40–45	0.1
		>45	0.1
7	Population density (ab/km^2)	>20,000	1
		18,000–20,000	0.91
		16,000–18,000	0.82
		14,000–16,000	0.73
		12,000–14,000	0.64
		10,000–12,000	0.55
		8,000–10,000	0.46
		6,000–8,000	0.36
		4,000–6,000	0.26
		2,000–4,000	0.15
		<2,000	0.07
8	Building height (m)	>20	1
		15–20	0.5
		10–15	0.38
		5–10	0.25
		<5	0.13
9	Street width (m)	>20	1
		15–20	0.5
		10–15	0.38
		5–10	0.25
		<5	0.13
10	Canyon orientation	South	1
		South–West	0.33
		South–East	0.33
		East–West	0.11
11	Irregularity of the city	High	1
		Medium	0.5
		Low	0.11

Table 4.35 UHII Calculation results for historic city center and suburban area

Parameter		Weight (v_i)	Weight (w_{ij})	Weight (w_{ij})	$v_i \cdot w_{ij}$	$v_i \cdot w_{ij}$
			Historic city center	Suburban area	Historic city center	Suburban area
1	Windless days	0.21	1	0.14	0.21	0.0294
2	Average maximum summer temperature	0.1	0.85	0.4	0.085	0.04
3	Average summer thermal excursion	0.1	0.7	0.25	0.07	0.025
4	Clear sky days	0.33	1	0.47	0.33	0.1551
5	Albedo of surfaces	3.47	1	0.41	3.47	1.4227
6	Vegetation coverage	2.62	0.5	0.3	1.31	0.786
7	Population density	0.75	0.55	0.26	0.4125	0.195
8	Building height	0.99	0.38	0.25	0.3762	0.2475
9	Street width	1.99	0.33	0.11	0.6567	0.2189
10	Canyon orientation	1.01	1	0.15	1.01	0.1515
11	Irregularity of the city	0.31	1	0.11	0.31	0.0341
	UHII Score				8.2404	3.3052

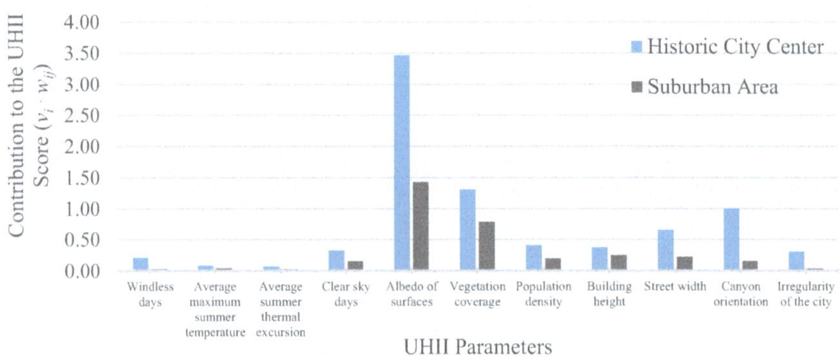

Fig. 4.33 Contribution of UHII Parameters in Different Urban Contexts

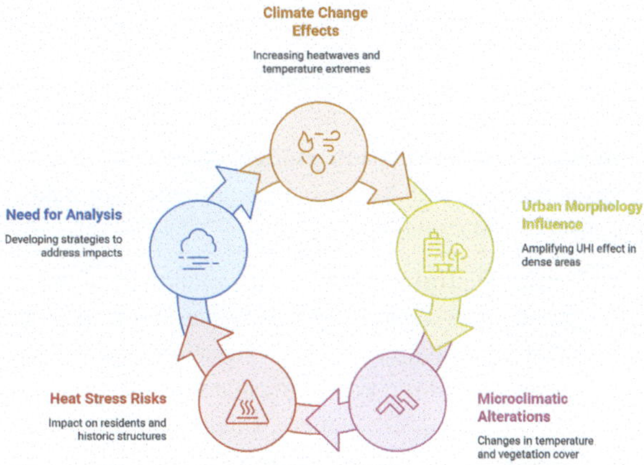

Fig. 4.34 Climate change and urbanization cycle

discomfort and **heat stress**, posing risks to both the well-being of residents and the **preservation of historic structures**, which may suffer from accelerated material degradation due to extreme thermal fluctuations.

The study of these phenomena requires robust microclimatic analysis techniques and advanced simulation tools to develop and assess mitigation strategies tailored to heritage contexts (Fig. 4.34).

A critical tool in assessing microclimatic conditions and the effectiveness of mitigation strategies in historic centers is **ENVI-met** (Huttner and Bruse 2009), a widely used computational model designed for high-resolution urban climate simulations. This model allows for the evaluation of how various environmental parameters—such as vegetation coverage, material reflectivity, street geometry, and atmospheric conditions—affect local temperature dynamics, humidity levels, and airflow patterns. ENVI-met utilizes **measured environmental data** as input to simulate microclimatic conditions in historic city centers. For specific urban contexts, researchers can calibrate their models comparing the model's outputs with field measurements.

The application of microclimate simulations in historic city centers is particularly relevant for testing **context-sensitive UHI mitigation strategies**, such as the integration of **green infrastructure** (e.g., urban gardens, vertical greenery, and shaded public spaces), the use of **high-albedo coatings and materials** that reduce heat absorption without altering the historic character of buildings or the optimization of **urban ventilation** by analyzing wind corridors and airflow dynamics in narrow, irregular street layouts.

Microclimate assessments are effective tools for ensuring the **long-term sustainability of historic centers** in a climate where the effects of climate change are increasing.

Digitization Notes

Digitalization plays a fundamental role in assessing and mitigating the **micro-climatic risks** affecting historic city centers. The integration of **digital tools, sensors, and simulation models** enhances the ability to monitor and predict climate impacts on built heritage, allowing for data-driven decision-making in conservation strategies.

- **Urban Digital Twins**: The development of **3D digital replicas** of historic city centers enables the simulation of microclimatic conditions under different climate scenarios. These models integrate **real-time sensor data** and historical records to evaluate the effectiveness of mitigation strategies such as increased greenery, shading structures, and material reflectivity.
- **GIS-Based Climate Mapping**: Geographic Information Systems (GIS) facilitate the digitization of urban heat maps, wind flow models, and solar radiation exposure assessments, providing valuable insights into localized climate conditions in heritage sites.
- **Remote Sensing and IoT-Based Monitoring**: The integration of **satellite imagery** and **ground-based sensor networks** enables continuous tracking of temperature variations, humidity levels, and air circulation within historic urban fabrics. These data sources enhance climate resilience planning by identifying high-risk areas prone to extreme temperature fluctuations and material degradation.
- **Heritage Building Information Modeling (HBIM)**: The application of **HBIM** to historic structures allows for the **digital preservation and simulation of environmental impacts** on built heritage. This approach integrates microclimate data with architectural and material properties to assess long-term climate risks and conservation needs.
- **Artificial Intelligence for Predictive Analysis**: Machine Learning (ML) algorithms analyze historical climate patterns and material deterioration data to forecast **future vulnerabilities of historic buildings** due to climate change. These predictive models support **proactive conservation planning**, reducing the risks associated with prolonged heat exposure, moisture infiltration, and pollutant-driven degradation.
- **Cloud-Based Climate Data Platforms**: The digitization of **climate records, sensor data, and simulation outputs** into centralized platforms allows for **open-access knowledge sharing** among researchers, conservationists, and policymakers. These databases facilitate the development of data-driven adaptation strategies tailored to heritage sites.

The integration of **digital tools into microclimatic analysis and heritage conservation frameworks** ensures a proactive approach to climate adaptation. By leveraging digital archives, real-time monitoring, and advanced simulations, historic city centers can **enhance their resilience against climate change** while preserving their **architectural, historical, and cultural significance** (Fig. 4.35).

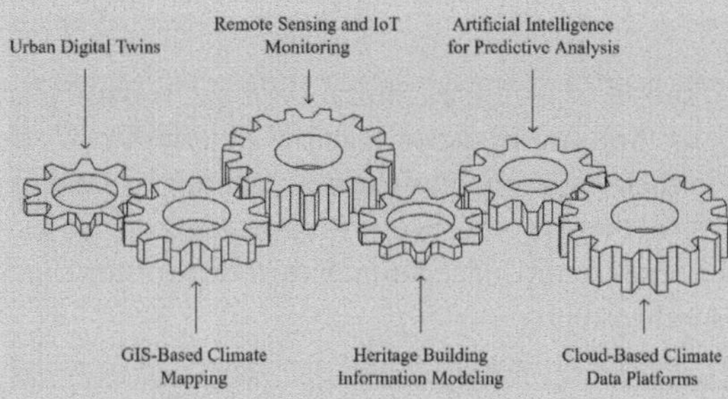

Fig. 4.35 Climate change resilience for heritage sites

An example of the implementation of such a process is the **INTERCONNECTING project** (Cultural Heritage Active Innovation for Next-Gen Sustainable Society, CHANGES project, under SPOKE 7—University of Florence. *Data, Analysis, and Digital Immersive Models for Sustainable Conservation of the Built Heritage: Risk Assessment and Proactive Strategies*). INTERCONNECTING is currently under development and operates within a collaborative framework involving several Italian universities and companies to develop an **integrated platform for multi-risk management and analysis**, including **seismic risk, climate change risk, and functional risk**.

By combining innovative digitization techniques with climate risk assessment, INTERCONNECTING aims to establish a framework for sustainable cultural heritage conservation, promoting climate change adaptation strategies for historic centers driven by scientific analysis and digital innovation.

4.6 Climate Change Vulnerability and Risk for Individual Heritage Buildings

4.6.1 Introduction

Cultural heritage includes tangible and intangible assets that define the identity and history of civilizations. Tangible cultural heritage, which includes individual monuments, architectural complexes, and sites of historical and aesthetic significance, interacts continuously with the surrounding environment. This interaction influences its natural degradation, a process accelerated by anthropogenic activities and, increasingly, by climate change (Bertolin 2019). In 2007 UNESCO (UNESCO 2007) identified the main climate parameters affecting cultural heritage, which include atmospheric moisture change, temperature change, sea-level rises, wind, desertification, climate and pollution acting together and climate and biological effects. These factors, alone or in combination, have the potential to alter the physical, chemical, and biological stability of heritage materials, impacting their conservation and structural integrity.

Built heritage, a subset of tangible cultural heritage, is particularly vulnerable to climatic variations as they contribute to material deterioration, affecting buildings differently based on their location, construction techniques, and material composition. The acceleration of these processes due to climate change poses a growing risk to historic structures, necessitating a deeper understanding of the specific mechanisms of degradation and vulnerability.

Since the 1990s, the scientific community has extensively studied the impacts of climate change on cultural heritage (Rowland 1992, 1999), highlighting the necessity for systematic evaluation methods and adaptive conservation strategies (Sesana et al. 2021). Understanding the relationship between climate factors and material deterioration is essential to develop tailored interventions that mitigate risks and extend the lifespan of heritage buildings. This section examines the specific vulnerabilities and risks posed by climate change to individual heritage buildings, focusing on the primary climatic stressors for built heritage: temperature, moisture, precipitation, wind, and solar radiation.

4.6.2 Temperature

Temperature fluctuations significantly impact the structural integrity and material durability of built heritage. Variations in temperature, particularly extreme heat events and freeze–thaw cycles, can accelerate the deterioration of historic materials, leading to irreversible damage over time.

4.6.2.1 Cryoclastim

Cryoclastism, also known as freeze–thaw weathering, is a physical process caused by the expansion of water that infiltrates the pores and cracks of materials and subsequently freezes (Fig. 4.36). The primary factors generating this phenomenon are water and freezing temperatures. As the water turns to ice, it expands, generating internal pressures that lead to microcracks and surface exfoliation. Repeated cycles of freezing and thawing increase these effects, weakening the structural cohesion of heritage buildings. This process is particularly problematic in regions subjected to frequent oscillations of temperature around the freezing point due to climate change (Grossi et al. 2007; Sabbioni et al. 2010; Sesana et al. 2021).

(a) (b)

Fig. 4.36 **a** Stone split by frost weathering in a mountainway to the tongue of Morteratsch glacier (Source: https://commons.wikimedia.org/wiki/File:Gletsjerpad_naar_de_Morteratschgletsjer_12-09-2019._(d.j.b)_28.jpg; Author: Dominicus Johannes Bergsma; License: Creative Commons Attribution-Share Alike 4.0; No changes were made); **b** Freeze–thaw weathering of a rock in southern Iceland (Source: https://commons.wikimedia.org/wiki/File:Weathering_freeze_thaw_action_iceland.jpg; Author: Till Niermann; Licence: Creative Commons Attribution-Share Alike 3.0; No changes were made)

4.6.2.2 Thermoclastism

Thermoclastism is a type of physical deterioration generated by rapid and intense temperature variations, which cause materials to expand and contract. The main factor generating this phenomenon is thermal variation. Thermoclastism affects various materials differently: stone and masonry structures may develop fissures, while wooden elements may warp or crack due to differential expansion (Bonazza et al. 2009a, b). The combination of solar exposure and temperature extremes can accelerate the breakdown of protective coatings, pigments, and surface treatments, further exposing the materials to weathering processes (Blavier et al. 2023).

4.6.2.3 Haloclasty

Haloclasty, also known as salt weathering, is a physical process of rock degradation driven by the crystallization and expansion of salts within porous materials. The primary factor that generates this phenomenon is the crystallization of salts. The process begins when saline water infiltrates rock crevices and, upon evaporation, leaves behind salt deposits. Over time, these crystals grow, and when subjected to temperature fluctuations, they expand, exerting stress on the surrounding material. This repeated cycle generates internal tensions, progressively weakening the rock structure and leading to fragmentation and disintegration (Goudie and Viles 1997; Mol and Viles 2012).

Fig. 4.37 Examples of water infiltration and capillary rise on the stone surfaces of the *Museo Domenico Ridola* in Matera. **a** Visible effects of rising damp on a masonry wall, with darkened areas and surface degradation due to prolonged moisture absorption. **b** Evidence of capillary rise along the base of the stone structure, leading to material deterioration, salt crystallization, and biological growth

4.6.3 Moisture

Moisture can accelerate the deterioration of built heritage by generating various degradation mechanisms in historic materials, including capillary rise, condensation, and biodeterioration. Additionally, fluctuations in soil moisture influence the stability of structures, particularly those built on expansive clay soils or near vegetation with high moisture absorption capacity.

4.6.3.1 Water Infiltration and Capillary Rise

Moisture infiltration occurs when water penetrates porous materials, either through direct exposure to precipitation or by capillary rise from the ground (Fig. 4.37). This phenomenon can lead to structural weakening, salt crystallization (haloclasty), and frost weathering (cryoclastism), particularly in masonry structures. Over time, repeated cycles of water absorption and evaporation contribute to the progressive degradation of building materials (Sabbioni et al. 2010; Curtis 2016).

4.6.3.2 Condensation and Internal Moisture Accumulation

Condensation forms when warm, moisture-laden air comes into contact with cooler surfaces, leading to the accumulation of water on walls, ceilings, and other building elements. This persistent moisture fosters microbial growth and promotes chemical reactions that deteriorate historical materials. The combination of high humidity and limited ventilation further increases the risk of moisture retention within built heritage (Brown and Rose 1996; Park 2009).

(a) (b)

Fig. 4.38 Examples of biodeterioration on architectural elements of the *Museo Domenico Ridola* in Matera. **a** Presence of biological patinas and microbial growth on the surface of a stone cornice, showing signs of flaking and material degradation. **b** Formation of mosses and lichens on a carved decorative element, highlighting the combined action of moisture and biological agents in the deterioration of the stone surface

4.6.3.3 Biodeterioration

Biodeterioration is the process by which biological agents, including microorganisms, fungi, lichens, and plants, accelerate material degradation (Fig. 4.38). High humidity levels and persistent moisture create ideal conditions for microbial colonization, leading to discoloration, biofilm formation, and surface erosion. Additionally, biological growth can introduce organic acids that chemically interact with construction materials, further weakening structural integrity (Ariño et al. 1995; Austigard and Mattsson 2020).

4.6.3.4 Soil Moisture Variations and Structural Stability

Changes in soil moisture can significantly impact the stability of heritage structures, particularly those built on expansive clay soils. These soils undergo cycles of shrinkage and swelling in response to moisture fluctuations, leading to differential foundation settlements and structural stress (Clarke and Smethurst 2008; Zumrawi et al. 2017). Similarly, buildings located near trees with high moisture absorption capacity may experience localized subsidence as roots extract water from the surrounding soil. In regions experiencing prolonged droughts followed by intense precipitation, rapid changes in soil saturation can further destabilize foundations, increasing the risk of cracking and deformation.

4.6.4 Precipitation

Precipitation directly impacts heritage buildings by introducing moisture into exposed surfaces, leading to material degradation and structural damage. Unlike general moisture-related mechanisms such as condensation and capillary rise, which result from prolonged humidity exposure, precipitation causes immediate and often severe deterioration, particularly during extreme weather events. Changes in rainfall patterns, including increased intensity and frequency of extreme precipitation events, can lead to severe damage in heritage structures (Dore 2005; Gobiet et al. 2014; Mo et al. 2019). Prolonged exposure to rainwater accelerates the deterioration of porous materials, while rapid shifts between dry and wet conditions exacerbate mechanical and chemical weathering processes. This section focuses on the direct impact of precipitation on historic buildings, while moisture-related deterioration mechanisms such as water infiltration and capillary rise have been discussed in the previous section.

4.6.4.1 Direct Impact of Rainwater

Rainwater infiltration is a primary cause of deterioration in heritage buildings, particularly those constructed with porous materials such as limestone, sandstone, and brick (Linden and Bossche 2022). Unlike the general moisture-related mechanisms discussed earlier, rainwater can cause immediate damage by saturating surfaces, washing away protective layers, and weakening mortar joints. The continuous absorption of rainwater fosters biological growth, salt crystallization, and freeze–thaw cycles. Furthermore, in the absence of adequate drainage, water pooling at the base of structures can accelerate erosion and foundation instability. In addition to these effects, **acid rain**, resulting from air pollution, significantly affects the preservation of historic buildings. When sulfur dioxide (SO_2), nitrogen oxides (NO_2) or other sources react with water in the atmosphere, they form acidic precipitation that dissolves carbonate-based materials such as limestone and marble (Fig. 4.39). This chemical weathering process leads to surface roughening, loss of carved details, and structural weakening. Acid rain also accelerates metal corrosion, particularly in architectural elements composed of iron, bronze, and copper, further endangering the integrity of built heritage.

4.6.4.2 Increased Rainfall and Erosion of Architectural Elements

Heavy rainfall and prolonged exposure to water and wind accelerate surface **erosion** in heritage structures (Fig. 4.38). Decorative carvings, reliefs, and inscriptions are particularly vulnerable to water-driven deterioration. Rainwater **runoff** (Markovi

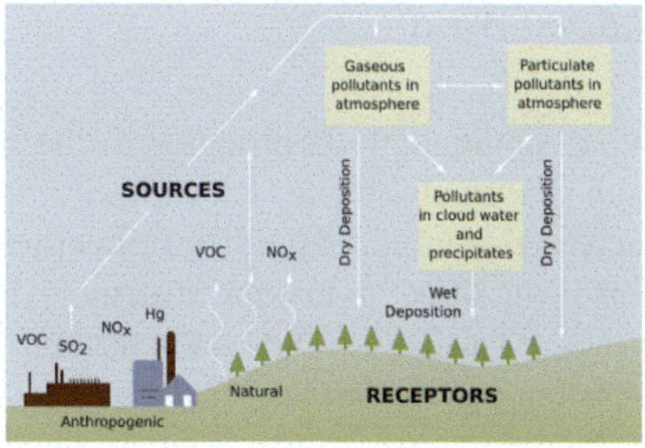

Fig. 4.39 Origin of the acid rain. *Source* https://commons.wikimedia.org/wiki/File:Origins_of_acid_rain.svg; License: Public Domain

Fig. 4.40 Examples of erosion caused by increased rainfall on architectural elements of the *Museo Domenico Ridola* in Matera. **a** Severe material loss and rounding of stone balustrades. **b** Surface erosion on the building façade, where water runoff has washed away the protective outer layer, exposing the underlying stone and leading to surface roughness and discoloration

et al. 2014), especially in areas with inadequate drainage, can remove protective layers from stone surfaces, leading to material loss and aesthetic degradation (Fig. 4.40).

4.6.4.3 Flooding and Rising Damp

In regions experiencing increased precipitation and extreme weather events, the risk of flooding poses a major threat to historic buildings (Arrighi et al. 2018). Floodwater can introduce contaminants, dissolve binding agents in masonry, and cause rapid structural deterioration. Additionally, prolonged water exposure can lead to rising damp, where moisture ascends through porous materials via capillary action, resulting in salt crystallization and internal degradation.

4.6.5 *Wind*

Wind significantly contributes to the deterioration of built heritage by affecting both structural integrity and surface materials. Changes in wind patterns, including increased wind speeds, shifting directions, and more frequent storm events, represent a growing threat to historic structures. Depending on its intensity and the nature of airborne particles, wind can contribute to structural damage, surface erosion, and moisture penetration.

4.6.5.1 Structural Effects of Wind Loads

Strong winds exert both static and dynamic loads on heritage buildings, which may result in progressive weakening of structural elements and, in extreme cases, partial or total collapse (Mosoarca et al. 2019; Hegyi et al. 2022). Historical buildings, particularly those with complex roof geometries, exposed façades, or lightweight timber structures, are particularly vulnerable (Fig. 4.39). In cases of windstorms and hurricanes, wind uplift forces can cause detachment of roofing elements, failure of connections, and cracking in masonry walls (Sesana et al. 2021).

4.6.5.2 Wind-Driven Erosion

In addition to structural risks, wind carries abrasive particles, including sand, dust, and pollutants, which gradually wear down building materials. This eolian erosion is especially pronounced in arid and coastal environments, where sand-laden winds can cause pitting, granular disintegration, and polishing of stone surfaces (Zhang et al. 2011; Wang et al. 2019). Wind erosion accelerates honeycomb weathering (**alveolarization or honeycomb weathering**), a process that results in cavity formation in calcareous stone (Fig. 4.42), a common material in historic buildings (Cardell et al. 2003).

4.6.5.3 Wind-Driven Rain (WDR) and Moisture Penetration

Wind-driven rain (WDR) facilitates the infiltration of moisture into porous materials, leading to accelerated deterioration (Blocken and Carmeliet 2004). When rain is driven at high velocities against façades, it penetrates cracks and capillaries, leading to moisture accumulation, salt crystallization, and biological growth (Orr et al. 2018). Increased rain intensity and wind pressure have been observed to accelerate decay processes in masonry and timber structures (Sesana et al. 2021).

4.6.6 Solar Radiation

Solar radiation is one of the primary factors influencing material deterioration in heritage buildings. It induces thermal stress, photodegradation, and microclimatic instability, which can accelerate the aging process of architectural surfaces.

4.6.6.1 Thermal Expansion and Contraction

Intense solar radiation leads to significant temperature fluctuations, which cause differential expansion and contraction in construction materials. This thermal stress contributes to microcracking, spalling, and material fatigue. In stone-built heritage, exposure to **infrared (IR) radiation** can increase thermal degradation and weaken structural cohesion (Sáez-Pérez and Rodríguez-Gordillo 2008).

4.6.6.2 Ultraviolet (UV) Degradation

Ultraviolet (UV) radiation is particularly damaging to organic materials such as wood, polymers, and paints (Andrady et al. 2011, 2019). It accelerates photo-oxidation, leading to color fading, embrittlement, and surface degradation. Historic wooden structures and decorative elements suffer from loss of mechanical resistance due to prolonged UV exposure (Blavier et al. 2023).

4.6.6.3 Microclimate Effects and Solar-Induced Pollutant Deposition

Solar radiation also influences the **microclimate** surrounding heritage buildings (Camuffo 2019). By increasing surface temperatures, it alters air circulation and humidity levels, which in turn impact **pollutant deposition** (Fig. 4.41) and biological colonization. Higher temperatures can promote the accumulation of dust and carbon particles on building surfaces, contributing to material discoloration and accelerating degradation processes (Blavier et al. 2023) (Figs. 4.42 and 4.43).

> ***Digitization Notes***
>
> The integration of digital technologies in heritage conservation is revolutionizing how climate-related vulnerabilities are assessed and managed at the building scale. High-resolution digital documentation methods, such as **Terrestrial Laser Scanning (TLS)** and **photogrammetry**, enable the precise recording of surface deterioration, structural deformations, and material decay

linked to temperature variations, moisture infiltration, and solar radiation exposure. These datasets serve as a fundamental reference for **long-term monitoring and comparative analyses** to track the progressive effects of climate change.

The use of **Heritage Building Information Modeling (HBIM)** allows for the aggregation of climate risk data into structured, **3D parametric models** that facilitate simulation and predictive analysis. By integrating **sensor-based monitoring systems** that track humidity fluctuations, thermal cycles, and pollutant deposition, HBIM provides a dynamic tool for informed decision-making in conservation strategies. Moreover, **AI-powered image analysis** is increasingly being applied to detect early signs of deterioration, such as micro-cracking, discoloration, and biological colonization, enabling preemptive interventions before significant damage occurs.

(a) (b)

Fig. 4.41 Aspen Hill, Tennessee, 1939. **a** Fiducia Church, wind damage (https://commons.wikimedia.org/wiki/File:Fiducia_Church,_wind_damage_-_NARA_-_280161.tif; Licence: Public Domain); **b** Kedron Church, wind damage (https://commons.wikimedia.org/wiki/File:Kedron_Church,_wind_damage_-_NARA_-_280159.jpg; Licence: Public Domain)

(a) (b)

Fig. 4.42 **a** Alveolization of Limestone, Viviers (Source: https://it.wikipedia.org/wiki/File:Limestone_alveolization_-_Viviers.jpg; Author: Coyau; Licence: CC BY-SA 3.0; No changes were made to the image); **b** Alveolization of sandstone, Strasbourg (Source: https://commons.wikimedia.org/wiki/File:Sandstone_alveolization_-_Strasbourg.jpg; Author: Coyau; Licence: CC BY-SA 3.0; No changes were made to the image)

(a)

(b)

Fig. 4.43 Monuments blackened by dirt and smog. **a** Milan Cathedral in 1962 (Source: https://commons.wikimedia.org/wiki/File:Milano_-_Duomo_-_1962.jpg; License: Public Domain). **b** Cologne (Koln) Cathedral in 1993 (Source: https://commons.wikimedia.org/wiki/File:Cologne_(Koln)_Cathedral_(9813075524).jpg; License: Public Domain)

References

Adamo C (2016) Impatti sulle coste italiane in seguito a eventi metereologici estremi. In: Zanchini E, Venneri S, Zampetti G (eds) Rapporto Ambiente Italia 2016. Presente e futuro nelle aree costrere in Italia. Edizioni Ambiente, Milano, pp 25–28

Andrady AL, Hamid H, Torikai A (2011) Effects of solar UV and climate change on materials. Photochem Photobiol Sci 10:292–300

Andrady AL, Pandey KK, Heikkilä AM (2019) Interactive effects of solar UV radiation and climate change on material damage. Photochem Photobiol Sci 18:804–825

Andreotti C, Liberatore D, Sorrentino L (2015) Identifying seismic local collapse mechanisms in unreinforced masonry buildings through 3D laser scanning. Key Eng Mater 628:79–84

Ariño X, Ortega-Calvo JJ, Gomez-Bolea A, Saiz-Jimenez C (1995) Lichen colonization of the Roman pavement at Baelo Claudia (Cadiz, Spain): biodeterioration versus bioprotection. Sci Total Environ 167:353–363

Arrighi C, Brugioni M, Castelli F, Franceschini S, Mazzanti B (2018) Flood risk assessment in art cities: the exemplary case of Florence (Italy). J Flood Risk Manag 11:S616–S631

Asteris PG, Sarhosis V, Mohebkhah A, Plevris V, Papaloizou L, Komodromos P, Lemos JV (2015) Numerical modeling of historic masonry structures. In: Handbook of research on seismic assessment and rehabilitation of historic structures. IGI Global, pp 213–256

Austigard MS, Mattsson J (2020) Monitoring climate change related biodeterioration of protected historic buildings. Int J Build Pathol Adapt 38:529–538

Benedetti D, Petrini V (1984) On seismic vulnerability of masonry buildings: proposal of an evaluation procedure. L'industria Delle Costruzioni 18:66–78

Bertolin C (2019) Preservation of cultural heritage and resources threatened by climate change. Geosciences (Basel) 9:250

Blavier CLS, Huerto-Cardenas HE, Aste N, Del Pero C, Leonforte F, Della Torre S (2023) Adaptive measures for preserving heritage buildings in the face of climate change: a review. Build Environ 110832

Blocken B, Carmeliet J (2004) A review of wind-driven rain research in building science. J Wind Eng Ind Aerodyn 92:1079–1130

Bonazza A, Messina P, Sabbioni C, Grossi CM, Brimblecombe P (2009a) Mapping the impact of climate change on surface recession of carbonate buildings in Europe. Sci Total Environ 407:2039–2050

Bonazza A, Sabbioni C, Messina P, Guaraldi C, De Nuntiis P (2009b) Climate change impact: mapping thermal stress on Carrara marble in Europe. Sci Total Environ 407:4506–4512

Borri A, Corradi M, Castori G, De Maria A (2015) A method for the analysis and classification of historic masonry. Bull Earthq Eng 13:2647–2665. https://doi.org/10.1007/s10518-015-9731-4

Borri A, Corradi M, De Maria A (2020) The failure of masonry walls by disaggregation and the masonry quality index. Heritage 3:1162–1198. https://doi.org/10.3390/heritage3040065

Borri A, Corradi M, De Maria A, Sisti R (2018) Calibration of a visual method for the analysis of the mechanical properties of historic masonry. In: Procedia structural integrity. Elsevier B.V., pp 418–427

Borri A, De Maria A (2009) Scheda di Valutazione dell'IQM (Indice di Qualità Muraria). Naples

Brando G, De Matteis G, Spacone E (2017) Predictive model for the seismic vulnerability assessment of small historic centres: application to the inner Abruzzi Region in Italy. Eng Struct 153:81–96. https://doi.org/10.1016/j.engstruct.2017.10.013

Brown JP, Rose WB (1996) Humidity and moisture in historic buildings: the origins of building and object conservation. APT Bull J Preserv Technol 27:12–23

Calvano M, Martinelli L, Calcerano F, Gigliarelli E (2022) Parametric processes for the implementation of HBIM—visual programming language for the digitisation of the index of masonry quality. ISPRS Int J Geoinf 11. https://doi.org/10.3390/ijgi11020093

Camuffo D (2019) Microclimate for cultural heritage: Measurement, risk assessment, conservation, restoration, and maintenance of indoor and outdoor monuments. Elsevier

Cantagallo C, Sangiorgio V, Verazzo C, Spacone E (2024) An analytical safety tree for heritage buildings: application to the archaeological museum "Domenico Ridola", Matera (Italy). Int J Architectural Heritage. https://doi.org/10.1080/15583058.2024.2383354

Cantagallo C, Spacone E, Perrucci D, Liguori N, Verazzo C (2020) A multilevel approach for the cultural heritage vulnerability and strengthening: application to the melfi castle. Buildings 10. https://doi.org/10.3390/BUILDINGS10090158

Cantagallo C, Terrenzi M, Barbagallo F, Di Domenico M, Ricci P, Camata G, Spacone E, Marino EM, Verderame GM (2023a) Effects of the extended N2 method on non-linear static procedures of reinforced concrete frame structures. Soil Dyn Earthq Eng 173:108144

Cantagallo C, Terrenzi M, Spacone E, Camata G (2023b) Effects of multi-directional seismic input on non-linear static analysis of existing reinforced concrete structures. Buildings 13:1656

Cara S, Aprile A, Pelà L, Roca P (2018) Seismic risk assessment and mitigation at emergency limit condition of historical buildings along strategic urban roadways. Application to the "Antiga Esquerra de L'Eixample" neighborhood of Barcelona. Int J Architectural Heritage 12:1055–1075

Cardell C, Delalieux F, Roumpopoulos K, Moropoulou A, Auger F, Van Grieken R (2003) Salt-induced decay in calcareous stone monuments and buildings in a marine environment in SW France. Constr Build Mater 17:165–179

Casapulla C (2021) Local out-of-plane failure modes in traditional block-masonry buildings. In: Masonry construction in active seismic regions. Elsevier, pp 289–322

Casapulla C, Argiento LU, Maione A (2018) Seismic safety assessment of a masonry building according to Italian Guidelines on cultural heritage: simplified mechanical-based approach and pushover analysis. Bull Earthq Eng 16:2809–2837. https://doi.org/10.1007/s10518-017-0281-9

Castori G, Corradi M, De Maria A, Sisti R, Borri A (2018) A numerical study on seismic damage of masonry fortresses. Bull Earthq Eng 16:4561–4580. https://doi.org/10.1007/s10518-018-0390-0

Circolare 7/2019 (2019) Istruzioni per l'applicazione dell'«Aggiornamento delle "Norme tecniche per le costruzioni"» di cui al D.M. 17 gennaio 2018 (in Italian)

Clarke D, Smethurst J (2008) Buildings on clay-the effects of geology, climate and vegetation on heave and settlement. Cathedral Communications Limited

Cosenza E, Del Vecchio C, Di Ludovico M, Dolce M, Moroni C, Prota A, Renzi E (2018) The Italian guidelines for seismic risk classification of constructions: technical principles and validation. Bull Earthq Eng 16:5905–5935. https://doi.org/10.1007/s10518-018-0431-8

Curti E, Lagomarsino S, Resemini S, Giovinazzi S (2007) Analisi non lineari di meccanismi locali di danno in strutture monumentali. Atti del XII convegno ANIDIS "L'ingegneria Sismica in Italia, pp 10–14

Curtis R (2016) Water management for traditional buildings: adaptation for a changing climate. APT Bull J Preserv Technol 47:8–14

Dalmastri C, Uccelli R (2024) Mortality from extreme meteorological and hydrogeological events in Italy: a rising health threat connected to climate change. Saf Extreme Environ 6:173–181

D'Ayala D, Novelli V (2010) Deliverable 8. Review and validation of existing vulnerability displacement-based models. PERPETUATE project

D'Ayala D, Speranza E (2003) Definition of collapse mechanisms and seismic vulnerability of historic masonry buildings. Earthq Spectra 19:479–509. https://doi.org/10.1193/1.1599896

De Matteis G, Brando G, Corlito V (2019) Predictive model for seismic vulnerability assessment of churches based on the 2009 L'Aquila earthquake. Bull Earthq Eng 17:4909–4936. https://doi.org/10.1007/s10518-019-00656-7

De Matteis G, Criber E, Brando G (2016) Damage probability matrices for three-nave masonry churches in abruzzi after the 2009 laquila earthquake. Int J Architectural Heritage 120–145. Taylor and Francis Inc.

DG ABAP (2017) http://www.benitutelati.it/. In: Italian ministry of culture

Diaferio M, Foti D, Sabbà MF, Lerna M (2021) A procedure for the seismic risk assessment of the cultural heritage. Bull Earthq Eng 19:1027–1050. https://doi.org/10.1007/s10518-020-01022-8

D.M. 02/07/1981 (1981) Normativa per le riparazioni ed il rafforzamento degli edifici danneggiati dal sisma nelle regioni Basilicata, Campania e Puglia

Dolce M, Manfredi G (2015) Libro bianco sulla ricostruzione privata fuori dai centri storici nei comuni colpiti dal sisma dell'Abruzzo del 6 aprile 2009. Doppiavoce

Dore MHI (2005) Climate change and changes in global precipitation patterns: what do we know? Environ Int 31:1167–1181

DPC, MiBAC (2001) Scheda per il Rilievo dei Beni Culturali—Danno alle Chiese

Eleftheriadou AK, Karabinis AI (2011) Development of damage probability matrices based on Greek earthquake damage data. Earthq Eng Eng Vib 10:129–141. https://doi.org/10.1007/s11803-011-0052-6

Eleftheriadou AK, Karabinis AI (2013) Evaluation of damage probability matrices from observational seismic damage data. Earthq Struct 4:299–324. https://doi.org/10.12989/eas.2013.4.3.299

EN 1998-1 Eurocode 8: Design of structures for earthquake resistance, Part 1: general rules, seismic actions and rules for buildings. Brussels, CEN

EN 1998-3 (2005) Eurocode 8: Design of structures for earthquake resistance—Part 3: assessment and retrofitting of buildings. Eurocode 8: design of structures for earthquake resistance—Part 3: assessment and retrofitting of buildings

Fabbrocino F, Vaiano G, Formisano A (2019) Parametric analysis on local collapse mechanisms of masonry churches. In: AIP conference proceedings. AIP Publishing

Fajfar P (1999) Capacity spectrum method based on inelastic demand spectra. Earthq Eng Struct Dyn 28:979–993

Fajfar P (2000) A nonlinear analysis method for performance-based seismic design. Earthq Spectra 16:573–592

FEMA P-1 (2012a) Seismic performance assessment of buildings. Volume 1-Methodology. Applied Technology Council California, Redwood City

FEMA P-2 (2012b) Seismic performance assessment of buildings. Volume 2—Implementation Guide. Federal Emergency Management Agency, Washington, DC, USA

Ferreira TM, Vicente R, Mendes da Silva JAR, Varum H, Costa A (2013) Seismic vulnerability assessment of historical urban centres: case study of the old city centre in Seixal, Portugal. Bull Earthq Eng 11:1753–1773. https://doi.org/10.1007/s10518-013-9447-2

Formisano A (2017a) Theoretical and numerical seismic analysis of masonry building aggregates: case studies in San Pio Delle Camere (L'Aquila, Italy). J Earthquake Eng 21:227–245. https://doi.org/10.1080/13632469.2016.1172376

Formisano A (2017b) Local- and global-scale seismic analyses of historical masonry compounds in San Pio delle Camere (L'Aquila, Italy). Nat Hazards 86:465–487. https://doi.org/10.1007/s11069-016-2694-1

GNDT (1993) Manuale Per Il Rilevamento Della Vulnerabilità Sismica Degli Edifici. Istruzione per la compilazione della scheda di 2° livello. Gruppo Nazionale per la Difesa dai Terremoti Rome, Roma

Gobiet A, Kotlarski S, Beniston M, Heinrich G, Rajczak J, Stoffel M (2014) 21st century climate change in the European Alps—a review. Sci Total Environ 493:1138–1151

Goudie A, Viles H (1997) Salt weathering hazards. Wiley, Chichester

Grossi CM, Brimblecombe P, Harris I (2007) Predicting long term freeze–thaw risks on Europe built heritage and archaeological sites in a changing climate. Sci Total Environ 377:273–281

Grünthal G (1998) European macroseismic scale 1998 (EMS-98). Centre Européen de Géodynamique et de Séismologie, Luxembourg

Hamidatou M, Sbartai B (2017) Deterministic assessment of seismic risk in Constantine city, Northeast Algeria. Nat Hazards 86:441–464. https://doi.org/10.1007/s11069-016-2693-2

Hegyi D, Armuth M, Halmos B, Marotzy K (2022) The effect of wind on historical timber towers analyzed by plastic limit analysis in the focus of a collapse. Eng Fail Anal 134:105852

Huttner S, Bruse M (2009) Numerical modeling of the urban climate–a preview on ENVI-met 4.0. In: 7th international conference on urban climate ICUC-7. Yokohama, Japan

Jacob D, Petersen J, Eggert B, Alias A, Christensen OB, Bouwer LM, Braun A, Colette A, Déqué M, Georgievski G, Georgopoulou E, Gobiet A, Menut L, Nikulin G, Haensler A, Hempelmann N, Jones C, Keuler K, Kovats S, Kröner N, Kotlarski S, Kriegsmann A, Martin E, van Meijgaard E, Moseley C, Pfeifer S, Preuschmann S, Radermacher C, Radtke K, Rechid D, Rounsevell M, Samuelsson P, Somot S, Soussana J-F, Teichmann C, Valentini R, Vautard R, Weber B, Yiou P (2014) EURO-CORDEX: new high-resolution climate change projections for European impact research. Reg Environ Change 14:563–578. https://doi.org/10.1007/s10113-013-0499-2

Kumar A, Pallav K (2020) Experimental and numerical investigation of old masonry wall using a macro-modeling approach. Open Civil Eng J 14

Lafrenz Samuels K, Platts EJ (2022) Global climate change and UNESCO World Heritage. Int J Cult Property 29:409–432. https://doi.org/10.1017/S0940739122000261

Lagomarsino S, Giovinazzi S (2006) Macroseismic and mechanical models for the vulnerability and damage assessment of current buildings. Bull Earthq Eng 4:415–443

Lagomarsino S, Podestà S (2004a) Seismic vulnerability of ancient churches: II. Statistical analysis of surveyed data and methods for risk analysis. Earthq Spectra 20:395–412. https://doi.org/10.1193/1.1737736

Lagomarsino S, Podestà S (2004b) Seismic vulnerability of ancient churches: I. Damage assessment and emergency planning. Earthq Spectra 20:377–394. https://doi.org/10.1193/1.1737735

Lagomarsino S, Podestà S, Lemme alberto (2005) Scheda Chiese di Secondo Livello per la Valutazione del Danno e della Vulnerabilità. In: Cigani G, Lemme A, Podestà S (eds) Beni Monumentali e Terremoto. Dall'emergenza alla ricostuzione. DEI, Tipografia del Genio Civile, Rome

Lang K, Bachmann H (2004) On the seismic vulnerability of existing buildings: a case study of the city of basel. Earthq Spectra 20:43–66. https://doi.org/10.1193/1.1648335

Lourenço PB (2015) Masonry modeling. Encyclopedia of earthquake engineering. Springer, Berlin Heidelberg, pp 1419–1431

Manigrasso M (2022) La città adattiva : il grado zero dell'urban design. Quodlibet, Macerata

Margottini C, Molin D, Serva L (1992) Intensity versus ground motion: a new approach using Italian data. Eng Geol 33:45–58

Marino M, Neri F, De Maria A, Borri A (2014) Experimental data of friction coeffcient for some types of masonry and its correlation with an index of quality masonry (IQM). In: 2nd European conference of earthquake engineering and seismology. Instanbul

Markovi G, Zele M, Káposztásová D, Hudáková G (2014) Rainwater infiltration in the urban areas. WIT Trans Ecol Environ 181:313–320

Mercogliano P, Montesarchio M, Barbato G (2017) Il Futuro Del Clima in Italia: Proiezioni Al 2050:219–225. https://doi.org/10.1406/88211

Meroni F, Petrini V, Zonno G (2000) Distribuzione nazionale della vulnerabilità media comunale. In: Bernardini A (ed) La vulnerabilità degli edifici: valutazione a scala nazionale della vulnerabilità sismica degli edifici ordinari. CNR-Gruppo Nazionale per la Difesa dai Terremoti Roma, Italy, Rome

MiBACT (2011) Guidelines for the seismic risk evaluation and reduction of the cultural heritage—alignment to the Technical Standards for Constructions (in Italian)

Ministry Decree n.58 28/02/2017 Guidelines for Seismic Risk Classification of Buildings

Mo C, Ruan Y, He J, Jin J, Liu P, Sun G (2019) Frequency analysis of precipitation extremes under climate change. Int J Climatol 39:1373–1387

Mol L, Viles H (2012) Conserving history in changing arid environments: a geomorphological approach. In: Mol L, Sternberg T (eds) Changing deserts: integrating people and their environment. White Horse Press, pp 215–236

Montesarchio M, Zollo AL, Bucchignani E, Mercogliano P, Castellari S (2014) Performance evaluation of high-resolution regional climate simulations in the Alpine space and analysis of extreme events. J Geophys Res Atmos 119:3222–3237. https://doi.org/10.1002/2013JD021105

Mosoarca M, Keller AI, Bocan C (2019) Failure analysis of church towers and roof structures due to high wind velocities. Eng Fail Anal 100:76–87

Müller SK, Pichelli E, Coppola E, Berthou S, Brienen S, Caillaud C, Demory M-E, Dobler A, Feldmann H, Mercogliano P, Tölle M, de Vries H (2024) The climate change response of alpine-mediterranean heavy precipitation events. Clim Dyn 62:165–186. https://doi.org/10.1007/s00382-023-06901-9

Munafò M, De Fioravante P, Cimini A, Riitano N, d'Antona M, Di Leginio M, Marinosci I, Assennato F, Luise A, Falconi M, Vecchio A, Jacomini C (2024) ISPRA land and soil monitoring, mapping, and assessment activities. In: Dazzi C, Benedetti A, Corti G, Costantini EAC (eds) Soil science in Italy: 1861 to 2024. Springer International Publishing, Cham, pp 607–619

Nguyen KN, Baker S (2023) Climate change and UNESCO world heritage-listed cultural properties: a systematic review, 2008–2021. Heritage 6:2394–2420

Nicolì D, Bellucci A, Ruggieri P, Athanasiadis PJ, Materia S, Peano D, Fedele G, Hénin R, Gualdi S (2023) The Euro-Mediterranean Center on Climate Change (CMCC) decadal prediction system. Geosci Model Dev 16:179–197. https://doi.org/10.5194/gmd-16-179-2023

NTC2018 (2018) D.M. 17/01/2018. Norme Tecniche per le Costruzioni (NTC2018). Gazzetta Ufficiale, n. 42 del 20/02/2018, Supplemento ordinario n.8 [in Italian]

Orr SA, Young M, Stelfox D, Curran J, Viles H (2018) Wind-driven rain and future risk to built heritage in the United Kingdom: novel metrics for characterising rain spells. Sci Total Environ 640:1098–1111

Pantò B, Raka E, Cannizzaro F, Camata G, Caddemi S, Spacone E, Caliò I (2015) Numerical macro-modeling of unreinforced masonry structures: a critical appraisal. In: Proceedings of the fifteenth international conference on civil, structural and environmental engineering computing. Civil-Comp Press Stirlingshire, UK, pp 85–86

Park SC (2009) Moisture in historic buildings and preservation guidance. ASTM International West Conshohocken, PA, USA

Petracca M, Camata G, Spacone E, Pelà L (2023) Efficient constitutive model for continuous micro-modeling of masonry structures. Int J Architectural Heritage 17:134–146

Petracca M, Pelà L, Rossi R, Zaghi S, Camata G, Spacone E (2017) Micro-scale continuous and discrete numerical models for nonlinear analysis of masonry shear walls. Constr Build Mater 149:296–314

Pietrapertosa F, Salvia M, De Gregorio Hurtado S, D'Alonzo V, Church JM, Geneletti D, Musco F, Reckien D (2019) Urban climate change mitigation and adaptation planning: are Italian cities ready? Cities 91:93–105. https://doi.org/10.1016/j.cities.2018.11.009

Podestà S, Curti E, Parodi S, D'Aristotile M (2010) Il sistema per la valutazione del rischio sismico (SIVARS). In: Editore G (ed) Linee Guida per la Valutazione e Riduzione del Rischio Sismico del Patrimonio Culturale. Allineamento alle nuove Norme Tecniche per le costruzioni. Ministero per i Beni e le Attività Culturali, pp 347–359

Putaud J-P, Bergamaschi P, Cavalli F, Dell'Acqua A, Douglas K, Goded I, Grassi F, Gruening C, Jensen N, Lagler F (2018) JRC-Ispra atmosphere-biosphere-climate integrated monitoring station: 2016 Report. Publications Office of the European Union

Ramírez Eudave R, Ferreira TM, Vicente R, Lourenco PB, Peña F (2024) Parametric and machine learning-based analysis of the seismic vulnerability of adobe historical buildings damaged after the september 2017 Mexico earthquakes. Int J Architectural Heritage 18:940–963. https://doi.org/10.1080/15583058.2023.2200739

Regione Molise, Consiglio Nazionale delle Ricerche (2005) Beni monumentali e terremoto. Dall'emergenza alla ricostruzione. DEI Tipografia del Genio Civile, Rome

Rosti A, Rota M, Penna A (2018) Damage classification and derivation of damage probability matrices from L'Aquila (2009) post-earthquake survey data. Bull Earthq Eng 16:3687–3720. https://doi.org/10.1007/s10518-018-0352-6

Rowland MJ (1999) Accelerated climate change and Australia's cultural heritage. Aust J Environ Manag 6:109–118

Rowland MJ (1992) Climate change, sea-level rise and the archaeological record. Aust Archaeol 34:29–33

Russo E, Kirchner I, Pfahl S, Schaap M, Cubasch U (2019) Sensitivity studies with the regional climate model COSMO-CLM 5.0 over the CORDEX Central Asia Domain. Geosci Model Dev 12:5229–5249. https://doi.org/10.5194/gmd-12-5229-2019

Saaty RW (1987) The analytic hierarchy process—what it is and how it is used. Math Model 9:161–176

Sabbioni C, Brimblecombe P, Cassar M (2010) The atlas of climate change impact on European cultural heritage. Scientific analysis and management strategies

Sáez-Pérez MP, Rodríguez-Gordillo J (2008) The influence of solar radiation on the deterioration of the marble columns in the courtyard of the lions in the Alhambra. Stud Conserv 53:145–157

Sandi H, Floricel I (1995) Analysis of seismic risk affecting the existing building stock. In: Proceedings of the 10th European conference on earthquake engineering, pp 1105–1110

Sangiorgio V, Fiorito F, Santamouris M (2020) Development of a holistic urban heat island evaluation methodology. Sci Rep 10:17913

Sangiorgio V, Uva G, Adam JM (2021) Integrated seismic vulnerability assessment of historical masonry churches including architectural and artistic assets based on macro-element approach. Int J Architectural Heritage 15:1609–1622. https://doi.org/10.1080/15583058.2019.1709916

Sbrogiò L, Valluzzi MR, Cardani G (2019) Recenti sviluppi sulle previsioni di vulnerabilità sismica di Campi Alto di Norcia alla luce degli interventi di riparazione e/o consolidamento e del sisma Centro Italia 2016. In: XVIII Convegno ANIDIS. L'Ingegneria Sismica in Italia. Ascoli Piceno

Sesana E, Gagnon AS, Ciantelli C, Cassar J, Hughes JJ (2021) Climate change impacts on cultural heritage: a literature review. Wiley Interdiscip Rev Clim Change 12:e710

Sisti R, Di Ludovico M, Borri A, Prota A (2019) Damage assessment and the effectiveness of prevention: the response of ordinary unreinforced masonry buildings in Norcia during the Central Italy 2016–2017 seismic sequence. Bull Earthq Eng 17:5609–5629. https://doi.org/10.1007/s10518-018-0448-z

Tavani C, Brando G, De Matteis G, Spacone E (2017) Damage probability matrices for eight historic centers after the 2009 L'Aquila Earthquake. In: 16th world conference on earthquake engineering, 16WCEE 2017. Santiago Chile

UNESCO (2007) Changement climatique et patrimoine mondial climate change and world heritage report on predicting and managing the impacts of climate change on World Heritage and Strategy to assist States Parties to implement appropriate management responses

Van Linden S, Van Den Bossche N (2022) Review of rainwater infiltration rates in wall assemblies. Build Environ 219:109213

Vicente R, Parodi S, Lagomarsino S, Varum H, Silva JARM (2011) Seismic vulnerability and risk assessment: case study of the historic city centre of Coimbra, Portugal. Bull Earthq Eng 9:1067–1096. https://doi.org/10.1007/s10518-010-9233-3

Wang X, Meng J, Zhu T, Zhang J (2019) Prediction of wind erosion over a heritage site: a case study of Yongling Mausoleum, China. Built Heritage 3:41–57

Whitman RV, Reed JW, Hong ST (1973) Earthquake damage probability matrices. In: Proceedings of the fifth world conference on earthquake engineering. Palazzo Dei Congresi, Rome, Italy, pp 2531–2540

Zhang Z, Wieland R, Reiche M, Funk R, Hoffmann C, Li Y, Sommer M (2011) Wind modelling for wind erosion research by open source computational fluid dynamics. Ecol Inform 6:316–324

Zollo AL, Rillo V, Bucchignani E, Montesarchio M, Mercogliano P (2016) Extreme temperature and precipitation events over Italy: assessment of high-resolution simulations with COSMO-CLM and future scenarios. Int J Climatol 36

Zumrawi MM, Abdelmarouf AO, Gameil AE (2017) Damages of buildings on expansive soils: diagnosis and avoidance. Int J Multi Sci Emerg Res 6:108–116

Chapter 5
Risk Mitigation Strategies: Conservation, Maintenance and Intervention

Abstract This chapter provides a comprehensive overview of risk mitigation strategies for built heritage, focusing on conservation, maintenance, structural interventions, and climate change actions. It begins by outlining the core principles of conservation and restoration, followed by an examination of monitoring systems for preventive maintenance, distinguishing between static and dynamic approaches and the use of specialized sensors for structural and environmental assessment. The chapter then presents strengthening interventions for both local and global collapse mechanisms, detailing traditional techniques such as tie rods, grout injection, and reinforced plaster, as well as innovative approaches like the application of composite materials. Additionally, the mitigation framework also considers actions to address the impacts of climate change, including the effects of temperature fluctuations, moisture, precipitation, wind, and solar radiation at both the urban and building scales. Strategies such as Urban Heat Island Intensity mitigation, bioprotection measures, moisture management, flood prevention, and UV protection are presented as part of an integrated approach. The chapter emphasizes the importance of combining traditional knowledge with innovative methods to ensure the long-term resilience and conservation of historic structures.

5.1 Conservation and Restoration Principles

Architectural heritage represents a pillar of societal identity and is one of the most valuable global economic assets, making its preservation a commitment to future generations. Public pressure demands that these structures be maintained in good condition, with interventions limited to the strictly necessary. However, over time, heritage buildings accumulate damage due to various factors, including material deterioration, repeated loading, and exceptional events.

In recent decades, extensive experience in consolidating and safeguarding historic buildings located in urban centers has underscored the importance of selecting techniques that are both compatible with and respectful of the existing architectural fabric. Over the past fifty years in Europe, numerous restoration and recovery projects

© The Author(s), under exclusive license to Springer Nature Switzerland AG 2025

C. Cantagallo et al., *Digitization of Built Heritage*, Digital Innovations in Architecture, Engineering and Construction, https://doi.org/10.1007/978-3-031-96598-2_5

have been carried out on masonry buildings. The lessons learned from unsuccessful projects have highlighted the need for thorough assessments before undertaking any restoration or rehabilitation work. When the extent of the damage or the effectiveness of repairs is not well understood, the outcome can be unpredictable, as demonstrated by cases where even advanced materials failed to deliver the desired results (Bertolesi et al. 2019).

Today, there is ample information supporting the choice of compatible materials and techniques for consolidation. To proceed in an informed manner, it is essential to have a deep understanding of the building's history and evolution, its geometry, structural details, crack patterns, material decay, construction techniques, and the properties of the existing materials, as indicated in Chapt. 3. This knowledge can be obtained through on-site experimental investigations, laboratory analyses, structural modelling, and a final vulnerability assessment (MiBACT 2011; Cantagallo et al. 2020, 2024). Consolidation interventions should only be designed after achieving this comprehensive understanding, recognizing that no single solution apply for every case.

Various intervention techniques can be discussed, and the most appropriate one should be selected based on its ability to respect the **authenticity** and **uniqueness** of the building, as well as its **compatibility** with existing materials. This also means that new materials used in repairs or consolidation must be chemically, physically, and mechanically compatible with the originals. However, it is impossible to expect these materials to be identical to those from centuries ago. Bricks, stones, or wood from the same quarry or source will not have the same characteristics as the original materials. For centuries, traditional materials and structural elements have been used in architectural heritage interventions, ensuring a natural compatibility with the existing fabric. It is also important to remember that for the sake of recognition and distinguishability, architects often incorporate new elements into existing buildings, ensuring material and structural compatibility (Bellini 1986; Carbonara 1997; Cvetanovic and Momcilovic-Petronijevic 2021).

Since no intervention is entirely permanent, the selected approach should ensure **durability** and be supported by a long-term **maintenance plan**.

Interventions on built heritage should be carefully assessed to ensure the **effectiveness** of the proposed solution while adhering to recognized **conservation principles**. These principles are outlined in international documents. Examples include the Venice Charter (1964), the ICOMOS/ISCARSAH Recommendations for the Analysis and Restoration of Architectural Heritage (ICOMOS/ISCARSAH 2005), and the ISO/FDIS 13822 guidelines (ISO/TC96/SC2 2010). While these criteria are not absolute, they provide valuable guidance in selecting the optimal repair or reinforcement solutions. In some cases, it may be impossible to meet all these criteria, and prioritization or engineering judgment is often necessary.

Adhering to these principles ensures that interventions are both effective and respectful of the building's historical and cultural significance. The most significant principles such as structural authenticity, reliability, minimum intervention, compatibility, durability, non-intrusiveness, reversibility and monitorability are further detailed in the following sections.

1. **Authenticity and Integrity**

The principles of authenticity and integrity form the cornerstone of heritage conservation. Authenticity refers to the preservation of the original characteristics of the structure, whether in terms of materials, design, or setting. Integrity, on the other hand, ensures that the essential features of the building or site remain intact and unaltered.

In practice, preserving authenticity means that every effort must be made to retain the original portions of a building, avoiding unnecessary replacement of historic elements with new materials. Even when intervention is necessary due to damage or decay, repairs should be minimal and should respect the original construction techniques. For example, where a building's façade has experienced material deterioration, the repair should use traditional materials and techniques rather than modern alternatives, unless these are demonstrably incompatible or unavailable.

Similarly, integrity requires a holistic approach to conservation. It is not enough to preserve isolated parts of a structure; instead, the building must be understood and preserved as a whole. This includes its relationship with its surrounding environment, historical context, and functional use. By maintaining both authenticity and integrity, the building retains its historical and cultural value, allowing it to continue conveying its historical significance to future generations. Thus, conservation should strive to ensure that any intervention respects the authenticity and integrity of the original structure, considering both visible and hidden elements of the heritage.

2. **Minimal Intervention**

The principle of minimal intervention advocates for doing only what is necessary to ensure the longevity and safety of a structure. It is rooted in the idea that any alteration or intervention inherently alters the historic value of a building. Therefore, the less intervention, the better. This philosophy ensures that the original materials and historical essence of the structure are preserved to the greatest extent possible.

Minimal intervention encourages a *conserve-as-found* approach, where elements are repaired rather than replaced. For instance, if structural problems arise, rather than rebuilding sections of the heritage building, conservation efforts should focus on reinforcing existing structures using methods that will least impact the original design. A typical application of this principle might involve the stabilization of a deteriorating stone wall by filling cracks with compatible, reversible mortar rather than replacing the stones.

Furthermore, this principle requires an understanding of the building's history and usage. Over time, buildings acquire new layers of history, and minimal intervention suggests that these layers should also be respected unless they are detrimental to the structure's overall integrity. Interventions should thus be conducted with careful research and a long-term view, preserving the building's historical and material authenticity.

3. **Reversibility**

Reversibility is a guiding principle in modern conservation. It underscores the importance of allowing future generations to undo interventions, should more advanced techniques or a deeper understanding of the structure arise. This concept plays a significant role in protecting the heritage value of a structure, ensuring that today's decisions do not permanently alter or damage the historical fabric.

In practical terms, reversibility involves using materials and methods that can be removed or replaced without compromising the structure's original elements. For instance, when injecting modern chemicals to consolidate decayed masonry, the materials chosen should not chemically react with the original stone, nor should they prevent future removal if superior restoration methods are developed later.

Additionally, the principle of reversibility demands that documentation be comprehensive and that interventions be designed in a way that allows for future inspection and, if necessary, removal or modification. It reflects a forward-thinking approach to conservation, ensuring that the structure remains flexible for future study, research, or treatment.

4. **Compatibility and Respect for Original Materials and Techniques**

Respect for original materials and construction techniques is fundamental to heritage building conservation, and any restoration interventions should ensure compatibility. Respecting original materials means that the authenticity of the structure is preserved by using, whenever possible, the same or closely similar materials and methods as those originally employed. This approach not only maintains the aesthetic and historical value of the building but also ensures that the intervention blends seamlessly with the original structure.

However, respect for original materials must go hand in hand with compatibility to prevent future degradation. Compatibility in restoration should be evaluated on several levels, including mechanical-structural, physical-chemical, thermal, and rheological compatibility (D'Ayala 2014; Quelhas et al. 2014). These aspects ensure that any new materials or techniques introduced do not inadvertently damage or destabilize the existing structure.

- **Mechanical-Structural Compatibility** refers to the use of materials that complement the original in terms of their mechanical properties, such as stiffness and load-bearing capacity. Using materials that are too rigid or too flexible compared to the original can result in mechanical stress and cause cracking or deformation under load. For instance, reinforcing a historic stone wall with modern materials requires careful consideration to avoid over-stiffening the structure, which might lead to stress concentrations and further damage.
- **Physical–Chemical Compatibility** involves ensuring that the interaction between new and original materials does not lead to harmful chemical reactions, such as the release of soluble salts or other substances that can deteriorate the structure over time. Materials used in restoration must be chosen to avoid long-term chemical incompatibilities, particularly in environments prone to moisture or pollutants, which could accelerate decay in historic materials.

– **Thermal and Rheological Compatibility** are crucial for managing how materials react to environmental changes and structural loads. New materials should have similar thermal expansion rates to the original materials to avoid stress from temperature fluctuations. Additionally, the rheological properties, which govern how materials deform under load, must be compatible to prevent differential movement that could cause cracks or other structural issues.

The **ICOMOS/ISCARSAH Recommendations** (ICOMOS/ISCARSAH 2005) emphasize that the materials used in any restoration project must be carefully studied for their long-term effects. Compatibility, while essential, is not sufficient on its own. Each intervention must be thoroughly evaluated not only for how well it integrates with the existing structure but also for its ability to protect and preserve the building's integrity over time.

The combination of respecting original materials and techniques with the requirement for compatibility ensures that any intervention enhances the building's stability and longevity without compromising its historical and architectural values. By prioritizing both authenticity and compatibility, conservators can strike a balance between preserving the past and securing the structure for the future.

5. **Durability**

The concept of durability is fundamental in restoration methodology, referring to the ability to preserve monuments and transmit them to future generations. This criterion emphasizes the necessity of interventions that not only endure over time but also maintain their material solidity and stability. Durability expresses the quality of an intervention, ensuring the structure's longevity, while minimizing the need for frequent or invasive maintenance.

It is essential that the materials and mechanical systems used in restoration and consolidation be highly durable and compatible with the original materials to prevent decay that could compromise both the overall safety of the structure and the durability of its original components. Furthermore, the choice of materials and techniques must focus on solutions that allow for future maintenance—both ordinary and extraordinary—without causing irreversible alterations to the monument. Conservation and restoration, therefore, aim to significantly extend the expected lifespan of heritage structures, ensuring that each action taken is beneficial not only in the short term but also in the long term.

According to the ICOMOS/ISCARSAH Recommendations (ICOMOS/ISCARSAH 2005), no intervention should be carried out without first thoroughly evaluating the probable benefits and potential damage to the architectural heritage, including an analysis of the long-term side effects. This assessment helps ensure that every decision is justified by its ability to add years of useful life to the building while respecting its historical and cultural integrity.

6. **Non-intrusiveness/Non-invasiveness**

The principle of non-intrusiveness or non-invasiveness is fundamental to the conservation of historic structures, advocating for interventions that preserve the integrity and authenticity of buildings without causing excessive alterations. This principle focuses on minimizing the impact of conservation measures on both the appearance and the structural elements of the monument, ensuring that the original character of the building remains intact.

This approach is based on the idea that interventions should not overwhelm or obscure the historical value of the structure. Instead, they should aim to stabilize, protect, and conserve the building while maintaining its aesthetic and cultural significance. By employing techniques that limit physical disruptions, the principle helps safeguard the monument's long-term historical integrity and its capacity to continue conveying its original significance.

- **Selection of Techniques**. According to the ICOMOS/ISCARSAH Recommendations (ICOMOS/ISCARSAH, 2005), the selection between "traditional" and "innovative" techniques should be made on a case-by-case basis, prioritizing those that are less invasive and more compatible with heritage values, while also ensuring safety and durability. Reinforcement interventions, particularly for buildings of significant historical or cultural value, should be as discreet as possible to minimize visible changes and maintain the building's authenticity.
- **Integration and Distinguishability**. The Venice Charter for the Conservation and Restoration of Monuments and Sites (1964) emphasizes that replacements for missing parts must blend harmoniously with the whole, while still being distinguishable from the original. This is crucial to avoid falsifying the artistic or historical value of the monument. Even when structural interventions are necessary, they must seamlessly integrate with the existing structure, preserving the overall appearance while remaining discernible as new and non-original additions.
- **Preference for Non-Invasive Techniques**. Non-invasive repair and reinforcement techniques should be favored over more intrusive alternatives. These techniques are crucial for preserving the material integrity of the existing structures. Among the possible solutions, preference should be given to methods that are the least invasive, ensuring respect for the monument's heritage values while meeting the necessary safety and durability requirements.

7. **Documentation and Research**

Before any conservation work begins, extensive documentation and research are critical components of the process. These activities help build a comprehensive understanding of the structure, ensuring that any interventions are informed by detailed knowledge of its history, current condition, and cultural significance. This research typically includes archival investigations, material testing, structural analysis, and condition surveys (see the guidance provided in Chap. 3).

The documentation process involves recording the building's current state through detailed drawings, photographs, and even 3D scans where appropriate. This information serves two primary purposes: first, it helps guide the conservation process by providing a baseline understanding of the building's historical and physical context. Second, it creates a lasting record for future researchers, enabling them to study past interventions and building conditions.

Effective documentation should be continuous throughout the conservation process. For instance, if a building is being restored, detailed records of the techniques and materials used during the intervention must be maintained. This not only serves as a guide for future conservators but also ensures accountability and transparency in conservation practices.

8. **Structural Stability and Functionality**

Structural stability is a non-negotiable aspect of any conservation project. The principle of structural stability demands that any conservation efforts ensure the building's continued safety and functionality while maintaining its historical significance. A balance must be struck between preserving the historical fabric and ensuring that the building meets modern safety standards, particularly in cases where the structure may be vulnerable to collapse or natural disasters like earthquakes.

Achieving this balance often requires sophisticated structural analyses to understand the existing load-bearing systems and their vulnerabilities. This can involve the use of modern engineering techniques like finite element modelling, combined with traditional approaches to reinforce or strengthen the structure where needed. However, such interventions must respect the historical fabric, making minimal visible changes to the building's appearance.

Functionality is also an important consideration. Historic buildings often need to accommodate modern functions without losing their historical character. For instance, adapting a medieval building to house modern offices might require changes to electrical systems or climate control. These upgrades must be done sensitively, ensuring they do not undermine the structural integrity or historical character of the building.

9. **International Collaboration and Professional Training**

In the field of heritage conservation, **international collaboration** and the **ongoing training of professionals** are essential components for the successful preservation of cultural assets. The complexity of conserving historical structures demands a multidisciplinary approach, where experts from different countries and fields share their knowledge and experiences. Organizations like **ICOMOS** and **UNESCO** play a pivotal role in fostering these collaborations, enabling the exchange of innovative techniques and best practices across borders.

Collaborative efforts between nations have become increasingly important as heritage sites face growing threats from environmental factors, natural disasters, and the effects of climate change. International cooperation ensures that research findings and technological advancements are widely disseminated, allowing professionals worldwide to adopt effective strategies tailored to the specific needs of individual sites. Furthermore, collaborative projects offer opportunities to address shared challenges, such as managing heritage in seismic zones or mitigating the impacts of urbanization on historical monuments.

Alongside international cooperation, the **continuous training and education** of professionals involved in heritage conservation are critical to maintaining high standards of practice. As new materials, methods, and technologies emerge, it is crucial that architects, engineers, conservators, and craftsmen remain informed and proficient in their application. Professional organizations regularly host workshops, conferences, and seminars aimed at keeping conservationists up to date with the latest advancements in the field. For example, the **ICOMOS International Training Committee** organizes activities that focus on the transfer of skills and knowledge necessary to deal with complex conservation issues.

Additionally, **public awareness and community involvement** are integral to the success of heritage preservation efforts. By engaging the local population and fostering a sense of ownership over cultural heritage, public outreach initiatives help build support for conservation projects. This is particularly important in ensuring that preservation is seen not only as a technical or academic endeavor but as a shared responsibility towards future generations.

Through the combination of **global collaboration, professional training, and public engagement**, heritage conservation efforts can be strengthened, ensuring that both modern techniques and traditional knowledge are effectively applied in preserving the world's cultural legacy.

Digitization Notes

The digitization of conservation processes is essential to ensuring accurate and long-lasting management of architectural heritage. Digital technologies enable the creation of detailed three-dimensional models that document the evolution of deterioration and structural changes over time, providing an objective basis for intervention decisions. The integration of these techniques into digital conservation platforms allows for centralized management of the collected information, which may include geometric surveys, material data, and structural models. Such a platform ensures not only secure and long-term storage of data but also its **easy access for future assessments and interventions**.

Moreover, the creation of digital archives linked to the interventions carried out allows for the **traceability of design decisions** and the **continuous monitoring of structures**. By recording detailed information about the interventions performed, future generations of experts an have a clear understanding of the conservation process. This enables them, if necessary, to make modifications or restorations based on precise and documented information.

5.2 Monitoring for Preventive Maintenance in Building Heritage

5.2.1 Key Concepts of Structural Monitoring Systems

Various tools can be employed to implement a structural monitoring system, each characterized with functionalities and applications. In the field of Structural Health Monitoring (SHM), the term *sensor* is often used to refer to the devices involved. However, they represent only one part of a more complex system, which consists of the following components:

- The **sensor**, which detects variations in a physical quantity, such as temperature, velocity, acceleration, force, displacement, or pressure.
- The **interface**, which converts the physical quantity detected by the sensor into a processable electrical signal (e.g., voltage, current, resistance, capacitance, or inductance), making the data suitable for processing and analysis.
- The **transducer**, which combines the sensor (that detects variations in the physical quantity) and the interface (that transforms the signal into a manipulatable electrical format), as illustrated in Fig. 5.1.

The system receives an input in the form of a physical quantity (e.g., temperature, force, light) and converts it into an electrical output signal (e.g., voltage, current, resistance, capacitance, or inductance) through the transducer.

5.2.2 Classification of Monitoring Systems

Monitoring systems can be classified into two main categories based on their function and the type of phenomena they are designed to detect: **static systems** and **dynamic systems**. The main characteristics of each of these categories are summarized in Table 5.1.

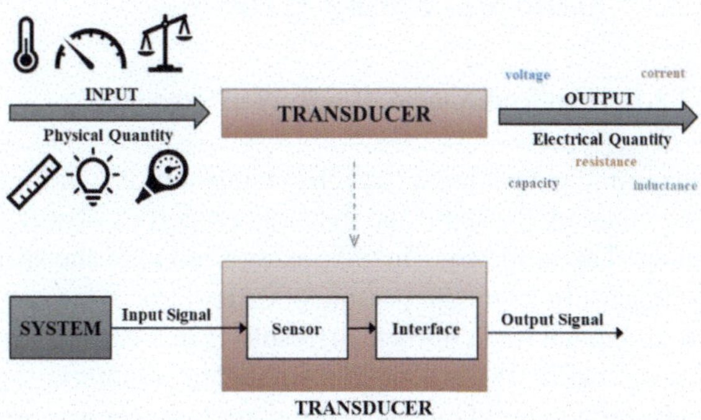

Fig. 5.1 Functional diagram of a transducer

Table 5.1 Comparison between static and dynamic monitoring systems

Characteristic	Static systems	Dynamic systems
Monitored phenomena	Slow and progressive	Rapid and transient
Acquisition frequency	Low or periodic	High and continuous
Examples of use	Settlements, subsidence, deformations	Vibrations, earthquakes, oscillations
Monitoring duration	Long-term	Short-term or continuous

Static systems are designed to measure slow and gradual variations of a monitored parameter over long periods. Generally, the number of measurements required is low (a few readings per minute or even per hour), making these systems well suited to monitor long-term phenomena. Among the most representative tools are **inclinometers**, used to measure the inclination or angular displacement of a structure or terrain relative to a predefined reference line. Another significant example is **crack meters**, which monitor the opening, closing, or evolution of cracks within structural elements, providing crucial information about the stability of the structure.

Dynamic systems are designed to record rapid and transient phenomena, such as **vibrations**, **oscillations**, or **seismic events**. Their ability to capture high-frequency data makes them indispensable for analysing structural behaviour during dynamic events or under variable loads. Among the most common tools in this category are **accelerometers**, used to measure accelerations induced by vibrations or earthquakes, and **velocimeters**, which measure velocity during dynamic events. Another example is **strain gauges** (dynamic extensometers), which measure dynamic variations in the deformation of a structural element. Dynamic systems can be installed in a continuous mode, for constant monitoring, or in a temporary mode, depending on the specific requirements of the project.

The aim of this chapter is to list, classify, and describe the different types of sensors, providing a comprehensive overview of the technologies available for static and dynamic monitoring of built heritage.

5.2.3 Static Monitoring Systems

5.2.3.1 Environmental Monitoring Sensors

Environmental parameters, such as temperature, wind, and humidity, are fundamental for evaluating the static and dynamic properties of structures. For instance, it is well known that an increase in temperature can alter the stiffness of a structure, leading to changes in its modal parameters. Similarly, wind can generate unforeseen forces not accounted for in the design, resulting in localized stress concentrations (Moreno-Gomez et al. 2018). Among environmental monitoring sensors, those with the widest application in building engineering are **temperature sensors**.

Temperature affects both the behavior of structures and the reliability of the sensors employed. Monitoring temperature allows the identification of potential structural issues, such as cracks or settlements. However, temperature can also negatively impact sensor performance, compromising the accuracy of measurements. Studies such as Gentile et al. (2019) demonstrate how temperature variations influence the natural frequencies of structures, emphasizing the need to consider these effects.

Temperature sensors can be divided into two main categories:

- **Contact Sensors**
  - **Thermocouples**: Cost-effective and widely used, thermocouples are suitable for long-term measurements but tend to be less precise and are prone to wear and tear (Moreno-Gomez et al. 2018).
  - **RTD (Resistance Temperature Detectors)**: Made of platinum wires, RTDs offer high accuracy due to the linear relationship between resistance and temperature. They are ideal for applications requiring reliable and stable measurements.
- **Non-contact Sensors**
  - **IR Sensors (Infrared)**: These sensors provide rapid measurements by detecting thermal radiation. They are particularly useful for assessing surface temperature variations without physically interfering with the monitored object. Tools like thermal cameras are widely applied to detect thermal anomalies, which can help identify cracks or voids.

In the case of historical buildings, **monitoring air humidity** is essential for assessing the conservation state of an asset and preserving its condition. The instrument used to measure this parameter is the **hygrometer**, which typically offers a measurement range of 0–100% with a resolution of 1%.

Wind monitoring involves the use of **anemometers** to measure wind speed. Two main types of anemometers are commercially available:

- **Mechanical Anemometers**. A mechanical anemometer consists of horizontal arms connected to a vertical shaft. The stronger the wind blows, the faster the shaft rotates. This movement is converted by a generator to provide an instant reading of wind speed. Other mechanical anemometers estimate wind speed by measuring the number of rotations of the arms.
- **Ultrasonic Anemometers**. Ultrasonic anemometers feature multiple pairs of transmitters and receivers positioned at 90-degree angles. They use differences in the speed of sound waves between components to calculate wind speed (Fig. 5.2).

5.2.3.2 Displacement Sensors

Some monitoring systems utilize displacement measurements to identify and analyze potential structural damage or to evaluate the effectiveness of the interventions

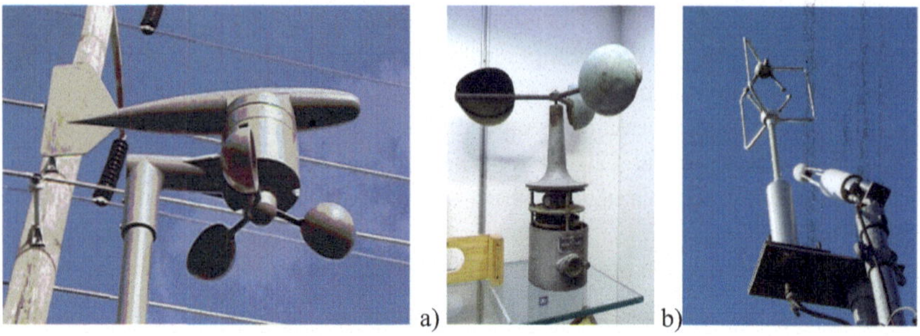

Fig. 5.2 Mechanical (**a** and **b**) and Ultrasonic (**c**) Anemometers. *Source* https://free-images.com/ Licence: Public Domain

on a structure (Saisi et al. 2016). To monitor structural movements, displacement transducers are employed, and the main instruments in this category include:

- **Electrical Crack Meters**. Displacement transducers (Fig. 5.3), such as **Linear Variable Differential Transformers** (LVDTs), are widely used to monitor variations in length, pressure, or force and are particularly effective in tracking the development of cracks over time. For a comprehensive assessment of both axial deformations and potential slippage phenomena, two strain gauges should be placed on the plane of the structural element under investigation. Alternatively, bi-axial or tri-axial crack meters can be used. A substantial number of measurements ensures the reliability of the results.
- **Vibrating Wire Crack Meters**. Vibrating wire crack meters are utilized to measure surface movements of **cracks**, **structural joints**, **rocks**, and **soil**. They consist of a stainless-steel cylindrical body housing a vibrating wire transducer and a sliding rod connected to the transducer through a spring. The two ends of the crack meter are equipped with anchoring devices and self-aligning spherical

Fig. 5.3 Electrical crack meters. *Source* https://commons.wikimedia.org/ Licence: CC0

joints, allowing the instrument to be fixed between the two parts of the crack whose movements are being measured.

- **Global Positioning Systems (GPS)**. Structural displacement measurements using GPS involve estimating the time required for signals from four connected satellites to reach the receiving station (Fig. 5.4). High-precision receivers are necessary to accurately measure displacements (Im et al. 2013).
- **Inclinometers**. An inclinometer is a device used to measure the angle of inclination of a structure relative to a reference plane. Various models are available, each with specific technical features that make them suitable for different applications (Gentile et al. 2019; Makoond et al. 2020; Sánchez et al. 2016). For example, the vertical inclinometer shown in Fig. 5.5 is used to measure the inclination of boreholes.

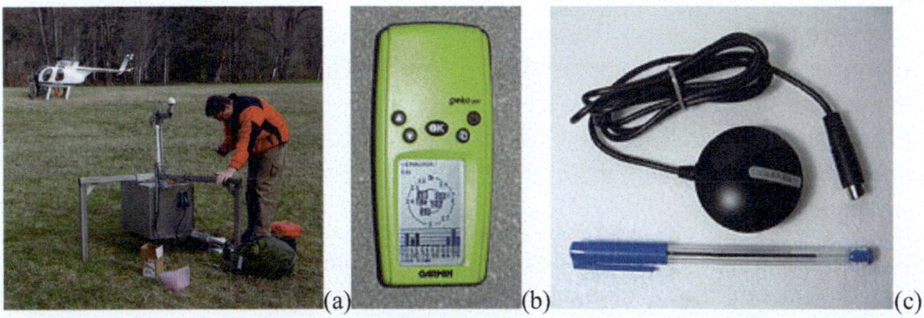

Fig. 5.4 Spider unit deployed to monitor landslide movement in NW Washington (March 22, 2014), equipped with GPS and geophones, transmitting real-time data to USGS (Source: https://www.flickr.com/photos/27784370@N05/13715290705/ Licence: Public Domain) (**a**); GNSS receivers (Source: https://commons.wikimedia.org. Licence: Public Domain) (**b** and **c**)

Fig. 5.5 Vertical inclinometer for measuring horizontal subsurface deformation in a borehole *Source* https://commons.wikimedia.org/wiki/Category:Inclinometers#/media/File:Vertical_inclinometert1.jpg. Author: Ekrem Canli. Licence: CC BY-SA 3.0. No changes have been made to the images

5.2.3.3 Strain and Mechanical Property Sensors

Strain gauges (or extensometers) are widely used to measure the **local deformation** (strain) of structural elements subjected to various loads. They are essential for analyzing the mechanical behavior of materials and structures under both static and dynamic conditions. These sensors detect minute changes in length relative to the original length of the element, providing data on stress and strain distribution. Strain gauges can be categorized as:

- **Mechanical Strain Gauges**. These instruments use mechanical components, such as levers and dials, to measure deformation (Fig. 5.6a). Although they are less common today due to advancements in electronic technologies, mechanical strain gauges are still used in specific applications where simplicity and robustness are required.
- **Resistive Strain Gauges**. Based on changes in electrical resistance when the sensor is stretched or compressed (Fig. 5.6b). These are commonly used for monitoring strains in structural elements like beams, columns, and bridges.
- **Fiber Optic Strain Gauges**. Offering higher precision and resistance to electromagnetic interference, these are used for long-term monitoring of sensitive structures.

In addition to strain gauges, other advanced techniques have emerged to detect structural vulnerabilities and quantify potential damage. These methods extend the range of mechanical property sensors and provide complementary information on the condition of structural elements. Below are some of the most widely used methods for crack detection and assessment.

- **Acoustic Emissions**. An innovative method for detecting damage in materials involves the use of sound waves. Acoustic emissions typically operate in the frequency range of 100 kHz to 1 MHz and are applied using specific ultrasonic devices. This technique leverages the propagation time of sound waves through

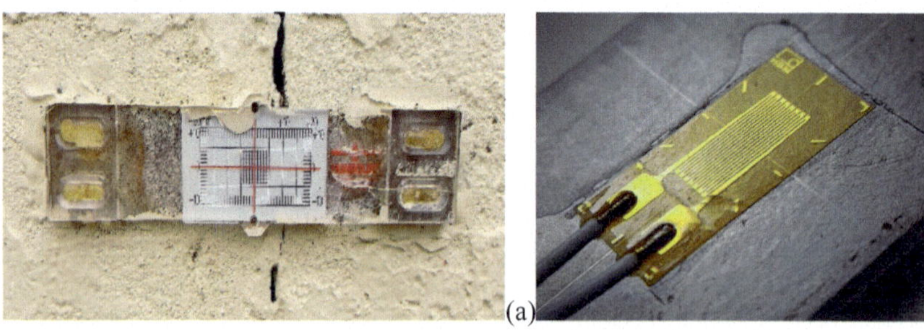
(a) (b)

Fig. 5.6 Mechanic Strain Gauges (Source: https://commons.wikimedia.org/ Licence: CC0) (**a**); Resistive Strain gauge (Source: https://commons.wikimedia.org/wiki/File:Strain_gauge_-.jpg. Author: Cristian V. Licence: CC-BY-4.0. No changes have been made to the images) (**b**)

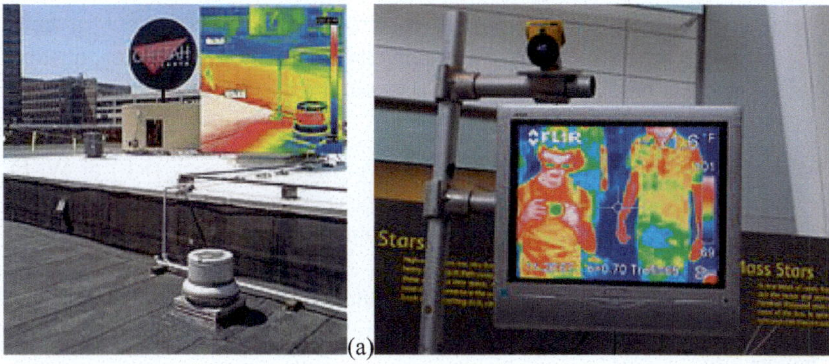

(a) (b)

Fig. 5.7 Infrared Thermography of Modified Bitumen Roof Source: https://commons.wikimedia.org/wiki/File:Infrared_Thermography_of_Modified_Bitumen_Roof.jpg Author: Alexander Cua. Licence: CC-BY-SA-3.0) (**a**); Thermal image of two people; one person holds a camera (Source: https://commons.wikimedia.org/ Licence: CC0) (**b**)

an intact or cracked body to identify potential areas of vulnerability (Ceravolo et al. 2016; Chik et al. 2024).

- **Thermography**. Thermographic imaging measures the surface temperature of masonry walls, allowing for the detection of cracks, voids, and changes in material composition (Fig. 5.7). This non-invasive technique is particularly valuable for heritage buildings, where it is often used to identify hidden structural issues and evaluate the homogeneity of building materials without damaging the original materials. For example, thermography could be employed in the restoration of historical churches to reveal areas where mortar had eroded over time, creating voids and weakening the structure. This information enables technicians to design the reinforcement on specific sections, minimizing unnecessary interventions and preserving the original fabric of the building. Thermography could also be used to map thermal bridges and assess energy efficiency in heritage buildings, providing information for improving insulation while respecting the historical authenticity of the structure.

5.2.4 *Dynamic Monitoring Systems*

5.2.4.1 Vibration Monitoring Sensors

Vibration monitoring is essential for understanding the dynamic behavior of structures. Sensors in this category measure oscillations induced by external forces, allowing engineers to evaluate the overall structural performance.

- **Accelerometers**. Accelerometers measure the acceleration of a structure caused by vibrations or dynamic loads. They are widely used in seismic monitoring and for detecting structural responses to external forces. Modern accelerometers can be piezoelectric, MEMS-based (Micro-Electro-Mechanical Systems), or fiber-optic, offering a range of sensitivity and precision depending on the application. Accelerometers can be classified into **wireless** and **wired (non-wireless)** systems based on how they transmit data:

- **Wireless Accelerometers** transmit data via radio frequency (RF) or Wi-Fi, eliminating the need for physical cables. These devices are particularly advantageous in applications where installing cables is impractical. They are especially useful for monitoring hard-to-reach areas or structures with extensive surfaces like bridges or large historical buildings. Wireless systems typically rely on high-capacity batteries and require periodic maintenance to ensure battery life.
- **Wired (Non-Wireless) Accelerometers** transmit data directly via coaxial or fiber-optic cables. These systems are often preferred in environments where signal reliability and consistency are paramount, such as laboratory testing or continuous long-term monitoring. Wired accelerometers can handle high-frequency sampling rates with minimal data loss and are less prone to interference compared to wireless systems.

- **Velocimeters**. Velocimeters, or velocity sensors, measure the speed of movement during oscillations. These are particularly useful for monitoring structures such as bridges, where dynamic loading due to wind or traffic can induce significant vibrations.
- **Strain Gauges**. Strain gauges can also be used in dynamic monitoring to measure rapid deformations in structural elements under fluctuating loads. Dynamic strain gauges are often characterized by higher sampling rates and greater sensitivity than static strain gauges.

5.2.4.2 Force and Load Sensors

Dynamic force and load sensors measure time-varying forces acting on structural elements, providing information about the load-bearing capacity and behavior of a structure under transient conditions.

- **Dynamic Load Sensors**. These sensors measure fluctuating loads, such as those caused by wind, traffic, or seismic activity. They are crucial for evaluating the distribution of forces in bridges, towers, and other load-bearing structures. For instance, on historical arch bridges, dynamic load sensors can track real-time force redistribution during seismic activity, providing data to prioritize intervention zones and ensure the preservation of the structure's integrity.
- **Pressure Sensors**. Pressure sensors are used to monitor transient pressure changes in structures like pipelines, tanks, and containment systems. These sensors can detect anomalies caused by dynamic loading, such as pressure surges or impacts.

5.2.4.3 Advanced Monitoring Sensors

Advanced sensors utilize emerging technologies to enhance dynamic monitoring systems, providing greater accuracy in data acquisition and analysis, along with a wider range of applications. These technologies are particularly suitable for monitoring the structural health of historic buildings, as they ensure effective and noninvasive assessments. Below are two key categories of sensors commonly employed in heritage structures:

- **Fiber Optic Sensors**. Fiber optic sensors are a non-invasive and highly sensitive solution for monitoring critical parameters in heritage buildings, such as strain, vibration, and temperature. A prominent technology in this category is Fiber Bragg Grating (FBG), which is widely used for measuring deformations and temperature variations. FBG sensors detect changes in the wavelength of light reflected within the fiber, which correspond to strain or temperature variations in the monitored structure (Fig. 5.8). The lightweight nature and the resistance to environmental factors of the fiber optic sensors make them particularly suitable for preserving the integrity of historical structures. These sensors can be discreetly embedded into walls, arches, or columns to detect deformations, monitor vibrations caused by traffic or wind, and assess temperature variations that could lead to material stress. In addition, fiber optic sensors are ideal for tracking the evolution of cracks and identifying changes in environmental conditions inside structures, such as humidity levels, which can impact decorative elements like frescoes. Their integration into remote monitoring systems enables continuous data collection, facilitating timely interventions and long-term preservation strategies.

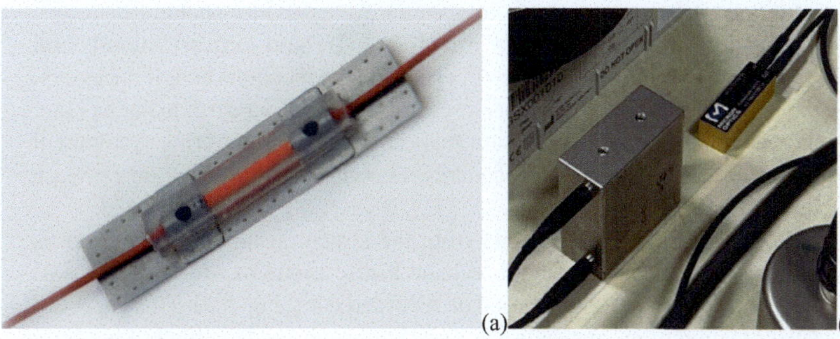

(a) (b)

Fig. 5.8 An example of a fiber-optic sensor based on Bragg grating technology (Source: https://commons.wikimedia.org/w/index.php?title=File:Fiber-optic_sensor.jpg&oldid=471511928. Author: Lena.lemon. Licence: CC-BY-SA-4.0 Self-published work. No changes have been made to the image) (**a**); Left, an accelerometer based on Bragg grating sensors; right, an accelerometer with an interferometric sensor (Source: https://it.wikipedia.org/wiki/Sensore_a_fibra_ottica#/media/File:Fiberopticsaccelerometer.png. Author: Ilmaggio98. Licence: CC BY-SA 4.0. No changes have been made to the image) (**b**)

- **Modal Analysis Sensors**. These sensors provide information on the natural frequencies, mode shapes, and damping properties of a structure by using a combination of accelerometers, velocimeters, and strain gauges to capture vibrations and deformations, along with technologies such as Laser Doppler Vibrometry (LDV) and fiber optic systems. Modal analysis systems also enable engineers to validate structural models by comparing analytical predictions (carried out using advanced structural models) with the actual behavior of the structure.

Digitization Notes

The monitoring and control system for built heritage can make use of innovative artificial intelligence (AI)-based systems. One of the most advanced approaches to using AI for the monitoring and management of heritage buildings is here outlined.

The large volume of data from sensor networks must be collected, cleaned, and transformed through a process of **data cleaning and feature engineering**. Once processed, the data feeds an AI-based model incorporating Artificial Neural Networks (ANNs) within sophisticated **control software**. This software enables the management and analysis of the complex data system generated by structural and environmental sensor networks, as well as a series of specific analyses applied to a predefined urban cluster.

The control software can operate using two parallel or alternative approaches, each characterized by different objectives and functionalities:

- **Data-Driven Approach**: Data is recorded in **real time** by the monitoring system, which trains the control software to detect anomalies indicating imminent risks or damage from specific events such as natural disasters. This approach allows for **early warnings to be sent to professionals and relevant authorities**, enabling timely on-site inspections and, if necessary, the suspension of infrastructure use. In the case of an earthquake or other natural calamities, real-time data analysis from the monitoring system will better inform emergency procedures, allowing for a distinction between damaged and intact buildings.
- **Model-Based Approach**: Data from the structural monitoring system is used to develop a reliable and realistic **Finite Element Modelling** (FEM) that updates automatically based on recorded data (Fig. 5.9). This approach enables simulations of events using real parameters and facilitates long-term statistical evaluations through the use of AI.

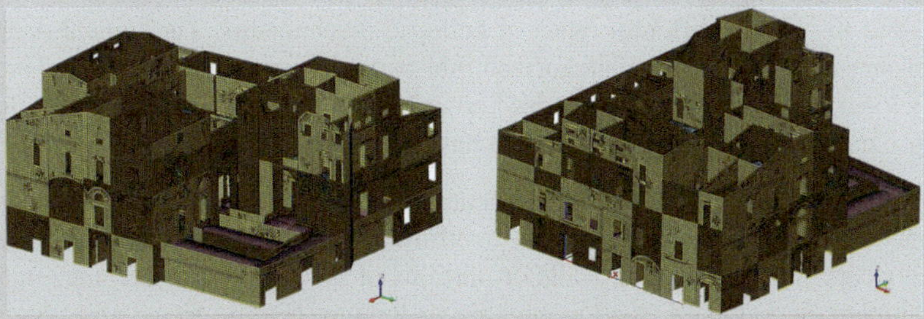

Fig. 5.9 FEM model of an historical palace of Central Italy

The data processed by the management software, using both the Data-Driven and Model-Based approaches, is displayed within a unified control platform ("**dashboard**"). This platform serves as the interface through which operators can visualize and manage data from both the sensors and structural and fluid dynamics models. The selected response parameters within this control platform allow for specific evaluations of urban risk, addressing both environmental well-being and structural safety.

From an environmental perspective, these evaluations enable: (a) predictions of specific urban pollution levels in certain city areas, (b) the design of interventions to mitigate pollution, (c) promotion of sustainable urban planning based on the built environment's impact on urban pollution, and (d) the definition of optimal temperature ranges during specific periods, contributing to reduced energy consumption in public infrastructure and enabling better energy consumption planning.

From a structural standpoint, the data visualized in the dashboard allows for the assessment of the vulnerability levels of selected infrastructures, identification of the necessary interventions to ensure their structural safety, and prioritization of the most urgent structural interventions and their corresponding costs. This facilitates the **planning of costs** for public spending management.

5.3 Strengthening Interventions

5.3.1 Overview

Masonry structures are a fundamental component of architectural heritage and require specialized intervention methods to ensure both their preservation and structural safety. This chapter introduces a series of advanced and traditional techniques aimed at enhancing the safety and conservation performance of masonry structures, and to prevent different types of collapse mechanisms. The typical damage in masonry buildings can be categorized into two primary collapse modes (Giuffrè 1996): local collapse mechanisms (also known as first-mode mechanisms) and global collapse mechanisms (second-mode mechanisms).

Local collapse mechanisms are associated with the out-of-plane behavior of masonry walls, resulting in flexural or overturning failure. These first-mode mechanisms occur when individual walls or portions of the structure experience out-of-plane kinematic behavior, such as rocking or flexure, leading to collapse due to a loss of equilibrium (Fig. 5.10). Such mechanisms are common in buildings where the connections between walls and floors, or between walls at intersections, are insufficient or ineffective, preventing the structure from behaving as a cohesive whole. In these cases, each wall tends to act independently, triggering out-of-plane collapse mechanisms. It is evident that, in such situations, a global verification using a three-dimensional structural model (such as pushover analysis) does not accurately reflect the seismic behavior of the building unless measures are first taken to prevent the activation of these local collapse mechanisms.

Global collapse mechanisms (second-mode mechanisms) involve the in-plane response of masonry walls, which lead to damage through flexure and shear (Fig. 5.11). The failure occurs when the material's ultimate strength is exceeded. Unlike first-mode mechanisms, second-mode mechanisms are activated in structures capable of developing a global response to seismic forces. This typically occurs when out-of-plane local collapse mechanisms are prevented or minimized through the use of tie rods, reinforced belts, or similar techniques. Adequate interlocking between walls and effective connections between walls and floors help prevent out-of-plane flexure and overturning. In cases where a building exhibits **box-like behavior**, the rigidity of the floors in their plane plays a crucial role. Rigid floors distribute seismic forces among the walls according to their in-plane stiffness and position, promoting in-plane collapse mechanisms. In contrast, if the floors are deformable (as is the case

(a)

(b)

Fig. 5.10 Typical examples of local collapse mechanisms

(a) (b)

Fig. 5.11 Typical examples of global collapse mechanisms

with traditional wooden floors), this beneficial redistribution does not occur. Each wall, regardless of its stiffness, is subjected to seismic forces proportional to the mass it supports, leading to an independent response based on the area of influence for vertical loads.

Given these behaviors, it becomes evident that the selection of interventions must be based on the specific weaknesses identified in the structure, ensuring that the solution is tailored to prevent the activation of either collapse mode.

Interventions on masonry structures can range from minimally invasive techniques to more radical solutions, depending on the specific condition of the building and the conservation goals. The primary objective is to **preserve the original integrity** of the masonry while simultaneously improving its structural performance to withstand current and future stresses.

The intervention techniques described in this chapter are characterized by: (a) varying effectiveness in addressing specific issues related to strength, durability, and material compatibility, which must be assessed in relation to the specific application context, and (b) differing degrees of reversibility, essential when dealing with architectural heritage.

Consolidation and reinforcement of masonry structures require a holistic and informed approach, grounded in an accurate assessment of the existing conditions and a deep understanding of the properties of the materials used. The selection of appropriate techniques is even more critical in historic buildings, where each intervention must not only respect the structure but also enhance its identity and integrity.

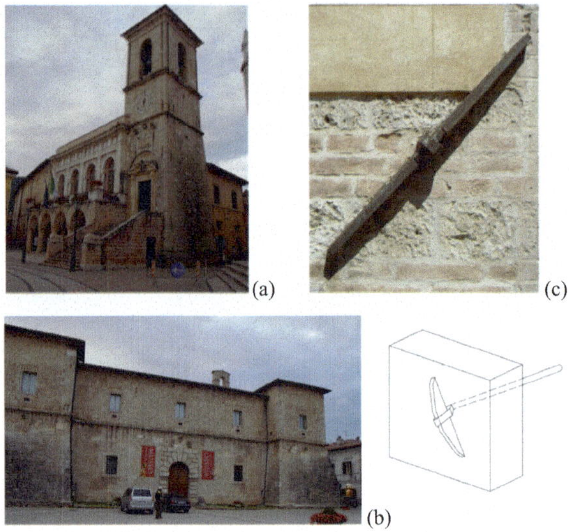

Fig. 5.12 Tie-rods on the civic tower (**a**)"Castellina" (**b**) of Norcia (PG, Italy) and details of the Steel T-rod anchorage (**c**)

5.3.2 Interventions for Local Collapse Mechanisms

5.3.2.1 Tie Rods

The use of tie rods is one of the oldest techniques applied to masonry buildings (Fig. 5.12), particularly for countering the horizontal thrust of arches and vaults. Structures with vaulted ceilings are considered flexible, and the installation of tie rods enhances the system's capacity to resist horizontal thrusts and out-of-plane rotations. This helps mitigate potential overturning mechanisms. This type of intervention is widely used due to its low cost, ease of implementation, and high effectiveness (NIKER Project 2010). This intervention is fully reversible, allowing the removal of the installed systems without causing significant damage to the masonry.

5.3.2.2 Tie Beams

This reinforcement technique involves constructing a structural element, typically made of reinforced concrete or steel, at the top of walls and at each floor level to function as a tie beam.

The primary functions of the tie beam are:

- To create a continuous connection between the vertical and horizontal structures, allowing for a better distribution of vertical loads.
- To improve the connection between orthogonal walls, enhancing the box-like behavior of the structure.

For reinforced concrete tie beams, holes are first drilled to connect to the underlying masonry, and connecting reinforcements are inserted. The tie beam reinforcement (longitudinal bars and stirrups) is then placed, ensuring proper cover and adherence to the specifications regarding the overlap of the bars. Finally, the formwork is installed, and the concrete is poured. Special care must be taken when working with masonry characterized by double-leaf walls with an internal core, as the holes need to be drilled alternately at an angle to ensure both masonry faces are engaged. According to scientific literature, the height of the reinforced concrete tie beam should not exceed the lesser value between the masonry thickness and 40 cm, and it should be constructed to span the entire width of the wall (Fig. 5.13).

In the case of steel tie beams, a structural element is placed around the perimeter of the floor, made of steel profiles appropriately connected to the masonry (Fig. 5.14). When working with hollow-tile concrete floors, the steel tie beam is connected to the existing masonry and a new reinforced concrete slab with shear connectors, which not only ensures the wall-to-floor connection but also prevents the slab from deforming out of its plane. For wooden floors, steel tie beams are connected to the load-bearing walls and a cross-laminated timber board, ensuring the diaphragm action of the floor.

Fig. 5.13 Reinforced concrete ties installed on the roof level

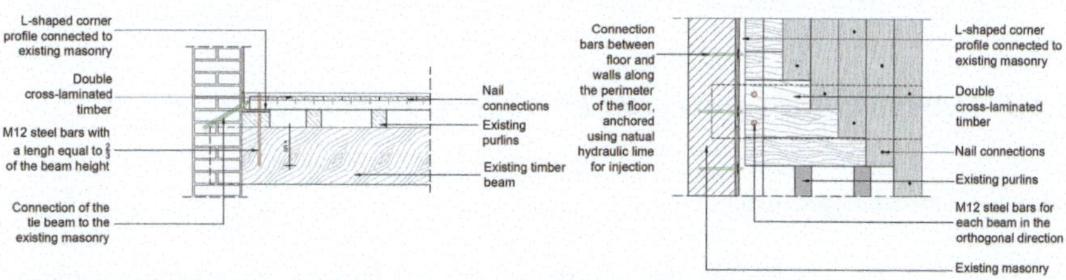

Fig. 5.14 Steel tie beams connecting existing masonry to wooden floor

5.3.2.3 External Ties

External ties are used to restore effective connections between orthogonal masonry walls. These ties can be made using either metallic elements or composite materials. In both cases, it is crucial to avoid creating stress concentration areas at the corners of the walls by using appropriate distribution elements. Metal ties are typically pre-tensioned to provide immediate confinement of the masonry box (NIKER Project 2010). The use of metallic materials on the exterior of the masonry does not create compatibility issues and ensures the complete reversibility of the intervention.

When using composite materials, the edges of the building must be chamfered before applying the bands. These materials offer extremely high mechanical strength while requiring very thin material layers (Mordanova et al. 2016; Cima et al. 2022). To ensure the compatibility of the intervention, it is advisable to use mortars that are compatible with the existing ones, such as natural hydraulic lime-based mortars, with low water-soluble salt content and high sulphate resistance.

5.3.2.4 Corner Stitching

This intervention is recommended to **restore the box behavior of a masonry building**. The reinforced injection creates a solid connection between the various masonry walls. The procedure involves drilling the masonry to create space for the steel bars, ensuring that the drilling does not penetrate all the way through to the opposite side. The holes are then washed with water at moderate pressure until saturation is achieved. Afterward, galvanized steel bars with improved adherence are inserted into the holes, which are subsequently sealed by injecting low-pressure mortar. It is important to note that the holes should be drilled along corner wall sections without windows, but only in solid masonry, to ensure the best possible performance of the intervention. The perforations must be staggered by a few centimetres to prevent them from intersecting. The stitching must penetrate the masonry to a depth equivalent to three times the thickness of the wall being stitched (Fig. 5.15). Special care should be taken during execution, as the success of the intervention depends on the proper adhesion between the bars, the injected mortar, and the masonry.

5.3.3 Interventions for Global Collapse Mechanisms

5.3.3.1 Grout Injection

The consolidation of masonry through the injection of binding mixtures is a widely used technique for filling voids and increasing the compactness of the walls (Valluzzi et al. 2004). Mechanically, this method enhances the strength of the masonry without significantly altering its stiffness (Valluzzi et al. 2002). The process involves drilling

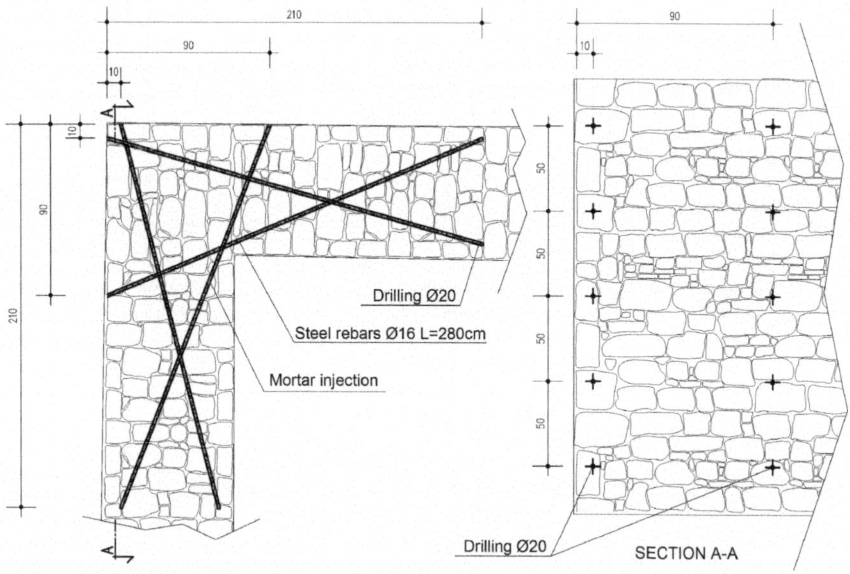

Fig. 5.15 Construction details of masonry corner stitching

holes, inserting injectors into these holes, and subsequently injecting the mixture until the masonry is fully saturated (NIKER Project 2010).

For optimal results, it is advisable to use materials that exhibit high breathability and are similar in form, size, stiffness, and strength to those used in the original masonry. In particular, the use of compatible mortars is recommended, especially those based on natural hydraulic lime, with a low content of water-soluble salts and high sulphate resistance. Such materials ensure not only **mechanical compatibility** but also **long-term durability**, reducing the risk of damaging the existing structure due to chemical reactions or differences in material properties.

5.3.3.2 Structural Repointing

This intervention is applied in cases where the mortar joints have suffered erosion, leading to the loss of the original mortar's surface binding function, while the original stone material remains in good condition. Repointing the joints can enhance the mechanical properties of the existing masonry by increasing its load-bearing capacity. The process involves the **removal of the deteriorated mortar** and the subsequent **repointing of the joints with new mortar** according to the phases specified in Fig. 5.16.

The repointing technique may involve the insertion of reinforcing bars into the joints, made of steel or fiber-reinforced polymer, on one or both sides of the wall, potentially connected transversely. In this case, it is referred to as **reinforced**

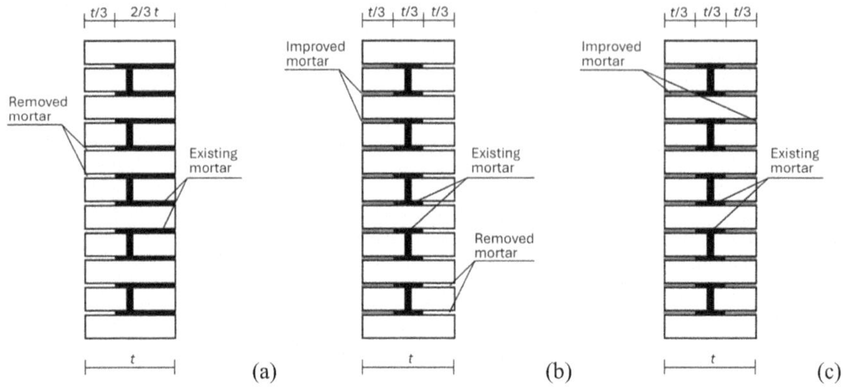

Fig. 5.16 Execution phases of the intervention of structural repointing. **a** Removal of mortar on the first side of the wall, **b** Replacement of mortar on the first side and removal of mortar on the second side, **c** Replacement of mortar on the second side. Image adapted from Tomazevic (1999)

repointing. This process can increase the strength of the masonry against both vertical and horizontal loads. Studies, such as those by Binda et al. (1999, 2001) and Valluzzi et al. (2005), have shown that the insertion of reinforcing bars into horizontal joints also **reduces lateral expansion of the masonry and the pattern of cracking**. However, reinforced repointing is not applicable when the joints are irregular, as is the case with most stone masonry structures.

When this strengthening technique is applied, it is highly recommended to use materials with high breathability and ensure compatibility with the existing ones. Specifically, the use of mortars based on natural hydraulic lime is suggested, as these mortars are characterized by low water-soluble salt content and high sulphate resistance. Such materials, not only provide durability but also prevent potential chemical incompatibilities that could harm the original masonry over time.

In heritage conservation, it is crucial to apply minimally invasive techniques like repointing, which improve the structure's mechanical performance without altering its historical value. Careful selection of materials and the precise application of the repointing process are key to ensuring the long-term preservation of both the visual and structural integrity of the building. Furthermore, ongoing maintenance and monitoring are essential to ensure that the restored joints continue to function effectively over time, particularly in environments exposed to weathering or pollutants.

5.3.3.3 Reinforced Plaster

The reinforced plaster technique involves applying a cement-based layer directly onto the masonry wall, reinforced with a **metal mesh** (Fig. 5.17). Specifically, as outlined by Bhattacharya et al. (2014) and Elgawady et al. (2004), a steel reinforcement mesh (with a diameter of 6.0–8.0 mm) is installed on both sides of the wall. The two meshes

Fig. 5.17 Metal mesh installed on an existing masonry building

are then connected using steel ties, and a cementitious plaster layer is applied to both surfaces. This intervention significantly increases the strength and stiffness of the masonry panel, while preventing lateral expansion (bulging) thanks to the partial confinement provided by the mesh and ties (Binda et al. 1994, 1997).

For the reinforced plaster technique to be effective, it must be applied to both sides of the wall, and adequate transverse ties must be installed. It is recommended to use steel bars anchored to the mesh and masonry to at least two-thirds of the wall thickness (NIKER Project 2010). Ensuring proper overlap of the welded wire mesh panels is also crucial, as it maintains the continuity of the reinforcement both vertically and horizontally. In addition, measures must be taken to ensure the durability of the reinforcement, ideally by using stainless steel mesh and ties. One of the primary vulnerabilities of this technique is the potential for corrosion of the reinforcement and poor anchorage to the masonry (Churilov and Dumova-Jovanoska 2012).

Nowadays, glass fiber-reinforced polymers (GFRP) or basalt fiber meshes, as well as stainless steel, are often used instead of welded wire mesh. These materials ensure **long-term effectiveness** by eliminating the risk of corrosion from environmental exposure, and they also allow the use of lime-based, gypsum, or pozzolan mortars, which are particularly suited for historic masonry.

The use of cement-based mortar, steel mesh, and connectors for reinforced plaster interventions is often incompatible with conservation principles. The reinforced concrete layer is virtually impossible to remove without causing significant damage to the underlying masonry. Additionally, it conceals the original masonry surfaces, altering the building's appearance and dimensions. This technique also changes the wall's physical properties, such as breathability, vapor permeability, and thermal insulation, as well as its deformability characteristics.

On the other hand, interventions utilizing composite materials for the mesh and connectors, combined with **compatible mortars** (such as **lime or pozzolanic mortars**), are less invasive. These materials do not alter the wall's physical properties or its deformability, making them a more suitable option for preserving the integrity of historic structures.

5.3.3.4 Anti-expulsion Tie-Rods

This intervention involves the insertion of metal rods through the masonry, anchored with small bolted plates. Anti-expulsion tie-rods are particularly suitable for walls with disconnected faces, such as "a sacco" masonry or walls with adjacent, unconnected layers. This technique is especially effective in cases where bulging or local instability is evident. In addition to improving the shear strength of the wall, the use of tie-rods helps to reduce deformation.

This method is best applied to masonry that does not contain small-sized stones, as in such cases, the action of the tie-rod would be limited to a small area, and the pre-tensioning applied during installation could induce cracks in the wall. For this reason, the choice of intervention should be carefully evaluated based on the specific characteristics of the masonry.

One of the main advantages of this reinforcement technique is the speed of execution and its relatively low cost, making it an efficient solution for stabilizing disconnected masonry (NIKER Project 2010).

To ensure compatibility with the historical fabric of the building, it is recommended that materials used in the intervention, such as the metal rods and plates, be selected based on their mechanical properties and their minimal visual impact. This approach aligns with the principles of minimal intervention, ensuring that structural reinforcement does not detract from the building's aesthetic and cultural value.

5.3.3.5 Composite Materials Strengthening

In recent decades, research in the field of composite systems has made significant strides, leading to the development of new technologies based on innovative materials. Regardless of the specific type of material, the composite materials are characterized by an adhesive component and by a matrix that facilitates the transfer of stresses, particularly tensile stresses, between the structure to be consolidated and the composite. This mechanism necessitates proper surface preparation, ensuring that the intervention is applied in direct contact with the load-bearing elements of the masonry structure (Cucco, 2008; Marino, 2023; Torsello & Musso, 2003; Vinci, 2022).

To provide a concise yet comprehensive overview, these systems can be categorized into three "macro-families" even though the range of possible interventions is often more complex, with cases of hybridization: FRP (Fiber Reinforced Polymers), a well-established technique over the past three decades; FRCM (Fiber Reinforced Cementitious Matrix), a more recent yet standardized system; and CRM (Composite Reinforced Mortar), a modern adaptation of traditional reinforced plaster:

- **FRP Strengthening**. The acronym FRP (Fiber Reinforced Polymers) refers to a wide range of composite systems combining fibers of carbon, glass, or aramid with a polymer matrix. The terminology for these composite materials varies depending on the type of fiber they contain: CFRP for **carbon fibers**, GFRP for

glass fibers and AFRP for **aramid fibers**. These composites employ fibers either in fabric and strip form (unidirectional, biaxial, and quadriaxial) or as preformed reinforcement systems (laminates, connectors, rods, and meshes), both in combination with **epoxy resins**. These technologies are characterized by low thickness (ranging from 2 to 5 mm), light weight (30–80 N/m^2), high elastic modulus, and quick application, although they require specialized labour. These materials are used in structural reinforcement to increase both the compressive and shear strength of masonry walls, but they can also be employed to enhance out-of-plane resistance in a wall. The application of fabric—whether unidirectional, bidirectional, or multiaxial—allows for **interventions on curved and irregular geometries**, typical of historic masonry (Torsello and Musso 2003; Cucco 2008; Vinci 2022; Liberotti 2023; Marino 2023). The primary advantages of this reinforcement system are its high mechanical properties, corrosion resistance, lightness, versatility, and quick application. However, careful attention must be paid to the physicochemical compatibility of these materials with historic masonry surfaces (Carbonara 2004).

- **FRCM Strengthening**. The acronym FRCM (Fiber Reinforced Cementitious Matrix) refers to composite materials used in situ for the structural consolidation of masonry, wood, or reinforced concrete structures. In international scientific literature, FRCMs are also referred to as TRC (Textile Reinforced Concrete), TRM (Textile Reinforced Mortars), and less commonly, FRM (Fabric Reinforced Mortar) or IMG (Inorganic Matrix-Grid Composites). These systems consist of two main components: fibers of various types (e.g., carbon, alkali-resistant glass, aramid, basalt, PBO, high-strength galvanized steel, polyparaphenylene benzobisoxazole) and **organic matrices**, such as **hydraulic lime** or **pozzolanic mortars** (Fig. 5.18). The most common method involves the widespread application of a bidirectional or unidirectional mesh embedded between two layers of mortar on one or both faces of the masonry. These systems are characterized by a reduced intervention thickness (generally around 10 mm), light weight (approximately 260 N/m^2), and significant elastic modulus—though lower than that of FRPs—along with easy and quick application (Liberotti 2023; Liberotti and Gusella 2021). These technologies are not limited to flexural and shear reinforcements of individual masonry panels; they can also be used for floor confinements, tie beams, and the confinement of pillars and columns. Furthermore, FRCMs are particularly well-suited for the strengthening of curved surfaces, such as columns and arch structures. Due to their organic matrix, these composites are **compatible with historic surfaces and align with the principles of restoration and conservation**. They do not alter the physicochemical properties of the masonry, such as breathability, vapor permeability, and thermal transmittance, nor do they compromise the mechanical functioning of the structure (Carbonara 2004). Unlike other composites, such as FRPs, FRCMs provide a **reversible consolidation strategy**. This is especially relevant given their material affinity with masonry, allowing for the intervention to be removed without causing damage to the surfaces. Compared to CRM techniques and other methodologies, FRCM applications are performed at low thicknesses, thus preventing any alteration to the formal relationships that

(a) (b)

Fig. 5.18 Application of FRCM systems on the extrados of historic masonry vaults

characterize the elevation development of heritage façades (e.g., pilasters, lesenes, cornices).

- **CRM Strengthening**. CRM (Composite Reinforced Mortar) systems are primarily used for the structural reinforcement of existing masonry elements. This acronym identifies an intervention that consists of a **mortar**, typically based on hydraulic lime or cement, into which a preformed bidirectional (flat or angular, depending on the location) **mesh made of various fiber materials** (e.g., glass, carbon, basalt, or aramid) is embedded. Regardless of the specific material combinations, CRM reinforcement systems generally range in thickness from 30 to 50 mm, excluding surface levelling. This system features **easy and quick application**, considerable thickness (minimum of 30 mm), a weight of 520 N/m^2, and a medium–low elastic modulus compared to other composite material families. CRM is a **modern evolution of traditional reinforced plaster**, differing primarily in the types of meshes and connectors used (Torsello and Musso 2003; Cucco 2008; Vinci 2022; Liberotti 2023; Marino 2023). CRM systems allow for uniform reinforcement across various masonry types and characteristics, improving both in-plane and out-of-plane mechanical performance. They are durable and offer some degree of compatibility and reversibility with historic masonry, especially when applied using lime-based mortars. However, CRM interventions do not fall within the category of low-thickness reinforcements, which can be a disadvantage. The **considerable thickness** of these systems not only reduces the usable space in interior rooms but also risks compromising the formal relationships that govern the decorative elements of heritage structures, especially those that protrude or recede from the main masonry surface.

5.3.3.6 Artificial Diatones

The reinforcement of masonry structures through the insertion of artificial diatones is a technique used to **consolidate degraded masonry by horizontally connecting the various layers that compose it** (Yavartanoo and Kang 2022). From a mechanical perspective, it increases compressive strength and shear stiffness, making the masonry less vulnerable to out-of-plane overturning. The intervention involves the drilling of holes, the insertion of the diatones, and the filling of the holes with mortar (Cattaneo and Vafa 2021). Depending on the material inserted into the holes, diatones can be classified into three types: (1) reinforced concrete diatones, (2) steel diatones, and (3) fiber diatones.

- **Reinforced concrete diatones**: In this method, a minimal reinforcement (typically φ 3–4 mm) bound by a spiral is inserted into a hole with a diameter of 15 cm, followed by the injection of mortar. The mortar should not be too fluid, as it could permeate the surrounding masonry (Mariani 2012). To ensure compatibility, natural hydraulic lime-based mortars with low water-soluble salt content and high sulphate resistance should be used.
- **Steel diatones**: A steel tube with a section of 3–5 mm and a diameter of 120 mm is inserted into the hole, and as with the reinforced concrete diatones, mortar is injected afterward, preferably with a low fluidity to prevent excessive permeation of the surrounding material.
- **Fiber diatones**: In this case, smaller diameter holes (8–15 cm) are used, and artificial diatones made of fabric strips are inserted. The ends of the fabric diatones are frayed to ensure adhesion to the masonry face, and low-pressure mortar injections are performed to consolidate the masonry and ensure collaboration with the frayed diatone. Once again, **natural hydraulic lime-based mortars** with low water-soluble salt content and high sulphate resistance are recommended to make the intervention compatible with the historic masonry materials.

Using appropriate mortars ensures the intervention remains compatible with the characteristics of the materials found in historic structures, thus maintaining the physical and mechanical integrity of the heritage building. These interventions provide enhanced structural stability while minimizing the impact on the original materials, aligning with the principles of conservation and restoration.

5.3.4 Local Repairs

5.3.4.1 Local Dismantling and Reconstruction

The Local Dismantling and Reconstruction technique ("unstitch-stitch") is a structural intervention aimed at restoring the continuity of masonry along fracture lines and rehabilitating severely damaged sections of walls. This method involves replacing deteriorated elements with new ones, playing a crucial role in maintaining or re-establishing the structural integrity of historic masonry. As highlighted by Modena et al. (2009) and based on Italy's post-earthquake experiences, this intervention is essential in achieving reinforcement practices that are not only technically sound but also respectful of the cultural and historical value of buildings.

The selection of materials for the stitching must be carefully considered, prioritizing those with similar shapes, dimensions, stiffness, and strength to the original masonry, in order to ensure compatibility and cohesive behaviour. Implementing this intervention requires a deep understanding of the type of masonry and the building's historical context, to avoid actions that may be ineffective or even harmful to the cultural heritage.

Noteworthy applications of the "unstitch-stitch" method can be found in works by Modena et al. (2011) and Orlando et al. (2023). This technique is typically employed on smaller, localized areas of damaged or deteriorated masonry, aiming to restore structural continuity by carefully removing damaged portions and inserting new materials without interrupting the overall masonry fabric. It is most suitable for masonry that exhibits a certain degree of regularity in its layout and texture.

Digitization Notes

The use of advanced digital technologies is transforming the landscape of building rehabilitation, particularly for cultural heritage. The integration of digital mapping methods, such as laser scanning and photogrammetry, with 3D printing (3DP) technologies is highly effective for interventions involving the strengthening, maintenance, and conservation of built heritage elements. These technologies enable the **accurate reproduction of historical building components** that have deteriorated or are missing, allowing for seamless integration with existing structures while preserving their original aesthetic and functional integrity.

Digital mapping techniques, specifically laser scanning and photogrammetry, play an essential role in accurately capturing the geometry and current condition of heritage structures. Laser scanning involves the use of LiDAR (Light Detection and Ranging) to create detailed point clouds that represent the surface geometry of a building in high resolution. Similarly, photogrammetry reconstructs three-dimensional spaces from a collection of high-resolution photographs taken from different perspectives. Both of these methods provide a comprehensive and precise digital representation of the building that serves as the foundation for further analysis and rehabilitation planning.

The use of 3DP, or additive manufacturing, in conjunction with these digital mapping methods, has introduced significant advantages to the field of heritage conservation. One of the key challenges in heritage rehabilitation is the **reproduction of intricate ornamental features** and s**tructural components** that may have deteriorated or are completely missing. Traditional manual reconstruction techniques require skilled artisans and can be labor-intensive, expensive, and time-consuming. On the other hand, 3DP offers a **faster and cost-effective alternative** by enabling the reproduction of these complex elements with a high degree of accuracy. Using compatible materials, such as lime mortars, ensures that the printed components maintain the physical properties of the original materials, allowing for a harmonious integration with the existing structure.

By leveraging 3D printing technology, custom elements, such as masonry units, decorative components, or structural supports, can be produced with exact specifications based on the digital models obtained through laser scanning or photogrammetry. This allows for the production of bespoke parts tailored to the unique characteristics of a specific building, which is especially important when dealing with irregular or intricately decorated architectural features (Jesus et al. 2023). Several rehabilitation case studies exemplify these advancements:

- **Restoration of the Grand Spada Palace Capitals**: in Ferrara, Italy, the restoration of five ornate capitals at the grand Spada Palace was completed using a Massivit 1800 3D printer, operated by Sismaitalia (MASSIVit 3D 2018). These capitals, integral to the palace's French balcony, had deteriorated significantly over the years. Traditional restoration methods were considered, but the time and cost involved were prohibitive. Sismaitalia used polystyrene to 3D print the capitals in two sizes: a smaller set measuring 480 mm high × 430 mm long × 215 mm thick and a larger set measuring 790 mm high × 790 mm long × 215 mm thick (Fig. 5.19). Using the printer's dual printhead, they were able to produce a pair of capitals in just 20 h. The hollow 3D printed structures were then filled with **polyurethane** foam to provide additional strength and finished with plaster and paint to ensure they matched the original counterparts. This approach provided a rapid, cost-effective solution while maintaining the aesthetic consistency of the palace.

(a)

(b)

Fig. 5.19 Restoration of the Grand Spada Palace Capitals with the use of 3DP. Lateral (**a**) and front (**b**) views of the capitals. *Source* © Google Earth

- **Reconstruction of Palmyra Arch of Triumph**: In collaboration with UNESCO, a 3D printed replica of the destroyed arch was created using 2D photographs of the original structure (Fig. 5.20). The Arch of Triumph in Palmyra was destroyed by ISIS militants in 2015 as part of a campaign to eradicate pre-Islamic cultural heritage sites, which were considered idolatrous. The digital reconstruction process required the use of advanced photogrammetry to achieve a high level of accuracy in replicating the intricate details of the historical arch. The reconstructed arch was first unveiled in Trafalgar Square, London, in 2016, before being displayed in other locations worldwide, including New York and Dubai, as a symbol of cultural resilience. The arch was two-thirds the size of the original monument, with a height of about 6 m, making it a manageable representation (Jalabi 2016). The replica served to draw international attention to the destruction of cultural heritage and the importance of preservation (Alexandrea 2017). However, it also opened a debate among archaeologists and cultural heritage experts. Some argued that creating replicas might trivialize the destruction of historical sites, posing questions about honesty and authenticity, while others believed that it provided a powerful tool to keep the memory of such structures alive and inspire restoration efforts (Jalabi 2016; Khunti 2018). The use of 3D printing allowed for the rapid and cost-effective reconstruction of the arch, showcasing the role of technology in filling the gap between destruction and cultural continuity (Turner 2016). This project demonstrated the potential of modern additive manufacturing to contribute to the preservation of cultural heritage, serving as both a tribute to history and an inspiration for future initiatives in digital conservation.

(a) (b)

Fig. 5.20 **a** The original Palmyra Arch of Triumph (source: https://www.flickr.com//photos/newpalmyra/33140885624. License: CC0 1.0); **b** the Arch re-built with the 3D-printed technology assembled in the City Hall Park, New York City (source: https://www.flickr.com/photos/bradhostetler/30180139473/. Title: Arch of Palmyra Replica—The Institute for Digital Archaeology. Author: Brad Hostetler. License: CC BY 2.0. No changes were made)

- **Notre-Dame Cathedral Gargoyle Replication**: After the devastating fire at Notre-Dame Cathedral in 2019, Concr3de proposed an innovative approach to replicate one of the cathedral's iconic gargoyles using 3D printing technology (Fig. 5.21a). Specifically, Concr3de proposed to create a replica that was not only visually accurate but also chemically similar to the original by utilizing a mixture of limestone and ash—materials reminiscent of those found in the original structure. The use of ash, sourced from the remnants of the fire, would allow the replica to embody a part of the cathedral's history, adding symbolic value to the reproduction (Dezeen 2019; France24 2019). A preliminary 3D printing process, conducted with high precision, enabled the production of the replica in a significantly shorter timeframe compared to traditional stone carving techniques, while also keeping costs low (Fig. 5.21b). This method demonstrated the potential of additive manufacturing to respond to urgent restoration needs effectively, while respecting the authenticity and materiality of the original structure.

(a) (b)

Fig. 5.21 Original Gargoyles of Notre-Dame Cathedral **a** source: https://www.flickr.com/photos/14824807@N00/8004584090/. Title: Gargoyles overlooking Paris. Author: Larry. License: CC BY 2.0. No changes were made; **b** source: https://www.flickr.com/photos/142463236@N07/39136545405/. Licence: PDM 1.0)

These case studies demonstrate the potential of **integrating digital fabrication into heritage conservation**, providing tailored, efficient, and high-quality solutions to the challenges of building rehabilitation.

Furthermore, digital tools and platforms are revolutionizing the way data related to heritage rehabilitation is managed and archived. By integrating high-resolution 3D models into cloud-based archives, it is possible to document the condition of heritage buildings before, during, and after interventions. This approach not only facilitates long-term monitoring and comparison but also aids in the **planning of future maintenance**. It creates a **comprehensive digital record** that can be accessed by researchers, conservators, and stakeholders to ensure that every aspect of the intervention is documented, promoting **transparency in conservation practices**. The archiving process is particularly significant for heritage buildings, where documentation of previous interventions can provide insights into the structure's evolution and inform future decisions.

The integration of these digital technologies into heritage conservation has also encouraged greater **collaboration among professionals** from different fields. Architects, engineers, conservators, and material scientists can work together using shared digital models to contribute their expertise and develop cohesive intervention strategies. This interdisciplinary approach is crucial for addressing the complex challenges associated with heritage rehabilitation, which often involve balancing structural integrity, aesthetic value, historical authenticity, and material compatibility.

5.4 Climate Change Actions for Heritage Buildings

Changes in the frequency, intensity, and patterns of climatic factors pose an increasing threat to cultural heritage worldwide, including built heritage (McIntyre-Tamwoy 2008; Markham et al. 2016).

At the urban district scale, various intervention strategies can help mitigate the Urban Heat Island Intensity (UHII). Moreover, at the scale of the single building, several studies and institutional declarations highlight effective mitigation approaches. Notably, documents such as the statement released by ICOMOS during COP27 emphasize the importance of integrating climate-responsive design, energy efficiency, and heritage conservation. The materials of historic buildings are subjected not only to increasingly frequent and destructive extreme weather events (Masson-Delmotte et al. 2021; Clarke et al. 2022) but also to **slow-onset processes**, which are cumulative, gradual, and widespread (Daly 2011; Sesana et al. 2021). Climate change accelerates existing deterioration mechanisms and may also generate new forms of decay, altering the preservation conditions of materials and structures (Bertolin 2019; ICOMOS 2019).

Despite extensive scientific research on the impact of climate change on the environment, studies focusing on adaptation strategies for built heritage remain relatively limited (Sesana et al. 2018; Blavier et al. 2023). The following subsections provide an overview of solutions aimed at protecting cultural heritage architecture from climate change as a function of 5 main climatological stresses: (1) temperature, (2) humidity, (3) precipitation, (4) wind, and (5) solar radiation, as suggested by Blavier et al. (2023).

5.4.1 Urban Scale Strategies for UHII Mitigation

Mitigating the Urban Heat Island (UHI) effect requires a combination of urban planning strategies and technological solutions. Key approaches include:

- Increasing urban greenery: Planting trees, creating green roofs, and expanding urban parks to enhance shading and evapotranspiration.
- Using reflective and cool materials: Implementing cool roofs and high-albedo pavements to reduce heat absorption.
- Enhancing urban ventilation: Designing open spaces and wind corridors to improve airflow and dissipate heat.
- Promoting water-based solutions: Integrating water bodies and fountains to leverage evaporative cooling effects.
- Reducing anthropogenic heat emissions: Improving energy efficiency, optimizing building materials, and promoting sustainable transportation.

These strategies, when combined, can significantly lower urban temperatures, improve thermal comfort, and reduce energy consumption.

5.4.2 Temperature

One of the most significant climatic factors affecting built heritage is temperature, which can cause various forms of damage to historic structures. Among the most concerning threats are material deterioration due to freeze–thaw cycles, thermal shocks, loss of bioprotection, foundation settlements, landslides, accelerated biological decomposition, and increased insect colonization in wooden buildings due to unstable frozen soils.

5.4.2.1 Mitigation Strategies for Freeze–Thaw Cycles and Thermal Shocks

The increasing frequency and intensity of freeze–thaw cycles and thermal shocks represent a major risk for historic materials. Thermal stress primarily depends on temperature fluctuations and increasing surface albedo through **light-colored paving**, **reflective coatings**, or regular facade cleaning can mitigate thermoclasty effects. Additionally, **urban vegetation** can contribute to regulating surface temperatures (Hayles 2022).

Reducing exposure to precipitation, extreme temperature fluctuations (both in winter and summer), and humidity variations is essential to mitigate the impact of freeze–thaw cycles. Various protective technologies have been implemented in the conservation of tangible heritage, and include:

- **Soft capping**, which utilizes plants, grass, and soil to cover horizontal masonry surfaces, protecting underlying structures especially in case of ruins (Fig. 5.22) or fragmented walls (Viles and Wood 2007; Lee et al. 2009; Hayles 2022).
- **Shading shelters**, designed to reduce thermal exposure and protect surfaces from excessive solar radiation (Grima et al. 2011; Becherini et al. 2016).

(a) (b)

Fig. 5.22 Examples of soft capping. **a** Ruins of the Clonmacnoise Castle; **b** Ruins of the Omberg Sweden Monastery. *Source* https://free-images.com/; License: Public Domain

- **Protective enclosures**, constructed with rigid materials such as wooden panels, rigid foam shells, or metal and plexiglass casings (Wölbert 2005).
- **Flexible protective covers**, consisting of membrane enclosures draped over vulnerable structures (Wölbert 2005).

Although these methods have proven effective in localized applications, their direct application on entire historic buildings has yet to be explored.

5.4.2.2 Mitigation Strategies for Bioprotection Loss

Microbial biofilms offer natural protection against water infiltration, extreme thermal events, and temperature fluctuations (Ariño et al. 1995; Chen et al. 2000; Sternberg et al. 2011; Concha-Lozano et al. 2012; Coombes et al. 2013; Cutler et al. 2013). However, rising temperatures may disrupt microbial ecosystems, preventing their regeneration and leading to increased material vulnerability (Carter and Viles 2004; Aptroot and Van Herk 2007).

A potential mitigation strategy involves **planting vegetation around historic buildings** (Fig. 5.23) to moderate surface temperatures, thus reducing both thermoclasty effects and microbial degradation. However, this approach requires careful assessment of potential disadvantages. Uncontrolled plant growth can cause physical damage to structures, increase pest infestations (e.g., insects and rodents), and, in some cases, obstruct visual appreciation of the building's original architectural features. Additionally, in clay-rich soils, vegetation-induced drying can lead to differential settlements, affecting the stability of historic foundations. The overall effectiveness of vegetation planting as a climate adaptation measure for bioprotection loss remains a topic requiring further investigation.

(a)

(b)

Fig. 5.23 Examples of planting vegetation around historic buildings. **a** Ruins of the Eldena Mona **b** Ruins of the Hohensyburg Castle, Germany. *Source* https://free-images.com/; License: Public Domain

5.4.2.3 Mitigation Strategies for Permafrost Thawing and Biodeterioration

Rising temperatures in Nordic and alpine regions may accelerate **biological decomposition** and **insect colonization**, as well as activate permafrost thawing, allowing water infiltration into historic wooden buildings, fostering **mold growth** and insect infestation. Few potential solutions have been identified for ancient buildings and include enhancing **drainage systems** and applying **protective treatments** and **biocides** (Koziróg et al. 2016; Ryszard and Małgorzata 2016; Flyen et al. 2020; Franco-Castillo et al. 2021).

5.4.2.4 Mitigation Strategies for Differential Settlements Due to Permafrost Thawing

Climate change poses a significant risk to the structural integrity of historic buildings in permafrost regions, leading to foundation subsidence, slope failures, and increased ground movement. Notable examples of affected buildings include the Blue Church in Sisimiut (Kaslegard 2011), protected structures in Svalbard (Mattsson et al. 2010), and the Church of Our Lady of Victory in Inuvik (Holubec 2008).

Mitigation strategies primarily aim to **reduce soil temperature** (Vyalov et al. 1993; Shur and Goering 2009). One proposed solution is **elevating buildings** using structural support systems, an approach already adopted in the United States for historic structures at risk of recurrent flooding [173].

Snow cover and vegetation play a crucial role in permafrost stability (Chang et al. 2012). In discontinuous permafrost areas, **vegetation shading** may help slow permafrost thawing, thus preserving the structural integrity of historic buildings (Perreault and Shur 2016). However, the impact of vegetation management is highly site-specific (Tan et al. 2022). In some cases, tree removal is recommended to prevent structural damage (Kaslegard 2011), while in others, deforestation can lead to deeper permafrost thawing (Holubec 2008). Therefore, any intervention related to vegetation management must be carefully evaluated on a case-by-case basis.

To mitigate permafrost degradation of foundations, **ventilated crawl spaces** and **cooling systems** have been proposed (Vyalov et al. 1993). Another potential measure is **seasonal insulation**, which, when applied during summer, can lower permafrost temperatures and improve foundation integrity in warming permafrost zones (Perreault and Shur 2016).

5.4.3 Moisture

Climate change is altering precipitation patterns and increasing humidity levels in several regions, increasing the risks associated with rising damp, condensation, biological colonization, and salt crystallization (Curtis 2016).

5.4.3.1 Mitigation Strategies for Ground Moisture-Related Degradation

Excessive ground moisture can cause various deterioration processes in traditional masonry walls, particularly in cases where improper drainage leads to capillary rise. Several preventive strategies can be implemented to minimize these effects:

- **Redesigning drainage systems** to improve water runoff and prevent accumulation near foundations.
- **Installing subsurface drainage systems** around buildings, where absent.
- **Lowering ground levels** when accumulated gardening materials contribute to moisture retention.
- **Converting surfaces into permeable landscaping**, reducing splash-back effects and preventing water retention.
- Installing or restoring a **Damp-Proof Course (DPC)** at the base of the building to create a moisture barrier and prevent capillary rise.
- **Using breathable paints**, which help regulate indoor humidity levels.

Promoting evaporation and controlling moisture levels inside historic buildings can significantly reduce rising damp issues. **Electro-osmotic** and **electromagnetic systems** have been used for moisture control, but studies report conflicting results, as in some cases no significant impact on water distribution or evaporation rates was observed (Melada et al. 2018; Julie et al. 2021).

Similarly, traditional air-trapping techniques such as **Knapen siphons** (Franzoni 2018) or **Schrijver stones** (Franzoni 2014) have been questioned for their impact on the historical and architectural integrity of heritage buildings. More conventional solutions, such as **mechanical moisture barriers**, remain among the most reliable

Fig. 5.24 **a** A Roman damp-proof course at Hardknott Roman Fort. The layer of slate in the wall blocks rising damp (Source: https://commons.wikimedia.org/w/index.php?curid=5442578; Author: Censorship Workaround—Own work; Licence: CC BY-SA 3.0. No changes have been made to the image). **b** Damp Proof Course (Source: https://commons.wikimedia.org/wiki/Category:Damp-proof_course#/media/File:Horizontalabdichtung_mit_schwarzer_Folie_in_einer_Ziegelwand.jpg. Author: HTL-Mödchen23. Licence: CC BY-SA 4.0. No changes have been made to the image)

interventions. These include **repairing, restoring, or replacing the Damp-Proof Course** (DPC) **at the base of the building** (Fig. 5.24). The DPC can be composed of traditional materials such as slate, as well as modern alternatives like PVC or metal sheets.

Moreover, increased humidity levels in many regions create favorable conditions for mold growth. Research has demonstrated that dirt accumulation on building surfaces intensifies this risk, allowing mold to develop even at relatively low humidity levels (Grant et al. 1989). Therefore, **regular facade cleaning** can be an effective measure to mitigate mold formation (Hayles 2022).

Lastly, moisture is a key factor in the transport of salts within building materials, leading to salt crystallization and associated deterioration. The use of **crystallization inhibitors** has been proposed as a sustainable solution to prevent salt-induced decay in porous materials. However, further studies are required to fully assess the long-term effectiveness of these treatments for heritage conservation (Grant et al. 1989; Bracciale et al. 2020).

Several **ventilation-based and clay-barrier techniques** have also been explored as potential solutions for mitigating rising damp in heritage buildings (Torres and de Freitas 2007; Franzoni 2014; Guimarães et al. 2017; Michette et al. 2017). However, their applicability in historic structures affected by climate change requires further investigation.

5.4.3.2 Mitigation Strategies for Air Humidity-Related Degradation

Airborne moisture can cause significant deterioration in heritage buildings, including leaching and corrosion in glass and metal components, as well as biodegradation. The increasing frequency of high-humidity events due to climate change highlights the urgent need for protective interventions.

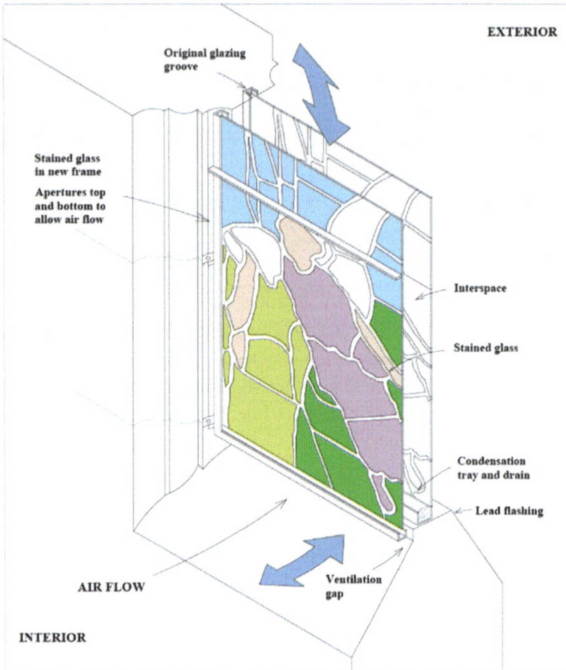

Fig. 5.25 Example of external protective glazing

For metal components, the frequent application of antioxidant protective coatings has been shown to prevent oxidation and corrosion. Similarly, **transparent waterproofing treatments** for stone materials can help shield vulnerable surfaces from prolonged moisture exposure.

Historic stained-glass windows are particularly susceptible to moisture-related damage, whether through condensation, splash water, or wind-driven rain. Acidic or alkaline pH conditions within water droplets can cause hydrated silica gels to form, leading to discoloration, microcracking, and long-term deterioration. In certain cases, installing **external protective glazing (EPG)** can be an effective conservation strategy (Fig. 5.25).

5.4.3.3 Mitigation Strategies for Clay Soil-Related Subsidence

Expansive clay soils can compromise the stability of historic buildings, particularly in regions where climate change leads to alternating periods of drought and heavy rainfall. Several soil reinforcement techniques have been developed to address this problem, including:

- Mini-piles and root piles (Fig. 5.26) and other traditional methods, which provide additional structural support (Cantagallo et al. 2020).

Fig. 5.26 (**a**) and (**b**) Root piles observed in the "Ridola" museum of Matera (MT, Italy); **c** Representative image of an reinforced system using root piles installed at opposing angles to stabilize a foundation on weak subsoil

- Lime columns, which stabilize the soil and reduce shrinkage effects (Hussein 2021; Kalita and Anitha Kumari 2021; Kumar and Thyagaraj 2021).
- Polyurethane resin injections, which enhance the load-bearing capacity of shallow foundations (Buzzi et al. 2008, 2010; Escolano Sánchez et al. 2017).

5.4.4 Precipitation

Climate change is causing significant changes in precipitation patterns, increasing the intensity and unpredictability of rain, snow and hail events. These changes can have a significant impact on historic buildings as many were not originally designed to cope with prolonged heavy rainfall. The resulting moisture infiltration, freeze–thaw cycles and increased surface runoff can accelerate structural and non-structural deterioration of the building, requiring targeted adaptation measures (Steenbergen et al. 2009).

5.4.4.1 Strategies to Manage Water Runoff on Gable Roofs

Several strategies can be adopted to minimize water infiltration and reduce the risk of wall saturation:

- Shape edge tiles at an angle to channel water flow to the center of adjacent tiles.
- Installing lead flashings or hidden gutters to direct water away from vulnerable spots, such as where roofs meet side walls.
- Adding all protective elements that can guide water away from weak structural points to drainage systems (Fig. 5.27).

Designing an efficient drainage system ensures that rainwater does not concentrate on the building's walls, preventing saturation damage over time.

Fig. 5.27 Strategy for manage water runoff adopted in the "Ridola" museum of Matera (MT, Italy)

5.4.4.2 Preventing Groundwater Accumulation and Structural Damage

Excessive moisture accumulation in the soil surrounding a historic building can cause serious structural problems, such as **foundation failure, cracks, and masonry deterioration**. Inadequate or obstructed drainage systems, along with fluctuating water tables, exacerbate these risks. Given the increasing frequency of rainfall events, it is essential to implement **systematic maintenance** and, where necessary, **improve drainage infrastructure**. Key preventive measures include:

- Clean roof drainage systems regularly to avoid blockages caused by debris and plants.
- Ensure that downspouts are properly secured and free of leaks to avoid localized water damage.
- Manage tree roots to prevent them from interfering with underground drainage systems.
- Improve drainage capacity by altering the size of gutters and conveying rainwater to appropriate drainage basins or catch basins (Sesana et al. 2021).

5.4.4.3 Strengthening Roofs Against Heavy Snowfall

In regions affected by heavy snowfall, shifting of tiles due to accumulated snow loads is a recurring problem. Several reinforcement techniques can help mitigate this risk, including:

- Fixing tiles with mortar (Fig. 5.28) to increase their resistance to displacement and wind uplift (Curtis and Snow 2016).
- Using additional fasteners to strengthen the tile attachment.

These interventions help maintain the integrity of historic roofing systems while reducing the likelihood of procuring personal injury caused by falling roofing elements.

Fig. 5.28 Tiles fixed with mortar in a historic building

5.4.4.4 Reducing Flood Risks in Historic Buildings

Flooding poses a serious risk to historic structures, not only through direct physical damage but also through **prolonged moisture retention**, which can lead to deterioration of materials (Fig. 5.29). Effective flood mitigation strategies aim to minimize water intrusion while ensuring efficient drainage to prevent long-term deterioration. The most common temporary flood mitigation solutions include:

- Displaceable **flood barriers**, such as gates and inflatable dams.
- Traditional **sandbags** or advanced high-expansion materials for temporary waterproofing (Snow 2014).

(a) (b)

Fig. 5.29 **a** Floods at Piazza Santa Croce in Florence September 13, 1557 (blue arrow) and November 4, 1966 (red arrow); **b** Carmine square after the flood in Florence 1966 *Source* https://commons.wikimedia.org/; Licenses: Public Domain, for both images

- **Flood-resistant coating systems**, often used in combination with water pumps for added protection (Eggleston et al. 2019). These include dry waterproofing systems such as waterproof sealants on exterior walls or removable anti-flood panels for doors (Pickles et al. 2015).

5.4.4.5 Prevention of Bulging in Historic Masonry

Many historic masonry buildings are built with double curtain walls and, during heavy and persistent rains, may be subject to water infiltration, swelling, and eventual collapse. To avoid this problem, reinforcement techniques commonly used in seismic retrofitting (such as mortar injections, as indicated in Fig. 5.30, or artificial diatons) can be adapted to protect historic masonry (Cascardi et al. 2020).

5.4.5 Wind

Strong winds and storm events are becoming increasingly frequent due to climate change. The effects of wind damage can be classified into two main categories: (1) **structural damage**, caused by wind-induced loads on buildings, and (b) **material deterioration**, resulting from wind-driven erosion, moisture infiltration, and chemical degradation [238]. In coastal and exposed areas, wind-driven salt deposition and water penetration further accelerate the decay of building materials. To safeguard cultural heritage structures, targeted mitigation strategies must address both direct structural risks and long-term material degradation.

Fig. 5.30 Grout injections carried out on Melfi Castle (PZ, Italy) from Prof. Arch. Paolo Rocchi in year 1999

5.4.5.1 Strategies to Protect Building Materials from Wind-Induced Decay

Wind, especially when carrying rain or salt, can contribute to **granular disintegration, erosion, and surface weakening** in historic materials. Buildings located in coastal or exposed environments are particularly vulnerable due to **salt crystallization, increased moisture penetration, and the gradual abrasion of surfaces** (Blavier et al. 2023). Two primary strategies can help mitigate these effects:

- **Modifying wind flow around the structures**. Installing **windbreaks**, such as trees, shrubs, or strategically placed barriers, can help diffuse wind velocity and reduce direct exposure to abrasive particles (Luo et al. 2020; Barčić et al. 2021).
- **Enhancing surface protection**. The application of **protective surface coatings** can create a barrier against wind-driven moisture and salt deposition.

Regular maintenance, including surface consolidation and protective treatments, helps **prevent wind-induced erosion and granular disintegration** (Chai et al. 2023).

5.4.5.2 Storm Protection for Windows and Openings

Historic buildings often lack modern storm protection systems, making them highly susceptible to **damage from windborne debris and high-speed gusts**. To safeguard openings without compromising the historical authenticity of a structure, several non-invasive strategies, including **removable storm panel,** can be installed when extreme weather events are forecasted. These can include **solid protective panels**, designed to fully cover windows and doors, **flexible wind-deflection screens**, which reduce direct pressure on openings or **perforated metal panels**, which allow airflow while providing impact resistance (1000 Friends of Florida 2008).

5.4.5.3 Reinforcement of Roofing Elements Against Wind Loads

Historic roofing systems are particularly vulnerable to **uplift forces and displacement** during windstorms. Ensuring that roof elements are securely anchored can significantly reduce the risk of damage. Recommended reinforcement measures include:

- **Mortar bedding** to fix tiles in place (Fig. 5.27), increasing their wind resistance (Curtis and Snow 2016).
- **Additional nailing techniques** to reinforce individual roof elements,
- **Adjusting the spacing of fasteners** for increased structural stability.
- **Ensuring gutters and drip edges** are securely fastened to prevent detachment during high-wind events.

5.4.5.4 Strengthening Structural Components to Withstand Wind Forces

Preventing **wind-induced structural failures** requires the creation of a **continuous load path** that transfers wind forces safely through the building structure (Chettri et al. 2022). Seismic retrofitting techniques are also generally effective against structural damage produced by wind action, since both forces are horizontal in nature (D'Ayala and Paganoni 2015). It is always necessary, however, in the case of locations where wind action is particularly significant, to adapt seismic strengthening solutions specifically to wind loading conditions.

5.4.6 Solar Radiation

Solar radiation can cause significant deterioration of historic buildings, primarily through **thermal stress and photodegradation**. Prolonged exposure to solar radiation can lead to severe material degradation, particularly in regions where climate change is increasing the frequency and intensity of extreme heat events.

The impact of solar radiation is not limited to direct heating; it also alters **humidity cycles and evaporation rates**, influencing moisture within porous materials. The combination of high solar exposure and poor ventilation can lead to persistent overheating, compromising both the structural integrity and the internal environmental conditions of heritage buildings.

5.4.6.1 Mitigation Strategies for Thermal Stress and Surface Overheating

One of the most significant risks associated with solar radiation is **thermal stress**, which arises from repeated expansion and contraction of materials due to temperature fluctuations. This phenomenon is particularly critical in **masonry structures, stone facades, and wooden elements**, where differential thermal movement can cause **cracking, detachment, and eventual structural failure**.

To mitigate these effects, several adaptive strategies can be employed:

- **Increasing surface reflectivity**. Applying **reflective coatings** or **light-colored finishes** can reduce heat absorption, minimizing temperature fluctuations and thermal-induced cracking. Regular maintenance, such as **surface cleaning**, can help maintain high albedo levels, preventing excessive solar heating (Hayles 2022; Fraleoni-Morgera et al. 2024).
- **Reducing direct solar exposure**. Installing **shading devices**, such as overhangs, screens, or retractable elements, can effectively decrease surface temperatures. Vegetation, when properly managed, can act as a **natural shading system**, helping to regulate microclimatic conditions around heritage structures.

- **Optimizing building ventilation**. Enhancing natural airflow can prevent excessive heat buildup in confined spaces. **Using passive ventilation techniques**, such as ventilated facades and cavity walls, helps dissipate heat and regulate interior temperatures (Lops et al. 2023).

5.4.6.2 Protecting Organic Materials from UV and Photodegradation

Ultraviolet (UV) radiation accelerates the deterioration of organic materials commonly found in historic buildings, including **wood, textiles, pigments, and varnishes**. Prolonged UV exposure can break down chemical bonds in these materials, leading to **fading, loss of cohesion, and structural weakening**.

To minimize UV-related damage, the following measures can be implemented:

- Applying **UV-filtering coatings** or treatments on exposed surfaces, particularly for wooden facades, painted decorations, and stained glass windows (Moretti et al. 2015).
- Installing **UV-protective glazing** for windows and display cases in heritage sites containing valuable artworks or artifacts (Blavier et al. 2023).
- **Limiting direct sunlight exposure** in interior spaces by using adjustable shading systems, such as curtains, blinds, or electrochromic glass panels.

5.4.6.3 Mitigation Strategies for Moisture-Related Solar Effects

Solar radiation indirectly influences moisture dynamics in historic buildings. **Increased evaporation rates** due to solar exposure can lead to **salt crystallization, material embrittlement, and façade discoloration** (Wallenhorst et al. 2018). The combination of **solar-driven drying and capillary rise** may also cause **cyclic damp-dry conditions**, favoring the formation of microcracks and reducing material cohesion.

To counteract these effects, heritage conservation strategies should focus on: (a) reducing excessive drying rates, as rapid evaporation can lead to salt crystallization inside porous materials, increasing mechanical stress; (b) ensuring that protective coatings and water-repellent treatments do not overly restrict vapor exchange, which could trap moisture within materials; (c) managing water runoff and drainage to prevent water accumulation at the base of walls, where solar-induced drying effects can intensify rising damp problems.

> ***Digitization Notes***
>
> The increasing impact of climate change on built heritage necessitates advanced digital solutions to assess, predict, and mitigate environmental stressors such as temperature fluctuations, solar radiation, and urban heat island effects. While digital tools are widely used in climatology and Computational Fluid Dynamics

(CFD), their integration within heritage conservation strategies remains underdeveloped. The following points present some innovative approaches leveraging digitalization to enhance the resilience of historic buildings to climate change.

- **Coupling Climate Modeling and CFD for Built Heritage**. Current climate analysis software provides large-scale meteorological data but lacks the resolution needed for building-specific assessments. CFD-based simulations offer detailed airflow modeling but do not incorporate climate variability over time. Integrating both approaches enables the simulation of seasonal temperature variations, solar exposure, and wind-driven heat dissipation at the building scale.
- **Heritage Digital Twins for Climate Risk Assessment, supported by AI-driven predictive models**. 3D digital twins can incorporate climate data to simulate how heritage structures respond to environmental changes, providing insights into material degradation, thermal stress, and moisture accumulation. AI-driven predictive models allow for real-time monitoring of heat loads and solar exposure, supporting conservation planning.
- **Urban-Scale Digital Mapping for Heritage Sites**. High-resolution satellite images and GIS-based radiation analysis help identify buildings most at risk from solar-induced degradation. Through spatial analysis, it becomes possible to determine which historic structures are most exposed to prolonged solar radiation, enabling conservation experts to anticipate risks such as accelerated surface deterioration, thermal expansion-induced cracking, and photodegradation of materials. Additionally, GIS-based models can simulate the effectiveness of various mitigation strategies before their actual implementation. For instance, if a historic building is highly exposed to solar radiation, GIS tools can virtually test the impact of urban vegetation, solar shading devices, or advanced surface treatments, helping to select the most effective and least intrusive solution for long-term conservation.
- **Intelligent conservation strategies**. The integration of meteorological datasets with digital simulations enables the development of data-driven climate adaptation approaches for heritage buildings. By leveraging real-time weather data, historical climate trends, and predictive models, it is possible to anticipate how environmental conditions will evolve and affect built heritage over time. This integration follows a structured process: first, climate datasets from meteorological stations, satellite observations, and historical records are collected and analyzed. These datasets, which include information on temperature fluctuations, humidity variations, wind patterns, and solar radiation levels, are then incorporated into CFD models and GIS-based tools to simulate the interaction between climate and built heritage at multiple scales. For instance, in areas suffering from urban heat island

(UHI) effects, the combined use of climate models and CFD simulations can assess how heat accumulates and dissipates around historic structures. These digital tools help conservationists test different mitigation strategies, such as increasing urban vegetation, improving ventilation through strategic airflows, or optimizing surface materials to enhance thermal reflectivity. If simulations show that a specific building retains excessive heat due to poor air circulation, CFD models can evaluate how adjusting airflow dynamics—by modifying street layouts, introducing passive cooling elements, or optimizing shading solutions—can contribute to more effective thermal management. Additionally, the use of digital twins allows conservation professionals to simulate climate change scenarios decades into the future, assessing how buildings will respond to projected temperature increases, changes in precipitation, or shifts in wind patterns. This predictive capacity makes it possible to develop conservation strategies that are not only reactive but also proactive, ensuring long-term resilience.

These advancements generated by the introduction of digitization in the **climate risk assessment** have the potential to **transform heritage conservation practices**, ensuring that historic buildings remain protected against the evolving challenges posed by climate change.

References

1000 Friends of Florida (2008) Disaster mitigation for historic structures: protection strategies. Division of Historical Resources, Tallahassee, FL

Alexandrea P (2017) 3D technology in the restoration of history. https://www.3dnatives.com/en/3d-technology-restoring-history-191020174/#!

Aptroot A, Van Herk CM (2007) Further evidence of the effects of global warming on lichens, particularly those with Trentepohlia phycobionts. Environ Pollut 146:293–298

Ariño X, Ortega-Calvo JJ, Gomez-Bolea A, Saiz-Jimenez C (1995) Lichen colonization of the Roman pavement at Baelo Claudia (Cadiz, Spain): biodeterioration vs. bioprotection. Sci Total Environ 167:353–363

Barčić D, Habjanec V, Španjol Ž, Šango M (2021) Analysis of raising windbreaks on the mediterranean Karst of Croatia. Sumar List 145:175–183

Becherini F, Cassar J, Galea M, Bernardi A (2016) Evaluation of the shelters over the prehistoric Megalithic Temples of Malta: environmental considerations. Environ Earth Sci 75:1–13

Bellini A (1986) Tecniche per la conservazione. Franco Angeli, Milano

Bertolesi E, Adam JM, Rinaudo P, Calderón PA (2019) Research and practice on masonry cross vaults—a review. Eng Struct 180:67–88

Bertolin C (2019) Preservation of cultural heritage and resources threatened by climate change. Geosciences (Basel) 9:250

Bhattacharya S, Nayak S, Dutta SC (2014) A critical review of retrofitting methods for unreinforced masonry structures. Int J Disaster Risk Reduction 7:51–67

Binda L, Modena C, Baronial G, Abbaneol S (1997) Repair and investigation techniques for stone masonry walls. Constr Build Mater 11:133–142

Binda L, Modena C, Baronio G, Gelmi A (1994) Experimental qualification of injection admixtures used for repair and strengthening of stone masonry walls. In: 10th international brick/block Masonry conference, pp 539–548

Binda L, Modena C, Saisi A, Tongini Folli R, Valluzzi MR (2001) Bed joints structural repointing of historic masonry structures. In: IX Canadian Masonry symposium spanning the centuries with Masonry. Fredericton (Canada)

Binda L, Modena C, Valluzzi MR, Zago R (1999) Mechanical effects of bed joint steel reinforcement in historic brick masonry structures. In: VIII international conference and exhibition, structural faults and repair. London

Blavier CLS, Huerto-Cardenas HE, Aste N, Del Pero C, Leonforte F, Della Torre S (2023) Adaptive measures for preserving heritage buildings in the face of climate change: a review. Build Environ 110832

Bracciale MP, Sammut S, Cassar J, Santarelli ML, Marrocchi A (2020) Molecular crystallization inhibitors for salt damage control in porous materials: an overview. Molecules 25:1873

Buzzi O, Fityus S, Sasaki Y, Sloan S (2008) Structure and properties of expanding polyurethane foam in the context of foundation remediation in expansive soil. Mech Mater 40:1012–1021

Buzzi O, Fityus S, Sloan SW (2010) Use of expanding polyurethane resin to remediate expansive soil foundations. Can Geotech J 47:623–634

Cantagallo C, Sangiorgio V, Verazzo C, Spacone E (2024) An analytical safety tree for heritage buildings: application to the archaeological museum "Domenico Ridola", Matera (Italy). Int J Architectural Heritage. https://doi.org/10.1080/15583058.2024.2383354

Cantagallo C, Spacone E, Perrucci D, Liguori N, Verazzo C (2020) A multilevel approach for the cultural heritage vulnerability and strengthening: application to the melfi castle. Buildings 10. https://doi.org/10.3390/BUILDINGS1009015

Carbonara G (1997) Avvicinamento al restauro. Teoria, storia, monumenti. Liguori, Napoli

Carbonara G (2004) Trattato di Restauro Architettonico. Torino

Carter NEA, Viles HA (2004) Lichen hotspots: raised rock temperatures beneath Verrucaria nigrescens on limestone. Geomorphology 62:1–16

Cascardi A, Leone M, Aiello MA (2020) Transversal joining of multi-leaf masonry through different types of connector: experimental and theoretical investigation. Constr Build Mater 265:120733. https://doi.org/10.1016/j.conbuildmat.2020.120733

Cattaneo S, Vafa N (2021) Tensile capacity of adhesive anchors in damaged masonry. Appl Sci (Switzerland) 11. https://doi.org/10.3390/app112110135

Ceravolo R, Pistone G, Fragonara LZ, Massetto S, Abbiati G (2016) Vibration-based monitoring and diagnosis of cultural heritage: a methodological discussion in three examples. Int J Architectural Heritage 10:375–395. https://doi.org/10.1080/15583058.2013.850554

Chai Y, Li Y, Wang W, Shi P, Luo H, Ma T, Zhao X, Zhang B, Zhu J (2023) Sandstone monuments still standing after storms: Waterproofing and durability protection. J Cult Herit 62:242–250

Chang X, Jin H, Wang Y, Zhang Y, Zhou G, Che F, Zhao Y (2012) Influences of vegetation on permafrost: a review. Acta Ecol Sin 32:7981–7990

Chen J, Blume H-P, Beyer L (2000) Weathering of rocks induced by lichen colonization—a review. Catena (Amst) 39:121–146

Chettri N, Dhaurali G, Neopaney T, Cheki T, Gautam D (2022) Wind vulnerability and strengthening of Bhutanese vernacular roofs. Geomat Nat Haz Risk 13:1511–1534

Chik TNT, Mahamood IA, Mohamad NAE, Yusoff NA, Rusidi MAY, Hakim SJS (2024) A review of structural health monitoring in heritage building due to cracking issues. In: IOP conference series: earth and environmental science. Institute of Physics

Churilov S, Dumova-Jovanoska E (2012) Analysis of masonry walls strengthened with RC jackets. In: 15 WCEE

Cima V, Bartolomeo C, Grande E, Imbimbo M (2022) Natural fibers for out-of-plane strengthening interventions of unreinforced Masonry buildings in aggregate configuration. Sustainability (Switzerland) 14. https://doi.org/10.3390/su14169967

Clarke B, Otto F, Stuart-Smith R, Harrington L (2022) Extreme weather impacts of climate change: an attribution perspective. Environ Res Clim 1:012001

Concha-Lozano N, Gaudon P, Pages J, De Billerbeck G, Lafon D, Eterradossi O (2012) Protective effect of endolithic fungal hyphae on oolitic limestone buildings. J Cult Herit 13:120–127

Coombes MA, Naylor LA, Viles HA, Thompson RC (2013) Bioprotection and disturbance: seaweed, microclimatic stability and conditions for mechanical weathering in the intertidal zone. Geomorphology 202:4–14

Cucco F (2008) Statica e consolidamento degli edifici storici. Palermo

Curtis R (2016) Water management for traditional buildings: adaptation for a changing climate. APT Bull J Preserv Technol 47:8–14

Curtis R, Snow JH (2016) Short guide-Climate change adaptation for traditional buildings. Historic Environment Scotland

Cutler NA, Viles HA, Ahmad S, McCabe S, Smith BJ (2013) Algal 'greening' and the conservation of stone heritage structures. Sci Total Environ 442:152–164

Cvetanovic A, Momcilovic-Petronijevic A (2021) Contemporary meets old in rehabilitating historic buildings. Facta Universitatis—Ser Architecture Civil Eng 19:1–13. https://doi.org/10.2298/fuace210301002c

Daly C (2011) Climate change and the conservation of archaeological sites: a review of impacts theory. Conserv Manag Archaeol Sites 13:293–310

D'Ayala (2014) Perspectives on European Earthquake Engineering and Seismology. In: Ansal A (ed) Perspectives on European earthquake engineering and seismology. Springer

D'Ayala D, Paganoni S (2015) Seismic strengthening strategies for heritage structures. In: Beer M, Kougioumtzoglou IA, Patelli E, Au S-K (eds) Encyclopedia of earthquake engineering. Springer, Berlin, pp 3090–3117

Dezeen (2019) Concr3de proposes using 3D printing to rebuild Notre-Dame. https://www.dezeen.com/2019/04/24/3d-printing-notre-dame-cathedral-concr3de/

Eggleston J, Parker J, Wellock J (2019) Guidelines on flood adaptation for rehabilitating historic buildings. US Department of the Interior, National Park Service, Cultural Resources, Partnerships & Science

Elgawady M, Elgawady M, Lestuzzi P, Badoux M (2004) A review of conventional seismic retrofitting techniques for URM. In: 13th international brick and block masonry conference

Escolano Sánchez F, Mazariegos De La Serna A, Sánchez Lavín JR, del Campo Yagüe JM (2017) Recalce de cimentaciones superficiales mediante la inyección de resina de poliuretano expandido. Caso de estudio: Palacio del Cardenal Diego Espinosa, en Segovia (España). Revista De La Construcción 16:420–430

Flyen A-C, Flyen C, Mattsson J (2020) Climate change impacts and fungal decay in vulnerable historic structures at Svalbard. In: E3S web of conferences. EDP Sciences, p 20006

Fraleoni-Morgera A, Afshani M, Montelpare S, Lops C (2024) Sustainable micro- and nanocomposites for thermal insulation in buildings. Adv Eng Mater 26. https://doi.org/10.1002/adem.202301064

France 24 (2019) Can Notre-Dame rise from the ashes with 3D printing? https://www.france24.com/en/20190427-paris-france-notre-dame-rebuild-reconstruction-concr3de-3d-printing-macronacron

Franco-Castillo I, Hierro L, de la Fuente JM, Seral-Ascaso A, Mitchell SG (2021) Perspectives for antimicrobial nanomaterials in cultural heritage conservation. Chem 7:629–669

Franzoni E (2018) State-of-the-art on methods for reducing rising damp in masonry. J Cult Herit 31:S3–S9

Franzoni E (2014) Rising damp removal from historical masonries: a still open challenge. Constr Build Mater 54:123–136

Gentile C, Ruccolo A, Canali F (2019) Continuous monitoring of the Milan Cathedral: dynamic characteristics and vibration-based SHM. J Civ Struct Health Monit 9:671–688. https://doi.org/10.1007/s13349-019-00361-8
Giuffrè A (1996) Sicurezza e conservazione dei centri storici. Il caso Ortigia
Grant C, Hunter CA, Flannigan B, Bravery AF (1989) The moisture requirements of moulds isolated from domestic dwellings. Int Biodeterior 25:259–284
Grima R, Stroud K, Torpiano A (2011) The protective shelters at Ħaġar Qim and Mnajdra: impacts, dilemmas and values. Challenges in conservation of architectural heritage Malta: Kamra tal-Periti, pp 3–14
Guimarães AS, Delgado JMPQ, De Freitas VP (2017) Case studies of rising damp treatment in historical buildings. Diffusion Found 10:107–119
Hayles C (2022) How resilient are buildings in the UK and Wales to the challenges associated with a changing climate? Practical recommendations for risk-based adaptation. Cardiff
Holubec I (2008) Flat loop thermosyphon foundations in warm permafrost
Hussein M (2021) Effect of sand and sand-lime piles on the behavior of expansive clay soil. Adv Civil Eng 2021:4927078
ICOMOS (2022) Statement to the United Nations Climate Change Conference (COP27). Sharm El Sheikh (Egypt)
ICOMOS (2019) The future of our pasts: engaging cultural heritage in climate action outline of climate change and cultural heritage. International Council on Monuments and Sites, Paris
ICOMOS/ISCARSAH (2005) Recommendations for the analysis, conservation and structural restoration of architectural heritage
Im SB, Hurlebaus S, Kang YJ (2013) Summary review of GPS technology for structural health monitoring. J Struct Eng 139:1653–1664. https://doi.org/10.1061/(asce)st.1943-541x.0000475
ISO/TC96/SC2 (2010) Bases for design of structures—assessment of existing structures
Jalabi R (2016) Replica of Syrian arch destroyed by Isis unveiled in New York City. https://www.theguardian.com/us-news/2016/sep/20/palmyra-arch-syria-new-york?CMP=fb_gu&utm_medium=website&utm_source=archdaily.com
Jesus M, Guimarães AS, Rangel B, Lino Alves J (2023) The potential of 3D printing in building pathology: rehabilitation of cultural heritage. Int J Build Pathol Adapt 41:647–674
Julie D, Vanhellemont Y, Lubelli B, de Bouw M (2021) Effectiveness of electromagnetic and electro-osmosis methods for the treatment of rising damp. In: 1st international conference on moisture in buildings (ICMB21) online
Kalita DM, Anitha Kumari SD (2021) Effect of composite pile foundation system on the behavior of soft ground. In: Seismic design and performance: select proceedings of 7th ICRAGEE 2020. Springer, pp 109–119
Kaslegard AS (2011) Climate change and cultural heritage in the Nordic countries. Nordic Council of Ministers
Khunti R (2018) The problem with printing palmyra: exploring the ethics of using 3D printing technology to reconstruct heritage. Stud Digital Heritage 2:1–12. https://doi.org/10.14434/sdh.v2i1.24590
Koziróg A, Rajkowska K, Otlewska A, Piotrowska M, Kunicka-Styczyńska A, Brycki B, Nowicka-Krawczyk P, Kościelniak M, Gutarowska B (2016) Protection of historical wood against microbial degradation—Selection and application of microbiocides. Int J Mol Sci 17:1364
Kumar KSR, Thyagaraj T (2021) Comparison of lime treatment techniques for deep stabilization of expansive soils. Int J Geotech Eng 15:1021–1039
Lee Z, Viles HA, Wood CH (2009) Soft capping historic walls: a better way of conserving ruins. English Heritage Research Project Report 1–69
Liberotti R (2023) Architectural and environmental sustainability of restoration interventions: innovative strategies using FRCM composites for cultural heritage safeguard. PhD Thesis, University of Perugia

Lops C, Di Loreto S, Pierantozzi M, Montelpare S (2023) Double-Skin Façades for building retrofitting and climate change: a case study in Central Italy. Appl Sci (Switzerland) 13. https://doi.org/10.3390/app13137629

Luo Q, Zhen L, Xiao Y, Wang H (2020) The effects of different types of vegetation restoration on wind erosion prevention: a case study in Yanchi. Environ Res Lett 15:115001

Makoond N, Pelà L, Molins C, Roca P, Alarcón D (2020) Automated data analysis for static structural health monitoring of masonry heritage structures. Struct Control Health Monit 27. https://doi.org/10.1002/stc.2581

Marino L (2023) Interventi di restauro sui ruderi. Didapress, Firenze, La protezione delle creste murarie

Markham A, Osipova E, Lafrenz Samuels K, Caldas A (2016) World heritage and tourism in a changing climate. UNESCO Publishing

MASSIVIt 3D (2018) Nineteenth century palace restored using MASSIVIT 3D printing. In: https://www.massivit3d.com/case-studies/nineteenth-century-palace-restored-using-massivit-3d-printing/

Masson-Delmotte V, Zhai P, Pirani A, Connors SL, Péan C, Berger S, Caud N, Chen Y, Goldfarb L, Gomis MI (2021) Climate change 2021: the physical science basis. Contribution of working group I to the sixth assessment report of the intergovernmental panel on climate change. 2:2391

Mattsson J, Flyen A-C, Nunez M (2010) Wood-decaying fungi in protected buildings and structures on Svalbard. Agarica 29:5–14

McIntyre-Tamwoy S (2008) The impact of global climate change and cultural heritage: grasping the issues and defining the problem. Hist Environ 21:2–9

Melada J, Gargano M, Veronese I, Ludwig N (2018) Does electro-osmosis work in moisture damage prevention? applicability of infrared-based methods to verify water distribution under electric fields. J Cult Herit 31:S38–S45

MiBACT (2011) Guidelines for the seismic risk evaluation and reduction of the cultural heritage—alignment to the Technical Standards for Constructions (in Italian)

Michette M, Lorenz R, Ziegert C (2017) Clay barriers for protecting historic buildings from ground moisture intrusion. Herit Sci 5:1–11

Modena C, Casarin F, Da Porto F, Garbin E, Mazzon N, Munari M, Panizza M, Valluzzi MR (2009) Structural interventions on historical masonry buildings: review of Eurocode 8 provisions in the light of the Italian experience. In: Cosenza E (ed) Eurocode 8: perspectives from the Italian standpoint workshop, pp 225–236

Modena C, Valluzzi MR, Da Porto F, Casarin F (2011) Structural aspects of the conservation of historic masonry constructions in seismic areas: remedial measures and emergency actions. Int J Architectural Heritage 5:539–558. https://doi.org/10.1080/15583058.2011.569632

Mordanova A, De Santis S, de Felice G (2016) State-of-the-art review of out-of-plane strengthening of masonry walls with mortar-based composites. In: Van Balen, Verstrynge (ed) Structural analysis of historical constructions—anamnesis, diagnosis, therapy, controls. Taylor & Francis Group, p 312

Moreno-Gomez A, Perez-Ramirez CA, Dominguez-Gonzalez A, Valtierra-Rodriguez M, Chavez-Alegria O, Amezquita-Sanchez JP (2018) Sensors used in structural health monitoring. Arch Comput Methods Eng 25:901–918. https://doi.org/10.1007/s11831-017-9217-4

Moretti E, Belloni E, Lascaro E (2015) The influence of solar control films on energy and daylighting performance by means of experimental data and preliminary unsteady simulations. Energy Procedia 78:340–345

NIKER Project (2010) Deliverable 3.1. Inventory of earthquake-induced failure mechanisms related to construction types, structural elements, and materials. Milan

Orlando M, Becattini G, Betti M (2023) Multilevel structural evaluation and rehabilitation design of an historic masonry fortress. J Build Eng 63. https://doi.org/10.1016/j.jobe.2022.105379

Perreault P, Shur Y (2016) Seasonal thermal insulation to mitigate climate change impacts on foundations in permafrost regions. Cold Reg Sci Technol 132:7–18

Pickles D, Rhodes P, Gooch J, Garlick J, SPAB Technical Panel, Kelly N, Hadley P, Berry S (2015) Flooding and historic buildings. English Heritage
Quelhas B, Cantini L, Guedes JM, da Porto F, Almeida C (2014) Characterization and reinforcement of stone masonry walls. In: Costa A et al (eds) Structural rehabilitation of old buildings. Springer, Berlin, pp 131–155
Ryszard K, Małgorzata M (2016) Smart environmentally friendly composite coatings for wood protection. In: Smart composite coatings and membranes. Elsevier, pp 293–325
Saisi A, Gentile C, Ruccolo A (2016) Pre-diagnostic prompt investigation and static monitoring of a historic bell-tower. Constr Build Mater 122:833–844. https://doi.org/10.1016/j.conbuildmat.2016.04.016
Sánchez AR, Meli R, Chávez MM (2016) Structural Monitoring of the Mexico City Cathedral (1990–2014). Int J Architectural Heritage pp 254–268. Taylor and Francis Inc.
Sesana E, Gagnon AS, Bertolin C, Hughes J (2018) Adapting cultural heritage to climate change risks: perspectives of cultural heritage experts in Europe. Geosciences (Basel) 8:305
Sesana E, Gagnon AS, Ciantelli C, Cassar J, Hughes JJ (2021) Climate change impacts on cultural heritage: a literature review. Wiley Interdiscip Rev Clim Change 12:e710
Shur Y, Goering DJ (2009) Climate change and foundations of buildings in permafrost regions. In: Permafrost soils. Springer, pp 251–260
Snow J (2014) Inform guide: flood damage to traditional buildings. Historic Scotland
Steenbergen RDJM, Geurts CPW, Van Bentum CA (2009) Climate change and its impact on structural safety. Heron 54:3–35
Sternberg T, Viles H, Cathersides A (2011) Evaluating the role of ivy (Hedera helix) in moderating wall surface microclimates and contributing to the bioprotection of historic buildings. Build Environ 46:293–297
Tan X, Luo S, Li H, Hao X, Wang J, Dong Q, Chen Z (2022) Investigating the effects of snow cover and vegetation on soil temperature using remote sensing indicators in the three river source region, China. Remote Sens (Basel) 14:4114
Tomazevic M (1999) Earthquake-resistant design of masonry buildings. World Scientific
Torres MIM, de Freitas VP (2007) Treatment of rising damp in historical buildings: wall base ventilation. Build Environ 42:424–435
Torsello BP, Musso SF (2003) Tecniche di restauro architettonico. Torino
Turner L (2016) Palmyra's Arch of Triumph recreated in London. https://www.bbc.com/news/uk-36070721
Valluzzi MR, Binda L, Modena C (2002) Experimental and analytical studies for the choice of repair techniques applied to historic buildings. Mater Struct 35:285–292
Valluzzi MR, Binda L, Modena C (2005) Mechanical behaviour of historic masonry structures strengthened by bed joints structural repointing. Constr Build Mater 19:63–73. https://doi.org/10.1016/j.conbuildmat.2004.04.036
Valluzzi MR, da Porto F, Modena C (2004) Behavior and modeling of strengthened three-leaf stone masonry walls. Mater Struct 37:184–192. https://doi.org/10.1007/bf02481618
Venice Charter (1964) International charter for the conservation and restoration of monuments and sites. Venice Charter
Viles HA, Wood C (2007) Green walls?: integrated laboratory and field testing of the effectiveness of soft wall capping in conserving ruins. Geol Soc Lond Spec Publ 271:309–322
Vinci M (2022) Metodi di calcolo e tecniche di consolidamento per edifici in Muratura. Palermo
Vyalov SS, Fotiev SM, Gerasimov AS, Zolotar AI (1993) Provision for bearing capacity of permafrost soils in conditions of climate warming. Soil Mech Found Eng 30:223–228

Wallenhorst L, Gurău L, Gellerich A, Militz H, Ohms G, Viöl W (2018) UV-blocking properties of Zn/ZnO coatings on wood deposited by cold plasma spraying at atmospheric pressure. Appl Surf Sci 434:1183–1192

Wölbert O (2005) Winterschutzverkleidungen für witterungsgefährdete Objekte. Klimastabilisierung und bauphysikalische Konzepte, ICOMOS Hefte des deutschen Nationalkomitees XLII

Yavartanoo F, Kang THK (2022) Retrofitting of unreinforced masonry structures and considerations for heritage-sensitive constructions. J Build Eng 49

Chapter 6
Automation of the Risk Analysis for Built Heritage

Abstract This chapter explores more effective and efficient methods for automating seismic risk analysis for built heritage. It first includes a review of major digital platforms developed for risk assessment. Then, it examines how the transition from datalakes to structured and interoperable databases enables advanced processing and integration of heterogeneous data. The chapter further explains how risk analysis automation can be efficiently performed using a multiscale digitalization framework, showing how data acquisition and management can be adapted to different levels of analysis. Expeditious approaches for large-scale vulnerability and risk assessments are presented at urban, compartment, and building levels, while detailed approaches focus on in-depth vulnerability evaluations of individual structures. Together, these methods support informed, data-driven decision-making and strengthen the resilience of heritage assets.

6.1 Objectives and Ambition

The most comprehensive digitization process of built heritage is obtained when all relevant information about the building and the artistic objects contained in it is collected within an integrated and easily accessible **digital platform**. This platform can be used as centralized system where data can be structured and processed, ensuring full availability to all stakeholders. By integrating a wide range of data—ranging from geometric and material information to structural behaviour and risk assessments—it creates a unified system that enhances decision-making in heritage conservation and risk management.

Currently, multiple digital platforms have been developed to facilitate the digitization, storage, and processing of data related to built heritage. A **comparative overview of existing platforms** is provided in this section based on a review of relevant scientific literature.

One of the most important advantages of digital platforms for built heritage consists in their ability to **automate large-scale seismic risk assessments**. A digital system can automatically integrate hazard, vulnerability, and exposure data to

© The Author(s), under exclusive license to Springer Nature Switzerland AG 2025
C. Cantagallo et al., *Digitization of Built Heritage*, Digital Innovations in Architecture, Engineering and Construction, https://doi.org/10.1007/978-3-031-96598-2_6

generate comprehensive risk maps and predictive models. Beyond risk assessment, a digital platform for heritage digitization provides key benefits in multiple areas:

- **Comprehensive Information Management**: The integration of geometric, historical, material, and structural data within a single system ensures accessibility and interoperability.
- **Automated Risk and Vulnerability Assessments**: Different methodological approaches for evaluating seismic vulnerability can be implemented within the system, allowing for real-time comparative analyses and greater accuracy in risk assessment.
- **Optimization of Resource Allocation**: The ability to compare vulnerability and risk indicators across different urban areas and historical centres allows for prioritizing conservation interventions at various administrative levels.

6.2 Overview of Existing Platforms for Risk Assessment

The assessment of seismic risk for built heritage has significantly advanced with the development of digital platforms designed to integrate hazard, vulnerability, and exposure data. These platforms enable real-time risk evaluation, providing decision-support tools for conservation planning, emergency response, and resource allocation. Below is an overview of some of the most relevant platforms currently used in risk assessment.

IRMA platform is a comprehensive seismic risk assessment tool developed by EUCENTRE with the support of the Italian Civil Protection Department (Borzi et al. 2021). Initially designed for residential buildings, it was extended to assess the seismic risk of school buildings and other critical infrastructures (Faravelli et al. 2022). IRMA integrates different exposure and vulnerability models with hazard data from the MPS04 seismic hazard map. The platform allows users to conduct probabilistic risk assessments over different time windows (e.g., 1 year, 50 years) and compare multiple fragility models developed by research institutions. Its flexibility makes it a valuable tool for national-scale risk assessments and policy planning.

SICURO + is a web tool (Faravelli et al. 2024) developed by EUCENTRE for the Italian Civil Protection Department. It translates complex risk assessment models into **accessible information for citizens** providing interactive maps that show seismic hazard, vulnerability, and exposure data at the urban level. The platform represents therefore an innovative approach to raising public awareness and promoting community resilience.

The **ARCHYTAS platform** has been developed within the EU-Greece funded ARCHYTAS project to serve as a **decision-support tool** for the seismic risk assessment of **monumental structures and archaeological sites** (Melissianos et al. 2024). It was specifically designed to support Greek authorities in the prioritization of **preservation and restoration** actions, as well as to manage resources before and after a seismic event. Specifically, the ARCHYTAS platform incorporates:

- **Sensor Networks**: On-site instrumentation, including accelerometers and telemetry systems, providing continuous monitoring of seismic activity and structural response.
- **Precomputed Seismic Hazard and Fragility Curves**: The platform utilizes probabilistic seismic hazard analysis (PSHA) and predefined fragility curves for different types of monumental structures based on existing numerical studies.
- **Risk Assessment**: A web-cloud system processes sensor data to identify damage levels and provide real-time alerts.
- **Decision-Support Interface**: Designed for government agencies and heritage conservation experts, the platform provides access to risk maps, real-time vulnerability assessments, and monitoring reports to support emergency planning and resource allocation.

ARCHYTAS has been tested on historical monuments, such as the Temple of Aphaia on Aegina Island, demonstrating its ability to prioritize conservation efforts based on real-time risk assessments. The platform enables authorities to (a) identify monuments at higher risk and plan preventive reinforcement strategies; (b) evaluate post-earthquake structural conditions based on monitoring data and (c) optimize intervention strategies by comparing risk levels of multiple heritage sites.

The **GENESIS platform** (Cantagallo and Sangiorgio 2024; Cantagallo et al. 2024a) is a web-based information system specifically designed to support the seismic risk assessment and management of **historic centers**, with a particular focus on Southern Italy. The platform is a **comprehensive digital tool** for the collection, integration, and analysis of structural and territorial data, enabling **multi-scale assessment** of seismic vulnerability and risk.

GENESIS integrates multiple datasets into a **unified web platform**, allowing users to access and manage information at various levels of detail. The system incorporates the following components:

- **Seismic Hazard**: The platform includes national-scale seismic hazard maps MPS04 (Meletti and Montaldo 2007) and can be supplemented with local microzonation studies to enhance accuracy in site-specific hazard assessments.
- **Exposure Data**: GENESIS contains information on building typologies, population density, and economic evaluations, which are essential for estimating potential seismic losses.
- **Survey Data Collection**: The platform includes two types of survey forms:
- *Expedited Survey Form*: Designed for **large-scale vulnerability assessments**, this form records basic building characteristics based on a rapid visual inspection.
- *Detailed Survey Form*: This form is used for comprehensive building assessments, incorporating HBIM models, material testing, and dynamic identification tests. It also requires input from detailed numerical models (FEM), including mode shapes, structural capacity, and comparisons between seismic demand and capacity. Additionally, users can input seismic risk evaluations conducted at an advanced level for individual buildings.
- **Seismic Vulnerability Assessment**: The platform automates large-scale vulnerability assessments based on the information provided in expedited surveys. It

generates **fragility curves**, which express the probability of exceeding a given damage threshold as a function of seismic intensity or Peak Ground Acceleration (PGA).
- **Seismic Risk Assessment**: Using the generated fragility curves, the platform computes expected loss curves associating potential economic losses with different damage levels.

The seismic risk assessment methodology implemented in GENESIS is based on expeditious techniques, enabling risk evaluations at different scales:

1. **Urban Level**: The entire municipality is analysed using average construction characteristics, provided by local engineers who classify the built environment into a maximum of three structural archetypes.
2. **Compartment Level**: The municipality is subdivided into homogeneous zones, each characterized by similar building features. Local experts define up to three representative building typologies (archetypes) per compartment.
3. **Building Level**: The platform allows individual building assessments by providing expeditious survey data, vulnerability estimates, and seismic risk calculations on a building-by-building basis. Additionally, by calculating the average vulnerability index for an entire compartment or municipality—provided that most of the buildings have been surveyed—it is possible to extrapolate the seismic vulnerability of broader urban areas with a more detailed accuracy level with respect to the urban or compartment approach.

Figure 6.1 illustrates the functional structure of the GENESIS platform. The diagram is divided into two main sections, each representing a different objective and corresponding methodological framework. On the left side, the focus is on seismic **risk assessment at a large scale**, which is essential for territorial planning of interventions and resource allocation. This approach integrates loss estimation and large-scale vulnerability assessment, which can be conducted at three levels: building-by-building knowledge (BBK), compartment knowledge (CK) and urban knowledge (UK). These assessments are based on a **Rapid Knowledge** (RK), as they rely on expedited survey methods to provide an overarching understanding of vulnerability across a given territory and provide a broad risk overview aimed to identify buildings that require more detailed assessments and potential interventions. On the right side, the platform addresses **accurate vulnerability assessment**, aiming at ensuring the structural safety of individual buildings through targeted strengthening and consolidation measures. This process involves a detailed analysis of failure mechanisms, distinguishing between local and global collapse modes. The assessment is based on in-depth investigations, including geometric surveys, material characterization, construction element studies, and soil and foundation surveys, classified under Accurate Knowledge (AK).

By integrating these two levels of analysis, GENESIS enables decision makers to efficiently prioritize resources and focus on the most vulnerable structures in need of

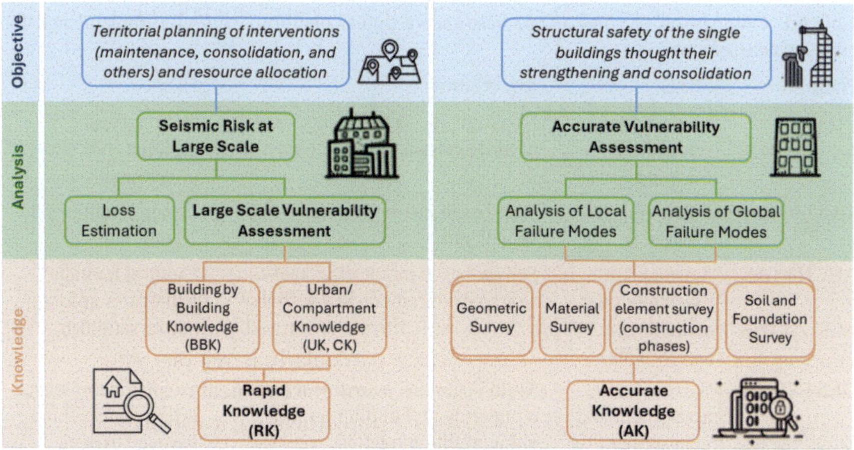

Fig. 6.1 Functional structure of the GENESIS platform, illustrating its dual approach to seismic risk assessment. The left side focuses on large-scale risk evaluation for territorial planning and resource allocation through rapid knowledge (RK), while the right side emphasizes accurate vulnerability assessment (AK) for individual buildings, supporting targeted strengthening and consolidation measures

in-depth assessments, as well as to understand in detail what vulnerabilities, interventions, and costs are needed to consolidate specific heritage buildings (Cantagallo et al. 2024b).

In the effort to identify the differences and strengths of the existing digital platforms for seismic risk assessment, it becomes evident that each system is designed with distinct objectives, functionalities, and target users. While all platforms aim to enhance risk evaluation and support decision-making, their approaches vary significantly depending on the scale of analysis, type of built environment considered, and methodological framework applied. Some platforms focus on broad territorial risk assessments, while others are tailored to the specific challenges of **historic centers and cultural heritage conservation** (Table 6.1).

IRMA provides a nationwide framework for seismic hazard and risk assessment, with a primary focus on large-scale mapping and vulnerability modelling. This platform is particularly suited for **policy planning** and the development of **regional risk mitigation strategies**, making it a valuable tool for national and regional governments responsible for disaster preparedness.

SICURO + plays a different role, as it is designed to bridge the gap between technical risk assessments and **public engagement**. Unlike other platforms that cater mainly to researchers and policymakers, SICURO + offers an **interactive environment** that translates seismic risk assessments into accessible information for citizens. By visualizing risk levels through web-based maps, it promotes **community awareness and preparedness**, empowering local populations to take preventive actions.

Table 6.1 Comparison of some of the best-known digital platforms currently implemented for seismic risk assessment

Platform	Focus	Key features	Users
IRMA	Large-scale seismic risk mapping	Exposure and vulnerability models, hazard assessment	National and regional governments
SICURO+	Public risk awareness	Web-based interactive maps, risk visualization	General public and communities
ARCHYTAS	Seismic risk assessment for ancient monuments	Sensor-based monitoring, real-time risk analysis, emergency response	Cultural heritage authorities and conservationists
GENESIS	Seismic risk assessment and management for historic centers	Multi-scale risk assessment (urban, compartment, building levels), vulnerability modeling, decision-support system	Local governments, urban planners and heritage experts

ARCHYTAS, in contrast, specializes in the seismic risk assessment of **ancient monuments**. The platform leverages **sensor-based monitoring** and integrates real-time seismic data to assess the vulnerability of heritage structures. A key strength of ARCHYTAS lies in its **emergency response capabilities**, which enable authorities to quickly evaluate post-earthquake damage, prioritize interventions, and develop effective conservation strategies for historic sites and archaeological monuments.

GENESIS, uniquely tailored for **historic centres**, implements a **multi-scale methodology** that allows seismic risk assessments to be conducted at urban, compartment, and individual building levels. By integrating seismic hazard models, exposure data, and survey-based vulnerability assessments, GENESIS provides a structured approach to both **expedited** and **detailed** evaluations. This comprehensive framework enables local and national authorities to optimize resource allocation and develop targeted conservation strategies, ensuring a more effective and data-driven approach to seismic risk management in heritage-rich urban areas.

As these platforms evolve, future advancements are expected to focus on AI-driven predictive analytics to refine risk models, improvement of the interoperability for data sharing, and the integration with satellite-based remote sensing technologies to enhance large-scale monitoring capabilities. These innovations will be instrumental in ensuring the long-term resilience of historic centres and cultural heritage assets in seismically active regions.

6.3 From Datalake to Database: Structuring and Interoperability of Data for Built Heritage Digitalization

The digitalization of built heritage requires an integrated and structured approach to manage data coming from heterogeneous sources, including geospatial analyses, HBIM models, survey campaigns, and sensor networks. In this context, the example of seismic risk analysis is proposed.

All raw data are collected in a **datalake**, a centralized repository that stores large amounts of raw, unstructured, or semi-structured data from multiple disciplines. The datalake enables the integration of diverse datasets from GIS maps, hazard models, HBIM, and in-field surveys without preliminary organization and processing.

The transition from datalake to database is essential to ensure interoperability between datasets and to enable advanced computational processes, such as vulnerability modeling, fragility curve derivation, and exposure-based risk assessments. The **database** acts as a structured and interoperable hub, transforming raw data into an organized system that can be queried and analyzed. Data accessibility and reusability are fostered by ensuring that data can be downloaded.

Once structured within the database, data becomes usable for the implementation of **algorithms** required for vulnerability assessment. A comprehensive digital platform should integrate multiple algorithms, each corresponding to a different large-scale assessment methodology. Each algorithm processes the data according to a specific methodology and generates distinct **fragility curves**, representing the probability of damage under varying hazard intensities.

After incorporating cultural, structural, and economic data about heritage buildings into the platform, exposure and vulnerability models are combined to estimate **expected loss curves**. These outputs support data-driven decision-making for conservation strategies and intervention planning.

Figure 6.2 visually represents the **data organization process** from raw data acquisition (datalake) to the structured database, which enables complete digital automation of vulnerability and risk analyses.

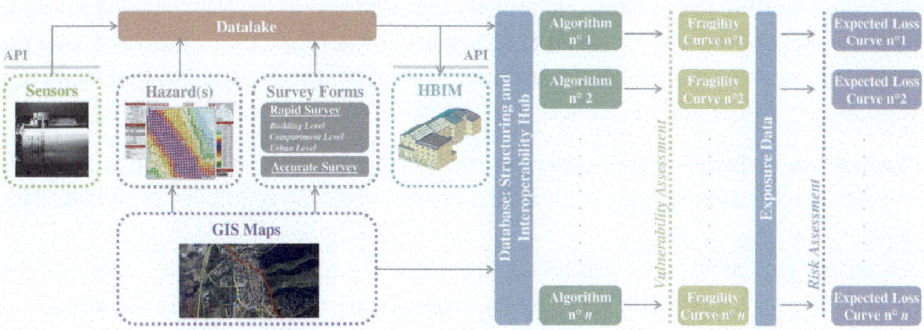

Fig. 6.2 Workflow of data structuring and interoperability for built heritage risk assessment

The workflow presented here highlights the importance of a structured approach in the digitalization of built heritage. By transitioning from a datalake-based raw data repository to a structured and interoperable database, this process enables:

- **Efficient data management**, reducing redundancy and improving data consistency.
- **Integration** of different data types (e.g., GIS, HBIM, and risk models).
- **Improved accessibility for stakeholders**, providing a unified and structured data environment for informed decision-making.

This approach ensures that built heritage assessments are based on a coherent, **data-driven methodology**, facilitating digitalization processes and risk-based conservation strategies.

6.4 Multiscale Digitalization: From Data Acquisition to Management

6.4.1 The Role of Multiscale Digitalization in Heritage Risk Management

If built heritage digitalization is carried out according to a **multiscale approach**, different **vulnerability and risk assessment methods** can be integrated within a single digital tool. The user can select the most appropriate vulnerability and risk assessment methodology depending on the objective to be achieved. This possibility can be a strong improvement in programming interventions of risk mitigation and conservation, since the results obtained from different methodologies can be compared and analyzed. Multiscale digitalization enables effective **management of built heritage**, ensuring that large-scale analyses and detailed building assessments complement each other, leading to more **informed and sustainable conservation actions**.

A comprehensive and inclusive digitalization of built heritage must therefore consider that heritage structures can be analyzed at different scales and levels of detail. This section proposes a **structured framework** for heritage digitalization based on a multiscale approach, ensuring that data acquisition, processing, and analysis are systematically integrated. Built heritage can be analyzed at two primary levels:

- **Expeditious (large-scale) analysis**—used for territorial vulnerability and risk assessments, often relying on rapid surveys without requiring complex data acquisition.
- **Detailed (or accurate) analysis**—used for in-depth structural vulnerability assessments, requiring high-precision data and advanced modeling techniques.

Figure 6.3 shows the multiscale approach by comparing expeditious and detailed methodologies. The expeditious approach is represented through large-scale

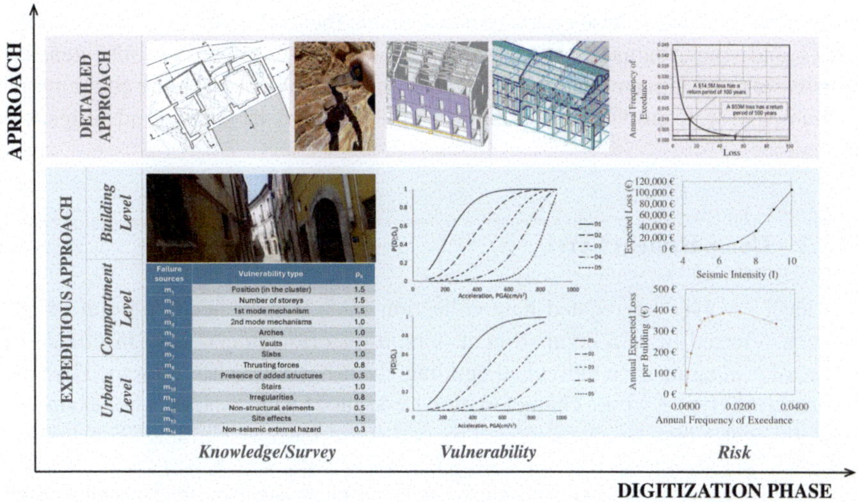

Fig. 6.3 Multiscale approach for heritage risk assessment

urban and territorial assessments, employing vulnerability indices and probabilistic fragility curves to estimate seismic risk at a broader level. The detailed approach, on the other hand, involves high-resolution digital documentation, structural analysis, and simulation models that allow for precise evaluation of the seismic behavior of individual buildings.

6.4.2 *Expeditious Approach: Large-Scale Vulnerability and Risk Assessment*

The **expeditious approach** is used for large-scale vulnerability and risk assessment, as discussed in Chap. 4. This methodology aims to identify high-risk areas, prioritize urgent interventions, and allocate resources efficiently for more detailed assessments.

Since this approach focuses on a territorial analysis, it is based on **rapid surveys** that do not require detailed geometric restitution, mechanical tests or data acquisition through sophisticated instrumentation. Expeditious surveys can collect visual inspection data structured at three levels: Urban Level, Compartment Level and Building Level.

6.4.2.1 Urban Level

At this level, the analyses generally refer to the entire municipality. The urban fabric is classified into two or three dominant archetypes, whose characteristics are defined by territorial experts. These characteristics enable the evaluation of a vulnerability index i_v and a mean damage level μ_D as a function of the considered seismic intensity. By assessing this damage level, fragility curves and loss estimates can be computed at the municipal scale, providing a valuable tool for disaster risk planning and mitigation strategies.

6.4.2.2 Compartment Level

This level involves aggregated data collection for a homogeneous portion of the territory, referred to as a compartment (e.g., a historic center). Also in this case, instead of evaluating each individual building, the urban fabric is represented by **two or three archetypes**, whose characteristics are defined by local experts. For example, a historic center may be characterized by two-story masonry buildings constructed in the late nineteenth century with irregular stone walls and flexible floors. Using average structural parameters, the mean damage level and fragility curves can be determined for the entire compartment.

6.4.2.3 Building Level

At this level, information is collected for each individual building within the study area. The collected data allows for the calculation of the vulnerability index i_v and the mean damage μ_D for each considered building. Using these parameters, an integrated, digitized and automatized system can calculate the fragility and loss curves for each building. Moreover, if data are available for multiple buildings, it is possible to compute the average vulnerability index i_v^*, calculating the average value of the single i_v. This enables the estimation of a territory-wide damage level and the derivation of fragility and loss curves for the entire area.

6.4.3 Detailed Approach: Single-Building Vulnerability and Risk Assessment

The detailed approach is used for in-depth vulnerability analysis of individual buildings, focusing on the identification of local and global collapse mechanisms as well as the most vulnerable components of the structure. This level of analysis is essential for designing tailored reinforcement and consolidation interventions. This approach is based on a **detailed (or accurate) survey**, which includes multiple aspects, such as:

- High-resolution geometric restitution using advanced instrumentation (e.g., point cloud data from laser scanning).
- Historical–critical analysis of construction phases and previous interventions.

- Documentation of architectural and structural details.
- Material investigations and on-site tests.
- Structural parameter acquisition through sensor-based monitoring.
- HBIM (Heritage Building Information Modeling) integration.

These highly detailed datasets allow professionals to develop numerical simulations using **Finite Element Models** (FEM), which provide an accurate assessment of the structural vulnerability of the building.

Currently, automated vulnerability analysis within a digital platform is not yet fully achievable due to interoperability limitations between finite element software and heritage databases. However, the results of these detailed analyses can be integrated into a centralized hub, helping stakeholders to understand the levels of vulnerability and risk of individual structures.

References

Borzi B, Onida M, Faravelli M, Polli D, Pagano M, Quaroni D, Cantoni A, Speranza E, Moroni C (2021) IRMA platform for the calculation of damages and risks of Italian residential buildings. Bull Earthq Eng 19:3033–3055. https://doi.org/10.1007/s10518-020-00924-x

Cantagallo C, Cianchino G, Sangiorgio V, Masciotta MG, Pierantozzi M, Lops C, Di Loreto S, Brando G, Spacone E (2024a) GENESIS: a web-based platform for managing the seismic risk of historic centres of Southern Italy. In: Boffill Y, Lombillo I, Blanco H (eds) REHABEND 2024. 10th Euro-American congress on construction pathology, rehabilitation technology and heritage management. University of Cantabria—Building Technology R&D Group, pp 2522–2530

Cantagallo C, Sangiorgio V (2024) An IT tool for managing seismic risk and energy performance of the building stock in Southern Italy. In: International conference of Ar. Tec. (Scientific Society of Architectural Engineering). Springer, pp 103–114

Cantagallo C, Sangiorgio V, Verazzo C, Spacone E (2024b) An analytical safety tree for heritage buildings: application to the archaeological museum "Domenico Ridola", Matera (Italy). Int J Architectural Heritage. https://doi.org/10.1080/15583058.2024.2383354

Faravelli M, Borzi B, Onida M, Cattari S, Alfano S, Masi A, Manfredi V, Lagomarsino S (2022) An Italian platform for the seismic risk assessment of school buildings. In: Procedia structural integrity. Elsevier B.V., pp 107–114

Faravelli M, Di Meo A, Borzi B, Cantoni A, Savadori L, Speranza E, Dolce M (2024) SICURO+: a web platform to raise awareness on seismic risk in Italy. Int J Disaster Risk Reduction 103. https://doi.org/10.1016/j.ijdrr.2024.104345

Meletti C, Montaldo V (2007) Stime di pericolosità sismica per diverse probabilità di superamento in 50 anni: valori di ag. Progetto DPC-INGV S1, Deliverable D2

Melissianos VE, Lachanas CG, Lignos XA, Vamvatsikos D, Chatzidaki A, Dasiou M-E, Manetas A (2024) A holistic platform for the seismic risk assessment of ancient monuments. In: 18th world conference on earthquake engineering (WCEE2024). English

Chapter 7
Conclusions

Abstract This chapter summarizes the main findings of the monograph, emphasizing the integration of advanced digital tools for the multi-scale and multi-risk assessment of built heritage. It presents the main aspects addressed, including the development of digital identity cards, automated risk analysis procedures, conservation and strengthening strategies, and integrated monitoring systems. The chapter also outlines future directions, focusing on artificial intelligence, interoperability, satellite-based monitoring, and enhanced public engagement. Together, these elements aim to advance sustainable and resilient conservation practices, supporting informed decision-making for the long-term protection of historic structures against seismic and climate-related threats.

The research presented in this book has explored the digitization of built heritage, emphasizing the integration of advanced technologies for data acquisition, analysis, and conservation. The work has provided a comprehensive examination of multi-scale and multi-risk methodologies for assessing the vulnerability of historic buildings, addressing both seismic hazards and climate change-related threats. Furthermore, it has proposed strategies for conservation, maintenance, and risk mitigation, supported by the implementation of digital platforms to enhance heritage management.

Specifically, the main points addressed in this work are as follows.

- **Digitalization and Heritage Knowledge**. The study has presented the fundamental role of digital tools in documenting, analyzing, and managing built heritage. Digital identity cards for historic buildings have been conceptualized to store and structure data, improving accessibility and usability. Technologies such as Heritage Building Information Modelling (HBIM), and Geographic Information Systems (GIS) have been shown to significantly enhance the accuracy and effectiveness of heritage conservation strategies.
- **Multi-Scale and Multi-Risk Assessment**. A structured four-level methodology for seismic vulnerability assessment has been outlined, distinguishing between preliminary, first-level, second-level, and third-level evaluations (each with an

© The Author(s), under exclusive license to Springer Nature Switzerland AG 2025

C. Cantagallo et al., *Digitization of Built Heritage*, Digital Innovations in Architecture, Engineering and Construction, https://doi.org/10.1007/978-3-031-96598-2_7

increasing level of detail in analysis). The approach facilitates scalable assessments, allowing for both territorial-scale risk mapping and detailed single-building analyses. This framework enables public authorities and conservation professionals to prioritize interventions and allocate resources more effectively.
- **Automation of Risk Analysis**. The implementation of digital platforms for risk assessment is illustrated through a nationally significant pilot case, the GENESIS system. These platforms integrate data from various sources, automating seismic risk assessments through the processing of hazard, vulnerability, and exposure parameters. The application of predictive analytics enables the proposed methodologies to enhance the accuracy of risk assessments and support well-informed decision-making in heritage conservation.
- **Conservation and Strengthening Strategies**. The monograph emphasized the importance of conservation principles such as authenticity, integrity, and compatibility of interventions. Strengthening strategies have been categorized into those addressing local collapse mechanisms (e.g., tie rods, external ties, and corner stitching) and those mitigating global collapse mechanisms (e.g., diaphragm strengthening and reinforced connections). The use of minimally invasive, reversible, and durable solutions is crucial in preserving the historical and cultural integrity of built heritage.
- **Monitoring and Preventive Maintenance**. This monograph presents a comprehensive dual approach, integrating structural monitoring with climate and environmental monitoring to enhance preventive conservation. Structural Health Monitoring (SHM) systems play a key role in safeguarding built heritage by providing real-time data on structural performance and deterioration. The deployment of static and dynamic monitoring sensors, such as accelerometers, strain gauges, and fiber optic sensors, enables data-driven maintenance strategies, reducing long-term conservation costs and mitigating structural risks. Similarly, climate and environmental monitoring systems provide important information about external stressors affecting heritage structures. Sensors measuring temperature, humidity, and wind conditions—including thermocouples, hygrometers, and anemometers—monitor environmental parameters that contribute to material degradation.

Future developments will focus on advancements in artificial intelligence to refine predictive models and enhance automation in risk assessment. The challenge of interoperability between different digital platforms must be addressed through the adoption of standardized data formats and open-source solutions. Additionally, the integration of satellite-based remote sensing and UAV technology will enable large-scale heritage monitoring, particularly in assessing climate change-related vulnerabilities. Expanding risk assessment frameworks to include hazards such as flooding, landslides, and urbanization effects is essential for comprehensive heritage protection. Furthermore, digital platforms should facilitate public engagement, allowing communities to actively participate in heritage preservation. Closer collaboration with policymakers and heritage institutions will be necessary to translate research findings into actionable conservation policies.

By embracing these future directions, the methodologies and tools presented in this work will continue to evolve, supporting more sustainable and resilient conservation strategies for built heritage.

The manufacturer's authorised representative in the EU is Springer Nature Customer Service Centre GmbH, Europaplatz 3, 69115 Heidelberg, Germany. If you have any concerns regarding our products, please contact ProductSafety@springernature.com

Printed and bound by CPI Group (UK) Ltd, Croydon, CR0 4YY
05/10/2026
02228129-0007